Building Performance Evaluation

Wolfgang F.E. Preiser · Andrea E. Hardy
Ulrich Schramm
Editors

Building Performance Evaluation

From Delivery Process to Life Cycle Phases

Second Edition

Springer

Editors
Wolfgang F.E. Preiser (deceased)
University of Cincinnati
Cincinnati, OH
USA

and

Arizona State University
Scottsdale, AZ
USA

Andrea E. Hardy
Creo Architects
Phoenix, AZ
USA

Ulrich Schramm
Bielefeld University of Applied Sciences
Minden, Bielefeld
Germany

ISBN 978-3-319-56861-4 ISBN 978-3-319-56862-1 (eBook)
DOI 10.1007/978-3-319-56862-1

Library of Congress Control Number: 2017939094

Printed on acid-free paper

This Springer imprint is published by Springer Nature
The registered company is Springer International Publishing AG
The registered company address is: Gewerbestrasse 11, 6330 Cham, Switzerland

In Memory of Wolfgang F.E. Preiser
colleague, teacher, mentor, and friend

Foreword

Why Building Performance Evaluation Matters

To understand the importance of this book, we need to stand back and look at the long-standing tension existing between those who design buildings and those who assess them after the fact. That tension rests, in part, on a false dichotomy between the sciences on the one hand and the humanities and arts on the other. Echoing the argument of the scientist and novelist C.P. Snow (1993) that these two cultures remain irreconcilable, some in the architectural profession act as if the art of architecture and the social science of environmental psychology occupy different worlds and will forever look at each other with suspicion.

This book offers ample evidence to refute that claim. The discipline of architecture, by its very nature, straddles the sciences and the arts, requiring calculation as much as creativity. And, the practice of architecture demands an understanding of the psychological, sociological, and anthropological implications of the built environment. Neglecting those social sciences, as some of the chapters here show, can produce highly dysfunctional, poorly performing, or decidedly inappropriate buildings, something few clients or occupants will tolerate for very long.

The real challenge for architecture lies not in bridging the two cultures of art and science, but in overcoming the poor communication between researchers and practitioners in this field. Unlike the sciences and to some extent the humanities, the architectural profession has yet to develop the communication networks and peer-reviewed journals that typically link research and practice in an effective knowledge loop, where problems confronted in practice get rigorously researched and the conclusions communicated back to practitioners in useful ways. This book helps close that loop with chapters full of useful information about the performance of buildings, but we need many more like this one to meet the real need for this knowledge.

One reason the profession has made so little progress on this front stems from a second and related fallacy to that of Snow's two cultures. The influential twentieth century architect and educator, Walter Gropius believed that architectural design took priority over and needed to precede social science (Alofsin 2013). In the curriculum he developed at Harvard, Gropius argued that schools of architecture should not hamper students' creativity by exposing them too early to history and other research-based social sciences. That suspicion of certain kinds of information coming too early in architectural education also affected practice, relegating the social sciences to the evaluation of buildings after their construction and occupancy.

At a superficial level, it may seem obvious that the design of a building precedes its evaluation and that its construction comes before the measurement of its success in meeting the needs of its owners and occupants. But at a deeper level, that temporal hierarchy makes no sense. An architect—or architectural student—cannot design a building without knowledge about how people occupy buildings, which turns Gropius's argument on its head: the evaluation and assessment of architecture must come before its design and construction, as well as afterward.

This may help explain the rise of building performance evaluation (BPE) over the last two decades. BPE represents an expansion of post-occupancy evaluation to encompass the entire process of pre-design, programming, design, construction, and occupancy in a non-dichotomous and non-hierarchical interweaving of architecture and social science. Clients should expect nothing less of the professionals they hire. The owners and occupants of buildings want—and deserve—to know that the environments they own and inhabit will meet their needs and that no part of the process fails in meeting that goal.

In some ways, BPE reflects a larger transition in the twenty-first century in many nations away from goods producing to service-based economies. Architecture stands in peculiar position in this shift. Buildings remain one of the most expensive goods that we produce, own, and use in our daily lives, and architects play a key role in the creation of that real estate. And yet, architects do not actually produce these goods; they produce the documents that specify materials and products and they observe the work of the contractors who build the structure. Architecture thus remains more of a service profession than a goods producing one.

Other members of the construction industry, like program managers and facilities managers, tend to understand this, but rarely do the keepers of the architectural culture. The American Institute of Architects (2016), like many architectural organizations around the world, awards buildings based on the design of the completed structure and the qualities of the finished product. While these awards programs acknowledge the teams of people involved in producing the building, the juries selecting these projects seldom take into account the quality of the design process or the assessment of the building's performance as part of their decision-making. The architectural culture, in other words, continues to focus on the goods that practitioners design rather than on the services they offer.

That divide between the reality of architectural practice as a service activity and the image of the architectural profession as a goods-producing discipline has presented profound problems for the field. Too many of the most recognized buildings have proven problematic as places to live or work, which has shaken the public's confidence that architects can produce a product that people can trust will meet their needs without encountering unwanted failures or creating unexpected costs. As other goods-producing industries have ratcheted up the performance and predictability of their products, the one-off, hand-made nature of architecture has not kept up.

BPE offers an opportunity for architecture to catch up. It represents a kind of continuous-improvement approach to the field, in which the quality and efficiency of every aspect of the design, construction, and occupancy process gets scrutinized and ideally enhanced. Larger architectural firms increasingly understand this as many have made great strides in integrating research—and conducting research themselves—as part of their services. To make this an expected part of every practice, the profession, again, needs to find ways to get this information into the hands—and the computers—of practitioners, regardless of the size and location of their offices. Maybe a book like this can continue to evolve as an ever-growing database of BPE knowledge, accessible not just to every architect, but also to everyone else involved in the design, construction, and management of buildings.

With that, the architectural profession might finally make the shift to thinking about itself as a service-design profession. Service-design emerged over the last two decades as practitioners began to apply the methods used in the creation of goods to the design of services. This has greatly expanded the demand for design and also the scope of its impact as people who might not need a new building, for example, recognize the many design-related challenges they face in their own organizations and communities.

Service-design also demonstrates that creative opportunities exist in all aspects of human activity, from the strategies we form to the assessments we make to the products and environments we create. And, BPE does the same for architecture. By engaging in conversations with and leveraging the ideas of the people who have the most at stake in a project—the owners and occupants of buildings—this approach shows how the creative process can improve every aspect of the design and building process. The social sciences do not interfere with our creativity; they are a source of it, as the chapters in this book amply show.

Thomas Fisher

References

American Institute of Architects (2016). AIA Honor Awards. *Architect*. Retrieved from http://www.architectmagazine.com/awards/aia-honor-awards. Accessed on January 15, 2017.

Alofsin, A. (2013). A Man of Parts: Walter Gropius in Five Acts—Teacher, *Architecture Boston*. Retrieved from https://www.architects.org/architectureboston/articles/man-parts. Accessed on January 15, 2017.

Snow, C.P. (1993). *The Two Cultures*. Cambridge: Cambridge University Press.

Author Biography

Thomas Fisher is Professor in the School of Architecture, the Dayton Hudson Chair in Urban Design at the University of Minnesota, and the Director of the Minnesota Design Center at the College of Design.

A graduate of Cornell University in architecture and Case Western Reserve University in intellectual history, he was recognized in 2005 as the fifth most published writer about architecture in the United States. He has written 9 books, over 50 book chapters or introductions, and over 400 articles in professional journals and major publications. Named a top-25 design educator four times by Design Intelligence, he has lectured at 36 universities and over 150 professional and public meetings in the U.S.

He has written extensively about architectural design, practice, and ethics. His books include *In the Scheme of Things, Alternative Thinking on the Practice of Architecture* (Minnesota), *Architectural Design and Ethics, Tools for Survival* (The Architectural Press), *Ethics for Architects, 50 Dilemmas of Professional Practice* (Princeton Architectural Press), two monographs on the work of architect David Salmela (Minnesota), a book on the work of Lake Flato (Rockport), and a book on system design entitled *Designing to Avoid Disaster, The Nature of Fracture-Critical Design* (Routledge). His latest book, *Designing our Way to a Better World* (Minnesota), was published in 2016, and he is currently at work on a collection of essays about design and ethics as well as a book manuscript on cities in the on-demand economy.

Preface

How did the book *Building Performance Evaluation* (formerly *Building Evaluation*) come to be?

In 1988, the International Association of People and Environment Studies (IAPS) organization in Europe organized one of their annual conferences at Delft University in the Netherlands. First Editor, Wolfgang Preiser organized symposia around the theme of Post-Occupancy Evaluation. These, in turn, became the basis for the original edition of *Building Evaluation*, published by Plenum Press in New York City in 1989 (Preiser 1989).

The process of this Second Edition began in 2015, with Wolfgang F.E. Preiser discovering that the previous publisher of this book was no longer in existence. With the original edition of this book still in print, the editors worked on bringing this Second Edition to fruition with our new publisher, Springer. From the very beginning, the concept was to select original chapters to be updated by their respective authors and to add new content from other professionals in the field.

New materials were drawn from the International Building Performance Evaluation (IBPE) consortium symposia and related paper sessions, which are held annually at the Environment Design Research Association (EDRA) conferences. There is ample material authored by practitioners, academics, and consultants, mostly in Europe, Asia, and North America. The resulting Second Edition, now called *Building Performance Evaluation*, thus presents an interesting contrast—namely a retrospective as to how the field of Post-Occupancy/Building Performance Evaluation (POE/BPE) evolved over the last 25 or so years. It also allows a look into the future by considering, for example, new workplace design concepts for offices, new opportunities utilizing digital media, and the like.

This book is seen as a continuation to *Assessing Building Performance* (Preiser and Vischer 2005) and *Enhancing Building Performance* (Mallory-Hill, Preiser, and Watson 2012). The first showcased the BPE conceptual framework for the first time in the context of real world applications. The second one modified and developed the conceptual framework to the BPE process model, together with an entirely new set of case studies from around the world. The present book once again is interdisciplinary and international as it brings together practitioners, administrators, academicians, as well as consultants from different disciplines and diverse countries. And with its unique balance of updated original chapters on the one hand and brand new contributions on the other, this book is considered to be intergenerational: it offers remarkable work from academics and practitioners being new in this field and complex theoretical approaches and analysis written by professionals well-known for many years.

Therefore, the audience of this book is envisioned as practitioners in the planning, design, and construction industries, facility managers, government organizations, academics, and students from various programs interested to learn from building performance evaluation.

By dividing the book into multiple sections, it is able to cover some of the topics in the original edition of the book, share advances within the field, and provide examples of the advances from within the field of building performance evaluation. The four book sections are: Introduction, Frontiers of Building Evaluation; Advances in Evaluation Knowledge; and, Advances in Evaluation Methods. During the process of editing and reviewing chapters, the editorial team and authors unfortunately lost their mentor, friend, and colleague Wolfgang Preiser. Thus, Epilogue, by Jacqueline C. Vischer, was redeveloped to be not only a summary of the history and future of the field of building evaluation, but is also a reflection of Wolf's life work.

Scottsdale, USA — Wolfgang F.E. Preiser
Phoenix, USA — Andrea E. Hardy
Minden, Germany — Ulrich Schramm
February 2017

References

Mallory-Hill, S., Preiser, W., & Watson, C. (Eds.). (2012). *Enhancing building performance*. Oxford: Wiley-Blackwell.

Preiser, W. (Ed.). (1989). *Building evaluation*. New York: Plenum Press.

Preiser, W., &. Vischer, J. (Eds.). (2005). *Assessing building performance*. Oxford: Elsevier.

Acknowledgements

This book largely would not have been possible without Wolfgang F.E. Preiser, who passed away before this book was able to be published. It was through his passion for publication and research in his last couple of years that he was able to get many of his books republished and discovered that there was an opportunity for a second edition to this book. Andrea and Uli want to extend a special thanks to Wolf for his knowledge, guidance, and optimism he shared with us. His mentorship will be greatly missed.

The editors thank all of the authors in this book, many of whom crossed language and geographic barriers to further the discourse on architectural design, construction, occupancy, and building evaluation. Chapter contributions share the authors' projects, research, and findings from both academia and practicing firms from around the World and reveal inspiring dedication to asking more of the built environment.

As the List of Contributors indicates authors of this volume come from Europe, North America, Australia, and Asia, bringing with them local research and experience in evaluating facilities at different scales. We thank all the organizations that were clients to the editors and authors, who provided releases of information, quoted text, illustrations, and photography.

We thank Natalie Jacobs, Senior Publishing Editor, and Johanna F.A. Pot, Editorial Assistant, at Springer for their continuing guidance and support in this project.

Thanks are due to our significant others, family, and friends who have supported us during the multiple year journey of developing, producing, and editing this book.

Contents

Contributors

MaryAnn Sorensen Allacci Rutgers University, New Brunswick, NJ, USA

James R. Anderson Champaign, IL, USA

Clinton J. Andrews Rutgers University, New Brunswick, NJ, USA

Azizan Aziz Carnegie Mellon University, Pittsburgh, PA, USA

Brodie Bain AICP, LEED AP, Perkins & Will, Seattle, WA, USA

Greg Allen Barker Needs Assessment - Facility Programming and Planning, San Luis Obispo, CA, USA

Franklin Becker Cornell University, Ithaca, NY, USA

Dominic Becking Bielefeld University of Applied Sciences, Minden, Germany

John Callewaert University of Michigan, Ann Arbor, MI, USA

Joon-Ho Choi University of Southern California Los Angeles, CA, USA

Ihab M. K. Elzeyadi University of Oregon, Eugene, OR, USA

Jay Farbstein Jay Farbstein & Associates, Pacific Palisades, CA, USA

Melissa Farling HDR, Phoenix, AZ, USA

Lindsey L. Fay University of Kentucky, Lexington, KY, USA

Sofia Fonseca HOK, Houston, TX, USA

Guido Francescato San Rafael, CA, USA

Mark Gorgolewski Ryerson University, Toronto, ON, Canada

Andrea E. Hardy Creo Architects, Phoenix, AZ, USA

Volker Hartkopf Carnegie Mellon University, Pittsburgh, PA, USA

Martin Hodulak Munich, Germany

Duncan Joiner Wellington, New Zealand

Akikazu Kato Mie University, Tsu, Japan

Masayuki Kato Naito Architects, Osaka, Japan

Vivian Loftness Carnegie Mellon University, Pittsburgh, PA, USA

Shauna Mallory-Hill University of Manitoba, Winnipeg, MB, Canada

Robert W. Marans University of Michigan, Ann Arbor, MI, USA

Shiho Mori Nihon Fukushi University, Mihama, Japan

Linda N. Nubani Michigan State University, East Lansing, MI, USA

Nigel Oseland Workplace Unlimited, Berkhamsted, UK

Jihyun Park Carnegie Mellon University, Pittsburgh, PA, USA

Steven Parshall HOK, Houston, TX, USA

Erin Persky Jay Farbstein & Associates, San Diego, CA, USA

John P. Petronis Architectural Research Consultants, Albuquerque, NM, USA

John W. Petronis Architectural Research Consultants, Albuquerque, NM, USA

Lexi Petronis Architectural Research Consultants, Albuquerque, NM, USA

Wolfgang F. E. Preiser Scottsdale, AZ, USA

Handi Chandra Putra Rutgers University, New Brunswick, NJ, USA

Sybille Reichart Bielefeld University of Applied Sciences, Bielefeld, Germany

Thierry Rosenheck US Government, Washington, DC, USA

Ulrich Schramm Bielefeld University of Applied Sciences, Minden, Germany

Jennifer A. Senick Rutgers University, New Brunswick, NJ, USA

Mille Sylvest Roskilde University and 3XN Architects, Copenhagen, Denmark

Ioanna Tsoulou Rutgers University, New Brunswick, NJ, USA

Jacqueline C. Vischer University of Montreal, Montreal, QC, Canada

Sue Weidemann Buffalo, NY, USA

Part I
Introduction

Chapter 1
From Linear Delivery Process to Life Cycle Phases: The Validity of the Concept of Building Performance Evaluation

Wolfgang F.E. Preiser, Andrea E. Hardy and Ulrich Schramm

1.1 Transcending the Meaning and Boundaries of POE

1.1.1 The Evolution of POE

The idea
The first precursors to post-occupancy evaluation (POE) started in the US with studies of dormitories in the late 1960s. Van der Ryn and Silverstein (1967) carried out case study evaluations of dormitories at the University of California, Berkley, while Hsia (1967) was doing the same at the University of Utah. At the time, they were not called post-occupancy evaluations, but they were attempts at assessing building performance from the building users' point of view (Connell and Ostrander 1976a). Wolfgang Preiser, first editor and author, was inspired by Van der Ryn and Hsia and developed his Master's thesis on evaluating dormitory performance at Virginia Tech (Preiser 1969). He used political science rating scales, with error rates of 3–5%, to create specific quality profiles of the three different housing types, as seen by the students.

Wolfgang F.E. Preiser passed away in August 2016 during the final phase of chapter editing.

W.F.E. Preiser
Scottsdale, USA

A.E. Hardy (✉)
Creo Architects, 2716 N 16th Street, Phoenix, AZ 85006, USA
e-mail: andrea.e.hardy@gmail.com

U. Schramm
Bielefeld University of Applied Sciences, Artilleriestrasse 9, 32427 Minden, Germany
e-mail: ulrich.schramm@fh-bielefeld.de

W.F.E. Preiser et al. (eds.), *Building Performance Evaluation*,
DOI 10.1007/978-3-319-56862-1_1

The term

The term 'post-occupancy evaluation' was used in a publication, for the first time in the *AIA Journal* issued in January 1975. Herb McLaughlin, author of that article, with a team of consultants performed POEs on hospitals in both Utah and California. At this point, information and studies were being mentioned in publications, but were also becoming in-house tools for designers and design firms (McLaughlin 1997). Table 1.1 gives a summary of milestones in the evolution of POE.

Table 1.1 Milestones in the evolution of POE and BPE

Year	Author(s)	Building type (s)	Contribution to the field
1967	Van der Ryn and Silverstein	Student dormitories	Environmental analysis; concept and methods
1968	Manning	Offices and schools	Comprehensive building appraisal
1968	Sanoff	Any facility type	"Evaluation Techniques for Designers"—first monograph on POE
1969	Preiser	Student dormitories	Environmental performance profiles; correlation of subjective and objective performance measures
1971	Field et al.	Hospital	Multi-method approach to data collection
1972	Markus et al.	Any facility type	Cost-based building performance evaluation model
1974	Becker	Public housing	Cross-sectional comparative approach to data collection and analysis
1975	General Services Administration (GSA)	US Courts Design Guide	Office system performance standards
1975	McLaughlin	Hospitals	"Evaluation of Hospitals"—first article published on POE
1975	Veterans Administration	Veterans Hospitals	POE of the Veterans Administration Hospital in San Diego
1976	U.S. Army Corps of Engineers	Military facilities	Design guide series with updatable, state-of-the-art criteria
1976	Goodrich	Public square	Observational POE methodology
1976	Connell and Ostrander	Government facilities	POEs of postal and enlisted housing
1978	Bechtel and Srivastava	Housing	Comprehensive review of POEs of housing
1979	Public works—Canada	Government facilities	POE incorporated into project delivery system
1980	Daish et al.	Military facilities	POE process guidelines
1980	Marans	Offices	Evaluation model linking perceptual and objective attributes
1981	Palmer	Any facility type	Programming linked to POE methodology
1982	Parshall and Pena	Any facility type	Simplified and standardized evaluation methodology for practitioners

(continued)

Table 1.1 (continued)

Year	Author(s)	Building type (s)	Contribution to the field
1983	Duffy and Chandor	Offices	Orbit 1: Systems design standards
1984	Brill et al.	Offices	Linking worker productivity and office design
1985	Davis et al.	Offices	Orbit 2: rating process on organizations, buildings and information technology
1987	Building Research Board	Any facility type	"POE practices in the building process"
1988	Preiser, Rabinowitz and White	Any facility type	"Post-Occupancy Evaluation"—first book on POE Methodology
1989	Farbstein	U.S. Postal Service	POE and organizational development
1989	Preiser	Any facility type	"Building Evaluation"—POE case studies from around the world
1992	Sanoff	Any facility type	Integrating programming, POE and user participation in design
1996	Baird et al.	Any facility type	"Building Evaluation Techniques"—first comprehensive methods book
1997	Preiser and Schramm	Any facility type	"Building Performance Evaluation"—conceptual BPE framework
2001	Federal Facilities Council	Any facility type	"Learning From Our Buildings"—Federal POE/BPE overview
2001	National Clearinghouse	Educational facilities	Feedback-based design standards for schools
2003	NCARB	Any facility type	"Improving Building Performance"—a study guide for architects
2005	Szigeti and Davis	Any facility type	Performance based building
2005	Preiser and Vischer	Any facility type	"Assessing Building Performance"—global BPE book
2006	Zeisel	Any facility type	Example POE—The Jerusalem Center for multi-handicapped visually impaired children
2007	Nasar, Preiser and Fisher	Any facility type	"Designing for designers: lessons learned from schools of architecture"
2007	Hartman	Any facility type	"Measuring a buildings success"—an article reviewing the potentials and fears of POEs for building designers
2008	Gonchar	Any facility type	"Looking back and moving forward"
2009	OECD	Educational facilities	First coordinated European effort
2009	Ireland	Any facility type	Importance of collaboration and building analysis—examples using information about LEED and intelligent building design
2010	Spataru	EON Research House	Research on 'Creative Energy Homes'

(continued)

Table 1.1 (continued)

Year	Author(s)	Building type (s)	Contribution to the field
2011	Borg	Any facility type	"A Dossier on Post-Occupancy Evaluation"
2012	Mallory-Hill, Preiser and Watson	Any facility type	"Enhancing Building Performance": State-of-the-art book on BPE process model, methodology and case studies
2012	Kampschroer	US Federal Buildings	Building requirements and evaluations for Federal Buildings
2012	Lenoir, Baird and Garde	Educational Facility	How to achieve thermal comfort through the use and efficiency of the ENERPOS building in La Reunion
2012	Newton et al.	Educational Facilities	Australia's "Building the Education Revolution"—template designs critiqued by an interdisciplinary team in POEs reviewing pedagogy, sustainability and life-cycle analysis
2014	Vischer and Malkoski	Workspace, Office	How and why certain work environments work for employees and their employers
2015	Preiser et al.	Any facility type	"Architecture Beyond Criticism": The paradigms of architectural criticism and performance evaluation and how they complement each other to give an overall result of how the building is used, observed, and appreciated
2015	Park	Any facility type	Concept and importance of using and understanding humans as sensors within the built environment
2015	Elzeyadi	Schools	"A Green Lesson: Measuring the Impacts of LEED Certification Credits on People, Planet, and Profit of K-12 Schools"
2017	NCARB	Any facility type	"Improving Building Performance"—a study guide for architects
2017	Fay et al.	Healthcare	Study of how a POE can be brought to actionary items through the use of charrettes

The methodology

Also in the late 1970s, the AIA Research Corporation commissioned a methodological review of POE techniques (Connell and Ostrander 1976b). And by the late 1980s, POEs were being carried out around the World, including the UK, Canada, New Zealand, Australia, and the US. At this time, public works projects, government buildings, airports, and other similar facility types were the main focus of POE studies. In the mid-1980s, the methodology of POE expanded when the National Academy of Sciences (Building Research Board 1987) established a committee to review the possible improvements within the practices of programming, POE, and database development.

Framework

The first books published on POE included *Post-Occupancy Evaluation* in 1988 by Preiser, Rabinowitz, and White and *Building Performance*, the predecessor of this present volume, in 1989 by Preiser. Both titles were considered as companion volumes: while the first book provided the framework and structure for completing an evaluation of a building, e.g., measurement techniques for the evaluation of the quality of the facility, the other one complemented the theoretical fundamentals with case studies from around the world.

Research, methodologies, and framework of POEs continue to evolve. POEs are one step, on the larger scale of building performance evaluation (BPE), in understanding how buildings function after they are occupied. This resource helps architects, building owners, and facility managers understand the implications and reactions to the facilities that they designed, built and/or commissioned. By considering the whole process from conception to future uses of the building, there can be a more holistic approach to the planning, programming, design, construction, occupancy, and future adaptability of the structure (see Fig. 1.1).

1.1.2 Holistic Thinking: The Building Life Cycle

Traditionally, building delivery was—and in parts of the building industry still is—considered as a linear process. For many architects, for example, the design of a building marks the beginning of a project and the hand-over of keys signifies the end. This perspective emphasizes the planning and construction of a facility with the product in mind, but omits the occupancy and future re-use phases. In reality,

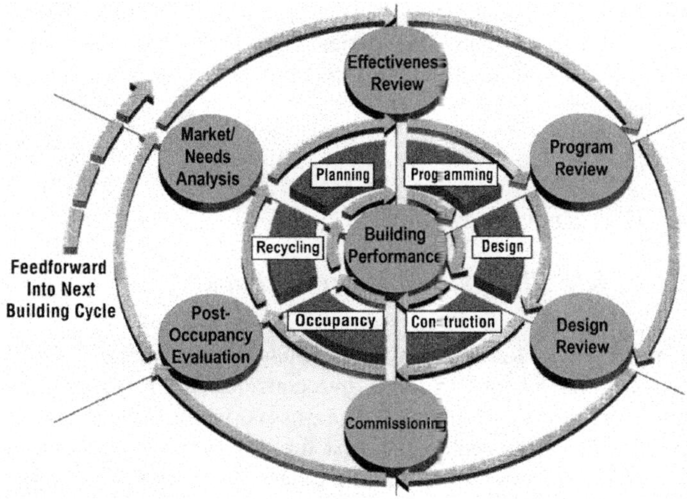

Fig. 1.1 Building performance evaluation (BPE) process model. *Source*: Authors

the client wants a building to be built in order to get a problem solved within their organization. They look forward to using the building for the given purpose. Therefore, from the client's perspective, the opening of a building does not mark the end of the process where the building's success can be determined. It is rather the starting point of building use, the longest period of time within a building's life cycle. So, compliance with construction cost limits and building time frames are only preliminary criteria to characterize its success. Although running costs, functionality, or user satisfaction and well-being are some other important factors to be considered when buildings are evaluated, for example in form of POEs. With a POE, important feedback is received about the building's strengths and weaknesses during occupancy allowing improvement in view of performance and quality. Finally, holistic thinking also includes a reflection about a building's re-use, transformation into a different building type, or even its potential future demolition. With this holistic perspective in mind, building life cycle developed consisting of six phases, covering all possible stages of a building's life: strategic planning, programming, design, construction, occupancy, and finally, adaptive re-use/recycling. Each of these phases is an indispensable component of the life cycle and has to be considered in a circular arrangement. Moreover, in order to get the complete picture, the evaluative stance, already implemented in form of POE in the phase of occupancy, was expanded to the other five phases as well with the goal to improve continuously the quality of buildings. While pre-design planning (Hershberger and Smith 2017), design, and building management all contribute to an economical and efficient building, 'feedforward' may be the more important, and unfortunately overlooked, step of BPE. By generating databases or allowing others to use your findings, the process of designing a building, similar to that of what has already been built and studied, will greatly help all steps of BPE, i.e., planning, programming, design, construction, and building management. Feeding forward information and knowledge helps to streamline all stages of building design and management, including avoiding costly mistakes. It is evident that BPE transcends the meaning and conventional boundaries of POE by focusing on evaluation throughout the building life cycle. This concept is discussed in several chapters throughout this book.

1.2 The BPE Process Model

1.2.1 Life Cycle Phases and Review Loops

BPE grew out of POE. Rather than focus solely on the phase of building occupancy, the process model for BPE takes into consideration performance evaluation over the entire life cycle of buildings. The phases, feedback loops, elements, and levels of BPE were introduced for the first time by Preiser and Schramm (1997). The model evolved over the years and the potential and benefits of applying BPE over the lifetime of a building is now discussed and researched regularly (Preiser and Schramm 2005, 2012; Preiser and Hardy 2015). This section summarizes the six phases and the respective internal feedback loops of BPE:

- Phase I: Strategic Planning—Effectiveness Review
- Phase II: Programming—Program Review
- Phase III: Design—Design Review
- Phase IV: Construction—Commissioning
- Phase V: Occupancy—Post-Occupancy Evaluation
- Phase VI: Re-use/Recycling—Market/Needs Analysis (see Fig. 1.1).

Phase I—Strategic Planning
The starting point of the building life cycle is the strategic plan, which establishes medium- and long-term needs of an organization through market/needs analysis, and, in turn, is based on mission and goals, as well as facility audits. If, for example, the statement, 'being close to the customer' is part of global organization's mission, the market for its products has to be analyzed in order to identify possible locations for regional headquarters or subsidiaries.

Feedback Loop I—Effectiveness Review
Outcomes of strategic planning are reviewed in terms of their effectiveness, relating to the specific 'big issue' categories of a given organization that match its mission and goals, such as corporate symbolism and image, visibility, innovative technology, flexibility and adaptive re-use, initial capital cost, operating and maintenance cost, and costs of replacement and recycling. Described by Schramm (2005), a possible outcome of Phase I and its feedback loop is the fact that a building may not be required at all.

Phase II—Programming
Once strategic planning, cost estimating and budgeting has occurred, a building project is a reality and programming can begin. Over the past 40 years, programming has become a required step in building delivery. It is outlined in the standard American Institute of Architects (AIA) Handbook (2013), is part of the Royal Institute of British Architects' (RIBA) 'Plan of Work' (2013), and described by the German Institute of Standardization (Deutsches Institut für Normung, DIN) in the new DIN 18205 'Brief for Building Design' (2015). A building program or brief documents outline the client's needs, aims, resources, and context for the project. Programming takes place between key stakeholders, including representatives of future building occupants in consultation with building specialists.

Feedback Loop II—Program Review
At the end of this phase, program review involves the client, the programmer, and representatives of occupant groups. This allows the project participants to reflect on the program document containing performance criteria and other outcomes of strategic planning. The review process allows the program to be evaluated step-by-step and to be modified in response to requirements or new priorities, which

might have emerged as part of the planning and programming process (Marmot et al. 2005).

Phase III—Design

This phase includes schematic design, design development, and construction documents. In the design process, schematic design is the initial phase of building design, during which a range of alternative solutions are developed, translating the programming parameters into one or more broad-brush building solutions. Design development is the second stage of building design, wherein one of the alternatives is chosen and elaborated on in order to address the program in more detail. Finally, construction contract documents are produced for the selected design.

Feedback Loop III—Design Review

The design phase has evaluative loops in the form of design review, or 'troubleshooting', involving the architect, the programmer, and client and/or user representative(s). The development of computer-aided design (CAD) techniques and building information modeling (BIM) makes it possible to evaluate solutions during the earliest phases of design. This allows designers to consider the effects of design decisions from various perspectives, while it is still not too late to make modifications. The goals of the organization, as well as its specific program requirements, provide evaluation criteria against which the programmer, client, and users can judge the building design as it develops. In areas where the design fails to meet program requirements, the client decides if the relevant program parameters, such as budget or functionality, need to be modified. Program tracking is therefore essential in order for subsequent, program-based design review to be realistic, useful, and effective (Vischer 2005).

Phase IV—Construction

Once design review has occurred with satisfactory outcomes, building construction can begin. The program, working drawings, and written construction documents are all part of the building contract, and they describe the expected performance of the future building in detail. In this phase, construction managers and architects share in construction administration and quality control to assure contractual compliance. In addition, national standards and codes, as well as local regulations need to be met, including quality standards or safety regulations. Failure to complete the previous phases can result in unforeseen change orders during construction, as some new requirement is identified or budgetary constraints imposed. Responding to change orders can substantially alter the cost of building construction.

Feedback Loop IV—Commissioning

At the end of the construction phase, inspections take place, which result in 'punch lists'; that is, items that need to be completed prior to acceptance and occupancy of the building by the client. As a formal and systematic review process, this loop is intended to insure that owners' expectations, as well as obligatory standards and norms, are met in the constructed building. This feedback loop is a 'reality check': it ensures that the builder fulfils his contract and that specific building performance criteria are made explicit, as well as compliance with relevant standards and norms is achieved (Holtz 2005). This process is specifically important for the building user or owner with smart buildings (see Sect. 1.3.3) to ensure that the facilities manager is properly informed on how to operate integrative and 'smart' building systems.

Phase V—Occupancy

In temporal terms, this phase is the longest of all those described. In fact, most analytic approaches to building delivery end at move-in. However, the BPE approach, with its reliance on feedback and evaluation, maintains a long-term perspective by including the period of occupancy in order to improve the quality of decisions made during earlier phases. While the earlier sub-phases normally last a couple of months, occupancy may last 10—50 years, depending on building type. To occupy a building is the original goal of a client when they decide on a building project. Although strategic planning, facility programming, and design are important phases in the quest to realize a building, it is only at move-in that the client obtains the architectural solution to the initial problem.

Feedback Loop V—Post-Occupancy Evaluation

During this phase, BPE is activated in the form of POEs that provide feedback on what works in the facility and what needs improvement. POEs also test some of the hypotheses behind key decisions made in the programming and design phases. Alternatively, POE results can be used to identify issues and problems in the performance of occupied buildings and to identify ways to solve these. Moreover, POEs are ideally carried out at regular intervals, that is, in 2- to 5-year cycles, especially in organizations with repetitive building programs, such as school districts and federal government agencies (Bordass and Leaman 2005).

Phase VI—Adaptive Re-use/Recycling

In many cases, recycling buildings for similar or quite different uses towards the end of their useful life has become quite common: warehouse lofts are often converted to artists' studios and apartments, and railway stations transformed into shopping centers, museums, and other functions of various kinds, for example. Such major use changes are as dramatic as constructing a new building. Even if building use does not change, building interiors are changing constantly throughout the lifetime of a building. The question of how well a building adapts and can be recycled is very important, not only in the sense of sustainable building practices, but also in the sense of adaptation to new uses (Preiser et al. 2017). The end of this phase constitutes the end of the useful life of a building, e.g., when the building is decommissioned, re-used, or demolished. In cases where construction and demolition waste reduction practices are in place, building materials with potential for re-use will be sorted and recycled into new products, and hazardous materials removed.

Feedback Loop VI—Market/Needs Analysis

This loop involves evaluating the market for the building type in question in terms of a prospective client organization's needs. It can mean assessing the rehabilitation potential of an abandoned or stripped-down building shell, or the potential of a prospective site in terms of future needs. Thus, in the BPE process model, the end point of this evolutionary cycle is also the beginning point of the next building life cycle.

1.2.2 The Performance Concept: Addressing Fitness for Use

As pointed out before, 'design' and 'construction' are important phases located between 'programming' and 'occupancy'. With their design, the architect gives an architectural solution to the problem analyzed during programming. Then, construction is the three-dimensional implementation of the architectural solution into a physical object. This object has to provide everything the client and users are looking for: its performance must be right!

The 'performance concept', focusing especially on the building industry, is the basis for the BPE process model. It evolved in the 1960s (Eberhard 1965), and became topic of the brand new international standard ISO 19208 *Framework for Specifying Performance in Buildings* (ISO 2016): "Over the last decade, this concept has been expanded to be used to address the beneficial or adverse impact of choices made regarding building materials, construction methods and resources, operating energy, water services and sanitary systems on economic conditions, the environment, a society or the quality of life, i.e., the contribution which a building makes to sustainable development" (p. V). The task of this standard is not only to present the necessary framework and principles to describe the performance of a building, but also "to provide the means of evaluation for solutions for all these applications" (ISO 2016, p. V).

As building performance evaluations evolved over the years, they continued to be evaluated in an informational manor. Due to their informal nature, the lessons learned were not necessarily applied in the next building or made to be public knowledge as a resource for similar building types. Although because of relatively slow change in the evolution of building types in the past, knowledge about their performance could be passed on from generation to generation of building specialists. Often individual craftspeople with multiple skills, i.e., artists, designers, draftsmen, and builders, had more control over the building delivery process than the members of multi-disciplinary teams of experts that are typically involved in buildings today. The emergence of new professions means increasing specialization; for example, project management focuses primarily on the building delivery process, and facility management focuses on operating the building over its lifetime. In order to achieve a successful building project and also a building object of high quality, project managers and facility managers may also take over the role of the evaluator, assessing the building performance throughout the building life cycle, thus assuring the expected quality of the facility.

The evaluation of performance in terms of quality assurance of a product is well-known in other industries. In the automotive industry, for example, every component, every sub-process, and every car is evaluated against specified performance criteria. The same is true for BPE. To be objective, actual performance throughout the building life cycle is measured against established performance criteria, taken from the literature, analogs and precedents, experts, codes, or other sources. The key aspect of performance criteria is that they constitute objective, quantifiable, and measurable 'hard' data, as opposed to 'soft' criteria, which derived from qualitative and often subjective assessments. Thus the BPE model is based on a feedback system comparing explicitly stated performance criteria with the actual, measured performance of a building.

Outcomes of BPEs vary according to short-, medium-, and long-term time-frames. Short-term outcomes include user feedback on problems in building performance within a specific sub-phase of the building life cycle, and identification of appropriate solutions. In order to assure the expected performance and quality, short-term outcomes are aimed at immediate implementation, with a limited budget and only minor adjustments. Medium-term outcomes include the application of the positive and negative lessons learned to inform subsequent phases within a building's life cycle, as well as the next building life cycle. This is especially useful for large organizations with recurring building types and programs. Long-term outcomes are aimed at the creation of databases, clearinghouses, and the generation of planning and design criteria for specific building types.

1.3 Validity of the BPE Concept

1.3.1 Applications Around the World

This book verifies the validity of the concept of BPE with examples from around the world. With contributions from eight countries and four continents, this volume expands the global perspective already characterized in the two former BPE books: *Assessing Building Performance* (Preiser and Vischer 2005) and *Enhancing Building Performance* (Mallory-Hill et al. 2012). All chapters are written by international experts to demonstrate the theories and advances in the field, as well as how to apply BPE in order to improve building performance. In addition, case studies relevant to a range of cultural contexts provide specific examples on how BPE methods and instruments are used in the field.

Many chapters are based on papers presented at IBPE-symposia (International Building Performance Evaluation). These symposia started in the 1990s at conferences of the Environmental Design Research Association (EDRA) and the International Association for People-Environment Studies (IAPS) in the United States and Europe, attended by a global audience.

1.3.2 Contributions to Sustainable Development

Around the early 2000s, amid growing concerns about the environment, a number of new voluntary third-party rating systems specifically aimed at green buildings emerged, including the Leadership in Energy and Environmental Design (LEED) system in the US, Building Research Establishment Environmental Assessment Method (BREEAM) in the UK, the National Australian Built Environment Rating System (NABERS) in Australia, and the German Seal of Quality for Sustainable Building of the German Sustainable Building Council (Deutsche Gesellschaft für Nachhaltiges Bauen, DGNB) in Germany. The World Green Building Council, formed in 2002, brings together member-based councils in over 70 countries "to create green buildings for everyone, everywhere—enabling people to thrive both today and tomorrow" (WorldGBC 2017). Green building councils promote sustainable design through green building rating systems that reward buildings different levels of certification when evaluated against an extensive set of sustainable design target performance criteria. Several contributions in this book focus on green buildings.

1.3.3 Improvements in Building Process and Quality

Integrative Planning

Integrative planning is basic for the success of a building project. It is only when all relevant participants consistently and comprehensively coordinate their work for a project, from the very beginning, that the process and the building itself will improve significantly. Therefore, an interdisciplinary team, together with the client and in participation with users and the public, works on a holistic building concept within an overarching planning strategy and on basis of a well-defined facility programme. These exact aspects, i.e., cooperative planning in a team, user participation, and facility programming, are mirrored in the results and recommendations for action the German Reform Commission 'Construction of Major Projects' highlighted in their final report in June 2015 (BMVI 2015). This commission was founded by the Federal Ministry of Building and Urban Development in view of major projects, such as the International Airport in Berlin, that are not delivered within budget or on schedule and exhibit distinct signs of procedures that do not represent value for money. In summary, the BPE process model emphasizes a process with interdisciplinary teams, it is user-oriented, and refers to programming as one of the major six phases of the building life cycle. Case studies in this volume cover these issues.

Building Information Modeling (BIM)

Many benefits and accomplishments of BPE are made possible and easier through the development of technology. Building information modeling (BIM) is now widely used in the architecture and building profession. This technology allows architects and designers to draw their buildings in 3D and to share these computer models with all members of the interdisciplinary team for coordination prior to construction. For example, client goals and user requirements translated by the programmer into performance criteria could be evaluated in the digital 3D-solution of BIM. Computer modeling highlights any conflicts prior to construction, which in turn streamlines the construction process by reducing the amount of questions in the field, e.g., requests for information, that typically pertain to details left off the drawings and/or conflicts shown between trades. Again, the German Reform Commission 'Construction of Major Projects' already mentioned above is recommending "the use of IT-based methods such as building information modeling (BIM), which can help to prevent such planning errors, [...] resulting in cost-intensive corrections" (BMVI 2015, p. 1). This issue is reflected in this book as well.

Smart Buildings
Other opportunities for feedback and feedforward information is through the development of smart buildings. These buildings contain highly advanced systems controlling heating, venting, air conditioning, shades, lighting, security systems, etc. On the one hand, they are able to collect data on how the building is used, contributing to the evaluation of the occupied building against planning parameters and simulation models. This information can then be used by architects and designers when developing similar structures. On the other hand, feedback by the occupants of smart buildings is most sensitive and critical, since they risk to be dominated by the building automation system—without any means of control. This issue is covered in this book as well.

1.4 Conclusion

The performance concept and framework for systematic evaluation of the built environment as outlined in this book is a much needed and timely methodological approach toward achieving higher quality in buildings, accountability in the building process, and ultimately, better building utilization and user satisfaction. Making explicit the performance requirements that are expected from a building, designing a building accordingly, and eventually comparing the actual performance of the building with that which was initially stated in the building program is the basis of the performance concept advocated for use in BPEs.

The goal of this second edition, of *Building Performance Evaluation*: *From Delivery Process to Life Cycle Phases*, is to update several original chapters with more recent findings to bring those theories up-to-date, while balancing those chapters with some new voices in the field, since the book's first publication. As such, the book is divided into Frontiers of Building Evaluation, Advances in Evaluation Knowledge, Advances in Evaluation Methods, and the Epilogue: Looking at the Influences of Wolfgang F.E. Preiser and the legacy that he's left behind. In summary, *Building Performance Evaluation* presents the past, the present, and the future.

Acknowledgements This chapter was updated from previously published versions: Preiser and Schramm (2005, 2012), and Preiser and Hardy (2015).

References

American Institute of Architects. (2013). *The Architect's handbook of professional practice*. Hoboken, NJ: Wiley.

Bordass, B., & Leaman, A. (2005). Phase 5: Occupancy—post-occupancy evaluation. In W. Preiser & J. Vischer (Eds.), *Assessing building performance* (pp. 72–79). Oxford: Elsevier.

Building Research Board. (1987). *Post-occupancy evaluation practices in the building process: Opportunities for improvement*. Washington, DC: National Academy Press.

Bundesministerium für Verkehr und digitale Infrastruktur (BMVI) (Ed.). (2015). *Reformkommission Bau von Großprojekten*. Berlin: BMVI.

Connell, B., & Ostrander, E. (1976a). *Methodological considerations in post-occupancy evaluation: An appraisal of the state of the art*. Washington, DC: AIA Research Corporation.

Connell, B., & Ostrander, E. (1976b). *Post-occupancy evaluation of mail delivery and communications in bachelor enlisted housing*. Washington, DC: AIA Research Corporation.

Deutsches Institut für Normung (DIN). (2016). *DIN 18205—Bedarfsplanung im Bauwesen—Brief for building design—Programme de conception dans l'industrie du bâtiment*. Berlin: Beuth.

Eberhard, J. (1965). Horizons for the performance concept in building. In P. Will (Ed.), *Proceedings of the symposium on the performance concept in building* (pp. 93–98). Washington, DC: Building Research Advisory Board, National Academy of Sciences, National Research Council.

Hershberger, B., & Smith, M. (2017). Predesign planning for flexibility. In W. F. E. Preiser, A. E. Hardy, & J. J. Wilhelm (Eds.), *Adaptive architecture: Changing parameters and practice*. London: Routledge.

Holtz, M. (2005). Phase 4: Construction—commissioning. In W. Preiser & J. Vischer (Eds.), *Assessing building performance* (pp. 62–71). Oxford: Elsevier.

Hsia, V. (1967). *Residence hall environments: An architectural psychology case study*. Salt Lake City, UT: University of Utah.

International Organization for Standardization (ISO). (2016). *ISO 19208 framework for specifying performance in buildings*. Geneva: ISO.

Mallory-Hill, S., Preiser, W., & Watson, C. (Eds.). (2012). *Enhancing building performance*. Oxford: Wiley-Blackwell.

Marmot, A., Eley, J., & Bradley, S. (2005). Phase 2: Programming/briefing—Programme review. In W. Preiser & J. Vischer (Eds.), *Assessing building performance* (pp. 39–51). Oxford: Elsevier.

McLaughlin, H. (1975). Post-occupancy evaluation of hospitals. *AIA Journal, 1,* 30–34.

McLaughlin, H. (1997). Post-occupancy evaluations: They show us what works, and what doesn't. *Architectural Record, 8,* 14.

Preiser, W. (1969). Behavioral design criteria in student housing. In H. Sanoff & S. Cohn (Eds.), *Proceedings of the first environmental design research association conference* (pp. 243–259). Raleigh, NC: North Carolina State University.

Preiser, W. (Ed.). (1989). *Building evaluation*. New York, NY: Plenum Press.

Preiser, W., & Hardy, A. (2015). Historical review of building performance evaluation. In W. Preiser, A. Davis, A. Salama, & A. Hardy (Eds.), *Architecture beyond criticism: Expert judgment and performance evaluation* (pp. 147–159). London: Routledge.

Preiser, W., & Schramm, U. (1997). Building performance evaluation. In D. Watson et al. (Eds.), *Time-saver standards for architectural data* (pp. 233–238). New York, NY: McGraw-Hill.
Preiser, W., & Schramm, U. (2005). A conceptual framework for building performance evaluation. In W. Preiser & J. Vischer (Eds.), *Assessing building performance* (pp. 15–26). Oxford: Elsevier.
Preiser, W., & Schramm, U. (2012). A process model for building performance evaluation (BPE). In S. Mallory-Hill, W. Preiser, & C. Watson (Eds.), *Enhancing building performance* (pp. 19–31). Oxford: Wiley-Blackwell.
Preiser, W., & Vischer, J. (Eds.). (2005). *Assessing building performance*. Oxford: Elsevier.
Preiser, W., Hardy, A., & Wilhelm, J. (Eds.). (2017). *Adaptive architecture: Changing parameters and practice*. London: Routledge.
Preiser, W., Rabinowitz, H., & White, E. (1988). *Post-occupancy evaluation*. New York, NY: Van Nostrand Reinhold. Republished in 2015, Routledge: London.
Royal Institute of British Architects (RIBA). (2013). *Plan of work*. London: RIBA Publications.
Schramm, U. (2005). Phase 1: Strategic planning—Effectiveness review. In W. Preiser & J. Vischer (Eds.), *Assessing building performance* (pp. 29–38). Oxford: Elsevier.
Van der Ryn, S., & Silverstein, M. (1967). *Dorms at Berkeley: An environmental analysis*. Berkeley, CA: Environmental Facilities Laboratories, University of California.
Vischer, J. (2005). Phase 3: Design—Design review. In W. Preiser & J. Vischer (Eds.), *Assessing building performance* (pp. 80–90). Oxford: Elsevier.
World Green Building Council (WorldGBC). (2017). Our mission. Retrieved from http://www.worldgbc.org/our-mission. Accessed on 8 Feb 2017.

Author Biographies

Wolfgang F.E. Preiser held a Ph.D. in Man-Environment Relations from Penn State, and several architecture degrees from Virginia Tech, Karlsruhe Tech (Germany) and Vienna Tech (Austria). He had over 40 years of experience in teaching, research and consulting in the evaluation and programming of environments, including health care facilities, public housing, public libraries, cross-cultural and universal design, as well as design research in general. He published 20 books and over 130 chapters, monographs, and articles. Most recent books include: *Architecture Beyond Criticism: Expert Judgment and Performance Evaluation* (Routledge, 2015); *Enhancing Building Performance* (Wiley, 2012); *Universal Design Handbook* (McGraw-Hill, 2010); *Designing for Designers: Lessons Learned From Schools of Architecture* (Fairchild, 2007). Preiser lectured worldwide at 69 venues and conferences in the United States and Canada, as well as 86 overseas. He served on national committees with the American Institute of Architects, the Building Research Board of the National Academy of Sciences, and the National Institute for Disability Rehabilitation Research. Preiser received many awards, including: 2 Progressive Architecture Awards; 2 Professional Fellowships from the National Endowment for the Arts; The Career Award from the Environmental Design Research Association (EDRA), 2 EDRA Achievement Awards, as well as other awards while at the University of Cincinnati.

Andrea E. Hardy Creo Architects, holds a Master's degree in Architecture from Arizona State University (2012), and a Bachelor of Science in Architectural Engineering Technology from Wentworth Institute of Technology (2007). She has been working in architecture offices for the past twelve years both during and between obtaining her degrees. After working professionally in Boston and Phoenix, and studying public architecture through Arizona State University in Buenos Aires, Argentina, Hardy is currently an Associate and Registered Architect working at Creo Architects in Phoenix, Arizona. While at Arizona State University, Hardy was a member of the American Institute of Architecture Students, served one term as Secretary for the American

Institute of Architecture Students, was a teaching assistant for a design studio and history class, received multiple scholarships, and participated in non-academic activities such as working on multiple design competitions, workshops, and volunteer work. Recently, Hardy was published as co-editor and co-author in *Architecture Beyond Criticism: Expert Judgment and Performance Evaluation* (2015), contributed to updating *Improving Building Performance* NCARB Monograph series (2016), and is co-editor and author in the forthcoming *Adaptive Architecture: Changing Parameters and Practice* (in print).

Ulrich Schramm is Professor in the Department of Architecture and Civil Engineering at the Bielefeld University of Applied Sciences, Minden Campus, in Germany. His appointment includes teaching and research responsibilities in the field of facility programming, building performance evaluation and building technologies. He is a trained architect and an experienced facility programmer. He received his doctorate in Architecture from the University of Stuttgart and a post-doctoral fellowship from the German Research Foundation (Deutsche Forschungsgemeinschaft, DFG) for his stay at the University of Cincinnati as Visiting Professor of Architecture. Results of his studies within the International Building Performance Evaluation (IBPE) consortium and the campus-based research project Intelligent Building Technologies (InteG-F) have been presented at conferences of the Environmental Design Research Association (EDRA) since 1995 and published in several articles and book chapters. Some recent chapters he has written were published in *Enhancing Building Performance* (Wiley-Blackwell, 2012) and *Architecture beyond Criticism: Expert Judgment and Performance Evaluation* (Routledge, 2015). He also co-authored the first book on Facility Programming in Germany *Nutzerorientierte Bedarfsplanung: Prozessqualität für nachhaltige Gebäude* (Springer, 2011). Currently, he is serving at the German Institute for Standardization (Deutsches Institut für Normung, DIN) in Berlin as expert and co-chair of the revised edition of the German building standard DIN 18205 *Brief for Building Design*.

Part II
Frontiers of Building Evaluation

Preamble

Nigel Oseland

With the second chapter, the author starts this section of the book by exploring how to overcome the barriers to uptake of Post-Occupancy Evaluation (POE). The author also explores how POE may be conducted in the future and become a shorter, ongoing, and integrated process rather than a one-off stand-alone exercise. Currently, POE appears to be more likely carried out pre-project as a part of the briefing process rather than just used post-project for understanding lessons learned; this represents a fundamental shift from POE to Building Performance Evaluation (BPE).

In the third chapter, Vivian Loftness and colleagues propose that in addition to using occupants as sensors, providing "integrated multi-sensory evaluation", field measurements are also made. Using feedback and measurements is referred to as POE+M and the chapter outlines the additional benefits of such an approach. Quite often studies adopting POE+M will focus on one environmental parameter but Loftness et al. suggest a more holistic approach and recommend that six environmental parameters are recorded: spatial, thermal, acoustic, visual, and air quality along with long-term building integrity against degradation. The field measurements must be considered in alongside psychological, psychological, and sociological variables and their combined effect on satisfaction, productivity, and health.

In the next chapter Persky, Farbstein and Farling cover the courthouse, an environment that tends to be evaluated less than other workplaces. The authors highlight the wide variety and number of stakeholders in courthouses, each with their own specific requirements: building owner, court staff, judges, litigants, attorneys, public visitors, jurors, and inmates. A toolkit for evaluating courthouses is proposed, one sufficiently flexible to apply to the different occupants. In addition to feedback the toolkit includes other pertinent measures such as maintenance requirements and technical performance plus safety and security. This chapter also highlights the importance of sharing the POE results.

The fifth chapter by Jennifer Senick and her team introduce the reader to the concept of synthetic populations. Traditionally synthetic populations, simulated people (or agents), are used in computer models in fields such as transportation planning and public health. Whereas computer models are increasingly used in the building design process, few incorporate the impact of occupant behaviour. The authors explain how small POE datasets can be combined to create larger synthetic populations. In turn, the synthetic populations may be used by researchers and building designers to develop more accurate models of performance and behaviour.

Schramm, Reichart, and Becking draw our attention to intelligent building technologies. They use a mix of pre-occupancy (pre-build) and post-occupancy evaluations to understand the lack of acceptance of smart technology in offices and lack of uptake in homes. The authors found that there is a high expectation of intelligent buildings, uncontrollable technology may decrease wellbeing, and lack of control is a cause of resistance to uptake. The chapter highlights the importance of BPE and evaluating the building from the planning to post-occupancy stages.

In chapter seven, Francescato, Weidemann, and Anderson revisit their original chapter in an earlier book edited by Wolfgang Preiser—*Building Evaluation* (1989). They propose that satisfaction scales alone should not be used for evaluating buildings. Satisfaction is multifactorial, it is an attitude and dependent upon who the respondent is, their relation to the building and the amount of time they spend in the building. "Environments are systems with multiple stakeholders, hence with multiple objectives … Therefore, they must be evaluated using multiple criteria".

In the final chapter of this section of the book, Wolfgang Preiser and the Petronis's revisit their chapter in *Building Evaluation* (1989). They introduce the reader to the concept of the Activation Process Model (APM) and demonstrate how it expands the scope of POE. Activation is a process of preparing people and a facility for move-in and operation. It is little understood but pertinent to the success of the buildings, in particular complex building such as hospitals.

Chapter 2
From POE to BPE: The Next Era

Nigel Oseland

2.1 The Future of POE

This second chapter in the second edition of *Building Performance Evaluation* provides an overview of how Post-Occupancy Evaluation (POE) evolved into building performance evaluation (BPE) and how it may evolve even further in the near future. The main focus of this chapter is on how to improve the recognition and uptake of BPE across the design and construction industry and, in particular, how to reposition POE and overcome the barriers to application. Whilst, (1) the benefits to conducting a POE are regularly demonstrated, (2) POE is required by some public institutions and (3) technology offers ease of collating feedback, there is nevertheless reticence in carrying out such evaluations post-project. POE now appears to be more likely carried out pre-project as a part of the briefing process, rather than used throughout the project, for building life cycle analysis or for understanding lessons learned. This represents a fundamental shift from POE to BPE. This chapter explores the development that BPE will need to make to encourage wider uptake, whether client portals and social media platforms make a difference, and are standards or best practice guidance driving POE uptake. In essence, it covers the future prospects for POE and BPE.

2.2 Barriers to Uptake of POE

In the workplace arena, it is not uncommon for conferences to include case studies of new office fit-outs. As a consequence, there are many presentations showcasing new office designs and the story behind how they were created. The case studies are

N. Oseland (✉)
Workplace Unlimited, PO Box 953, Berkhamsted, Hertfordshire HP4 IZN, UK
e-mail: oseland@workplaceunlimited.com

W.F.E. Preiser et al. (eds.), *Building Performance Evaluation*,
DOI 10.1007/978-3-319-56862-1_2

usually presented as success stories; and thus, any lessons learned are usually not shared, and typically there is little systematic data to substantiate their success.

For example, in a two-day conference on smart working in mainland Europe, some 12 case studies of new workplaces were presented but only two of them included occupant feedback or any robust post-project data. Whilst this evidence may be considered anecdotal, this level of reporting is fairly representative of the current understanding and uptake of POE across Europe.

Bearing in mind the importance of POE addressed in the following chapters, a key question is why there is still minimal uptake of POE by the workplace community, and what are the barriers to uptake? In the POE training courses regularly delivered by this chapter's author, the three barriers most often identified are:

1. *Payment*—Quite often it is perceived, particularly with a one-off project, that the designer rather than the occupier benefits from carrying out a POE. Hence, the occupier is not inclined to pay for such analysis. In reality, the occupier benefits by knowing whether they met the project objectives, fulfilled the needs of their workforce, and delivered value. In most organisations today, there is an expectation that the project lead and design team deliver value and prove it, sometimes referred to as benefits realisation. The many benefits of conducting a POE need to be clearly explained to the occupier but the designer does, of course, benefit hugely from the POE. The chapter author worked for an architectural practice many years ago and offered POE as a value-add service. The architects found that their clients not only appreciated their interest but then also commissioned them to carry out more design work. Furthermore, the post-project evaluation was a small expense, as usually an occupant survey was carried as part of the briefing process and a similar survey was used post-project.
2. *Reputation*—Designers and architects may be concerned that the POE will raise issues with the design that may in turn affect their reputation. If the project genuinely fails, then it needs to be flagged as the occupants' lives will be affected indefinitely, not just the reputation of the designer. However, the POE is usually carried out six–twelve months after project completion and after the defect liability period and the commissioning phase. The POE relates more to how the building supports the occupying business and depends on the brief, how the business uses the building and other factors outside the control of the designer. As mentioned above, the authors own experience is that a POE is usually greatly appreciated by the occupier and seen as a joint learning, rather than blaming, opportunity.
3. *Expectation*—It is often voiced that POE may lead to an expectation that further changes will be made after the project budget is spent. Again, the author's own experience is that POE can identify behavioural changes with no additional capital costs or identify other quick wins with minimal costs. It is more important to share the findings of the POE and if any issues are uncovered then explain how they may be resolved in the future. Also, it is important to provide a balanced POE reporting the successes and positive benefits of the project and phrasing criticism as recommendations.

The main point is that regardless of the barriers to uptake there is always a means of overcoming them. The lack of uptake is more likely to be a lack of understanding and awareness of the benefits and importance of POE. As ambassadors of POE that action lies with the workplace strategy and design community.

2.3 POE Influencers

In contrast to the above lack of reporting of POEs, there appears to be a good uptake on training in it. Back in 2007, the author of this chapter was the lead author of the *British Council for Offices Guide to Post-Occupancy Evaluation* (Oseland 2007), and since then, he has run one-day courses on how to conduct POEs. If anything, the course is more popular now than ever and attended by designers, engineers, facilities managers, and representatives from educational institutions. Quite often the participants attend the course to verify that they are doing their own POEs properly rather than to learn how to conduct a POE *per se*.

In the UK, the uptake of "underground" (in-house) POE is partly due to a number of influencing bodies that are raising awareness and making POE more accessible:

- *RIBA*—The Royal Institute of British Architects (RIBA) original *Plan of Work*, dating back to 1963, included *Stage M Feedback* but it was later dropped. However, the new *Plan of Work 2013* includes *Stage 7 In Use,* which suggests "Conduct activities listed in Handover Strategy including Post-occupancy Evaluation, review of Project Performance, Project Outcomes and Research and Development aspects" (RIBA 2013).
- *Government*—The Office of Government Commerce (OGC), part of HM Treasury, introduced annual evaluations of all central government buildings along with Post-Implementation Reviews of all new building projects as part of the *OGC Gateway Process Review 5*. More recently, the Cabinet Office introduced the *Government Soft Landings* (GSL) policy alongside *Building Information Modelling* (BIM) for all central government departments. GSL includes annual POEs for 3 years to ensure the facility meets performance targets.
- *Education*—The Higher Education Funding Council for England (HEFCE) and Association of University Directors of Estates (AUDE) introduced their best practice *Guide to Post Occupancy Evaluation* in 2006. Since then the Skills Funding Agency (SFA) requires a post-occupancy evaluation as a condition of consent following a capital development.
- *World Green Building Council (WGBC)*—A campaign for health, wellbeing, and productivity in offices. The WGBC are promoting a framework for collating evidence and data on productivity gained through POE.
- *Leesman Index*—There are many standardised POE methodologies in the marketplace, but since its launch in 2010, the Leesman Index is the one growing

most rapidly. The Leesman team have now collated over 262,000 responses from more than 2,000 workplaces in 67 different countries. Not only is the database growing but the founder Tim Oldman is a regular and passionate speaker on the importance of occupant feedback in creating better workplaces.

So based on the author's experience of POE training and based on the awareness raised by the above bodies, it appears that POE is still very much on the agenda of many organisations in the UK. The conclusion is therefore that POEs are being carried out, but just for "personal consumption" and not shared amongst the workplace community. It is possible that POE offers some form of competitive advantage and perhaps the instigators simply do not want to dilute that advantage? Whilst it is good that POEs are being carried out, the workplace industry nevertheless needs to share results, and they need to know the lessons learned and not just hear a design sales pitch.

2.4 POE Throughout the Project and Building Life-Cycle

The focus of the government, education, and RIBA guidance highlighted above is on post-project reviews. However, the Leesman team explain that "many of our clients use the survey findings to establish the business case for a major strategic project" (Leesman 2015), i.e., conduct pre-project and post-project evaluations. As mentioned under the barriers to uptake, the chapter author and his architectural colleagues used feedback surveys to readily gather information to inform the design brief, but then repeated the same survey post-project. Pre-project evaluation often includes space analysis and utilisation surveys as well as occupant feedback. The key is to offer the pre- and post-project reviews as a package with the expectation that there will be a follow up survey to conclude the design process.

So whilst many organisations will associate "post-occupancy evaluation" as a post-project assessment of a newly occupied workplace, it is the systematic evaluation of an existing workplace followed by the new workplace that is becoming more prevalent, and offers most value. This is a significant change in the primary use of POE and supports the reported shift from POE to BPE where the building is assessed throughout its life cycle. This approach also offers a more robust analysis and more meaningful comparison to an isolated one-off post-project review.

The Leesman team also reported that "once a project has started, the data finds new value, evaluating readiness for change, engaging employees in the change management process" (Leesman 2015). In the workplace arena, change management is quite often seen as a bolt-on service, starting once the design has been agreed, meaning it is then considered the time to convince staff of the benefits of their new workplace. But of course, the change process actually commences with the initial engagement and briefing stage of the project. Back in the 2007, in the *BCO Guide to Post Occupancy Evaluation,* it was advised that POE is used not only for (1) measuring project success, and (2) setting a baseline for measurement, but also for (3) informing the design process, and (4) inputting to change management and communications. Standard POE methodologies, e.g., on-line surveys, interviews, and workshops, are all great tools for gathering occupant insight and initiating the change process.

There is less evidence to support that POE is used for ongoing, such as annual, assessments. Nevertheless, there are occasional reports of a few key questions on the workplace being added to the annual staff satisfaction and culture surveys. This may change as the nature of how the design industry collects feedback and evaluates the ever-changing workplace.

2.5 New Means of Data Collation

Ongoing and real-time feedback is becoming more prevalent in, for example, coffee shops, doctor's waiting rooms, hotels, and even public toilets. The feedback may be gathered through cards/forms, electronically through kiosks, or push-button systems. In 2005, the *Bop*! Project (Wilson 2006) took the idea of instant feedback one step further. *Bop*! was a Department of Trade and Industry (DTI) funded project in which Central Saint Martins, Imperial College, Arup, and others studied how ubiquitous wireless sensors could be deployed to monitor the use of and satisfaction with buildings. The project developed an array of sensors and quirky interactive devices to collect movement around and feedback on the workplace. The author's favourite notable technique was two pressure sensitive mats, one with "yes" and one with "no" printed on them. A different question was posed above the mats each day and people voted as they entered or left the building. Whilst *Bop*! itself may not have progressed, this initiative did lay the foundation for using technology to collect instant feedback. Whilst such data is less detailed, it is quick, accessible, in real-time, and likely to improve response rates. It does mean collating lots of small bits of feedback rather than one-off surveys with detailed feedback, so a slightly different approach to analysis is required.

More progress has been developed in the use of sensors to monitor the utilisation of the workspace. Utilisation studies use observers to provide a one or two week in-depth study of how space is utilised over time. Embedded passive infrared (PIR) sensors allow the occupancy levels of the various spaces to be continuously monitored over time. There are now several cost-effective systems in the market place. Some of the larger corporate occupiers have even adopted them to monitor their space in real-time to help ensure maximum utilisation. Sensors to monitor the environmental conditions, e.g., temperature, humidity, and sound levels, are also becoming more common place. Notably *The Edge*, the new Deloitte Amsterdam headquarters, is seamlessly fitted throughout with sensors to monitor and adjust the environmental conditions and occupancy for both comfort and sustainability (Randall 2015). In due course, as technology progresses, real-time monitoring may replace pre- and post-project measurements for those organisations who wish to proactively manage their workplaces.

Users of on-line and app based companies such as Uber and Airbnb are expected to make reviews after using the service; and the data is then used to build cross-platform customer profiles. When purchasing on-line or downloading apps or documents it is not uncommon to answer questions to unlock the download. *CAPTCHA* (Completely Automated Public Turing test to tell Computers and Humans Apart) is an on-line device for checking human interaction and deterring automated inputs (robots). However, *CAPTCHA* is also collating data on behalf of Google and, for example, is serving as a benchmark task for artificial intelligence technologies. This all indicates that there has been a shift towards requesting real-time feedback in return for securing or using a service.

Embedded sensors and on-line feedback indicate that the next step is incorporating embedded monitoring connected to real-time feedback of workplaces. Of course there will still be need for independent expert opinion and critiques, and the set-piece pre- and post-project evaluations will be carried out when more detail is required. However, it is feasible that occupiers looking for more proactive and ongoing feedback will replace lengthy on-line questionnaire surveys with quick automated real-time reviews made by the occupants, those customers experiencing the workplace on a daily basis.

2.6 Conclusion

In the meantime, the workplace design community has a responsibility to promote POE, obtain occupant feedback, and regularly share our results and lessons learned. Back to the Leesman Index, and their survey of over 262,000 occupants, they report that "only 54% of those we've surveyed agree that their workplace design allows them to be productive" (Leesman 2015). Cooper's (2001) comments that without a feedback loop every building, to some extent is a prototype – spaces and systems put together in new ways, with potentially unpredictable outcomes. So whilst architectural creativity should not be stifled, the workplace community also needs to ensure that they do not build workplaces that are reported to not benefit the occupants. The first step to this is, and always has been, honest feedback and considered evaluation integrated into the design process.

References

AUDE/HEFCE. (2006). *Guide to post occupancy evaluation.* The Higher Education Funding Council for England and Association of University Directors of Estates.

Cabinet Office. (2013). *Government Soft Landings (GSL): Section 1—Introduction*. London: Cabinet Office.

Cooper, I. (2001). Post-occupancy evaluation—Where are you? *Building Research and Information, 29*(2), 158–163.

Imperial College. (2005). *Bop!—Making sense of space*. http://www.doc.ic.ac.uk/~asher/ubi/bop/. Accessed on 1 Nov 2016.

Leesman. (2015). *Leesman Review, Issue 19*. http://leesmanindex.com/wp-content/uploads/LEESMAN_Review_19_Interactive-SP.pdf. Accessed on 1 Nov 2016.

OGC. (2007). *OGC Gateway™ process review 5: Operations review and benefits realisation*. London: Office of Government Commerce.

Oseland, N. A. (2007). *British Council for offices guide to post-occupancy evaluation*. London: British Council for Offices.

Preiser, W. F. E. (Ed.). (1989). *Building evaluation*. New York: Plenum Press.

Randall, T. (2015). The smartest building in the world. Bloomberg Businessweek, September 23. www.bloomberg.com/features/2015-the-edge-the-worlds-greenest-building/. Accessed on 1 Nov 2016.

RIBA. (2013). *RIBA Plan of Work 2013*. London: The Royal Institute of British Architects. https://www.ribaplanofwork.com. Accessed on 1 Nov 2016.

SFA. (2014). *Post-occupancy Evaluation Guide*. London: Skills Funding Agency. https://www.gov.uk/government/publications/sfa-capital-funding-evaluation-and-use-of-capital-grant-forms. Accessed on 1 Nov 2016.

WGBC. (2014). *Health, wellbeing & productivity in offices: The next chapter for Green Building*. UK World Green Building Council.

Wilson, D. (2006). BOP: Making sense of space. In *Proceedings of IET wireless sensor networks conference*. London, 4 December 2006.

Author Biography

Dr Nigel Oseland is an environmental psychologist, workplace strategist, researcher, change manager and author with 11 years research and 19 years consulting experience. Nigel is an internationally recognised expert in occupant feedback methods, performance and productivity, agile working, environmental conditions and psychophysics in the workplace.

Nigel authored the British Council for Offices (2007) *Guide to Post-Occupancy Evaluation*. He continues to write articles and guidance, present regularly at international conferences, and he organises the biannual *Workplace Trends* conference and annual *Learning Environments* conference.

Chapter 3
Critical Frameworks for Building Evaluation: User Satisfaction, Environmental Measurements and the Technical Attributes of Building Systems (POE + M)

Vivian Loftness, Volker Hartkopf, Azizan Aziz, Joon-Ho Choi and Jihyun Park

3.1 Synopsis

An integrated approach to building performance evaluation mandates that post-occupancy evaluation subjective tools be matched by metrics (POE + M). While leveraging occupants as sensors to quickly capture indoor environmental quality or IEQ conditions in a work environment is valuable the addition of measured environmental conditions across all variables, and of carefully captured records of critical

V. Loftness (✉)
School of Architecture, Carnegie Mellon University,
Pittsburgh, PA 15213, USA
e-mail: loftness@cmu.edu

J.-H. Choi
Human-Building Integration Research Lab, University of Southern California,
Los Angeles, CA 90089-0291, USA
e-mail: joonhoch@usc.edu

A. Aziz · J. Park
Center for Building Performance and Diagnostics, School of Architecture,
Carnegie Mellon University, Pittsburgh, PA 15213, USA
e-mail: azizan@cmu.edu

J. Park
e-mail: jihp@cmu.edu

V. Hartkopf
Center for Building Performance, Carnegie Mellon University,
Pittsburgh, PA 15213, USA
e-mail: Vh02@andrew.cmu.edu

W.F.E. Preiser et al. (eds.), *Building Performance Evaluation*,
DOI 10.1007/978-3-319-56862-1_3

workplace attributes that define their physical environments, are equally critical to understanding building occupant comfort, satisfaction, health, and performance. With over 15 years of POE + M field measurements in office workplaces, the Center for Building Performance and Diagnostics (CBPD) at Carnegie Mellon University (CMU) has a database of over 1600 workstations with statistically significant findings about the measured and user-perceived quality of the indoor environment, as well as the technical attributes of building systems that contribute to successful, high performance buildings. This chapter provides an overview of the National Environmental Assessment Toolkit (NEAT) developed with the US General Services Administration (GSA), and an array of findings that will catalyze future indoor environmental standards and improve building enclosure, mechanical, lighting, and interior design.

3.2 Total Building Performance Is Critical to Building Evaluation

The need for a manageable yet comprehensive list of performance mandates for designing or evaluating buildings is critical to improving overall environmental quality in today's work environments (BRAB 1988). An emphasis on one performance area, such as energy, can lead to the discovery of failures in other performance areas, such as serious air quality and façade degradation failures, resulting from rapid investments in energy efficient enclosures. Yet building evaluations continue in singular performance areas, e.g., acoustic studies in factories, lighting in offices, and heat loss in old buildings, with recommendations for action that will solve that performance problem—and create three more (Preiser et al. 1988).

Instead, the workplace evaluation community must begin with a comprehensive outline of "total building performance", which is finite enough to be manageable in the field, yet developed enough to represent the "integrated multi-sensory evaluator" known as a human being. The authors of this chapter have built on the efforts of the International Organization for Standardization (ISO), the Chartered Institution of Bulding Services Engineers (CIBSE), the National Institutes of Standards and Technology (NIST), the American Society of Heating, Refrigerating, and Air-Conditioning Engineers (ASHRAE), the Illuminating Engineering Society of North America (IESNA) and others to develop a manageable yet comprehensive list of six performance mandates for the built environment: spatial quality, thermal quality, acoustic quality, visual quality, air quality, and long-term building integrity against degradation (see Fig. 3.1) (Hartkopf et al. 1986).

3.2.1 Defining Total Building Performance

Field evaluation must cover these six performance areas with adequate depth, measuring against physiological, as well as psychological, sociological, and

DEFINING TOTAL BUILDING PERFORMANCE

I. SPATIAL QUALITY = satisfactory:
based on knowledge of the building occupancies, occupancy functions, and organizational structures
   A. Individual Space Layout Quality
   Useable space, furnishings, layout efficiency, access, ergonomics, image, flexiblity/growth
   B. Aggregate Space Layout Quality
   Proximities, access/ wayfinding, layout efficiency, image, amenities, flexibility/growth
   C. Building Siting Layout Quality
   Access, public interface/image, indoor-outdoor relationships,
   outdoor space layout, flexiblity/growth
   D. Quality of Conveniences and Services
   E. Occupancy Factors and Controls

II. THERMAL QUALITY = satisfactory:
   A. Air Temperature
   B. Mean Radiant and extreme radiant temperature
   C. Humidity
   D. Air Speed
   E. Occupancy Factors and Controls

III. AIR QUALITY = satisfactory:
   A. Fresh Air
   B. Fresh Air Distribution
   C. Restriction of Mass Pollution - gases, vapors,
   micro-organisms, fumes, smokes, dusts
   D. Restriction of Energy Pollution - ionizing radiation,
   microwaves, radio waves, light waves, infrared
   E. Occupancy Factors and Controls

IV. ACOUSTIC QUALITY = satisfactory:
   A. Sound Source - Sound Pressure Levels and Frequency
   B. Sound Source - Background Noise
   C. Sound Path - Noise Isolation (air and structureborne)
   D. Sound Path - Sound Distribution; absorption, reflection,
   uniformity, reverberation
   E. Occupancy Factors and Controls

V. VISUAL QUALITY = satisfactory:
   A. Ambient Light Levels - artificial and daylight
   B. Task Light Levels - artificial and daylight
   C. Contrast and Brightness Ratios
   D. Color Rendition
   E. View, visual information
   F. Occupancy Factors and Controls

VI. BUILDING INTEGRITY = satisfactory:
based on knowledge of loads, moisture conditions, temperature
shifts, air movement, radiation conditions, biological attack, manmade and natural disasters
   A. Quality of Mechanical/Structural Properties
   compression, tension, shear, abuse
   B. Quality of Physical/Chemical Properties
   watertightness, airtightness
   transmission, reflection, absorption of heat, light
   and sound energy, fire safety
   C. Visible Properties
   color, texture, finish, form; durability, maintainability

Fig. 3.1 Defining total building performance. *Source* Authors

economic limits of acceptability (see Fig. 3.2). Research efforts, and consequently codes and standards, tend to be focused on the physiological limits of acceptable performance, aimed at ensuring the physical health and safety of the building's

	Physiological Needs	Psychological Needs	Sociological Needs	Economic Needs
	Performance Criteria Specific to Certain Human Scenes, in the Integrated System			
1 Spatial	Ergonomic Comfort Handicap Access Functional Serving	Habitability Beauty, Calm; Excitement, View	Wayfinding, Functional Adjacencies	Space Conservation
2 Thermal	No Numbness, Frostbite; No Drowsiness, Heat Stroke	Healthy Plants Sense of Warmth Individual Control	Flexibility to Dress with the Custom	Energy Conservation
3 Air	Air Purity; No Lung Problems; No Rashes, Cancers	Healthy Plants, Not Closed in, Stuffy No Synthetics	No Irritation From Neighbors; No Smoke, Smells	Energy Conservation
4 Acoustic	No Hearing Damage; Musical Enjoyment, Speech Clarity	Quiet, Soothing; Activity, Excitement "Alive"	Privacy Communication	First-cost Conservation
5 Visual	No Glare, Good Task Illuminance, Wayfinding, No Fatigue	Orientation, Cheerfulness, Calm, Intimate, Spacious, Alive	Status of Window Daylit Office "Sense of Territory"	Energy Conservation
6 Building Integrity	Fire Safety; Struct, Strength + Stability; Weather tightness	Durability, Sense of Stability Image	Status/Appearance Quality of Const. "Craftsmanship"	Material/Labor Conservation
	Performance Criteria General to All Human Sense, in the Integrated System			
	Physical Comfort Health Safety Functional Appropriateness	Psych. Comfort Mental Health Psych. Safety Esthetics Delight	Privacy Security Community Image/Status	Space Conservation Material Conservation Time Conservation Energy Conservation Money/Investment Conservation

Fig. 3.2 Physiological, psychological, sociological and economic assessments of building performance. *Source* Authors

occupants by sheltering basic bodily functions—sight, hearing, breathing, touch, and movement, from wear or destruction over time.

3.2.2 Physiological, Psychological, Sociological, and Economic Assessments of Total Building Performance

In the evaluation of occupied environments, it is also necessary to establish the length of time during which the building or space must perform (Hartkopf et al. 1985). The building evaluator must establish the level of suitability, reliability, and flexibility expected by the clients for the time they expect to remain in the spaces to be evaluated. This framework for evaluating six performance mandates, and their physiological, psychological, sociological, and economic acceptability over time, might be entitled the "Field Evaluation of Total Building Performance."

3.3 Expertise on the Integrated and Occupied Settings Is Critical to Building Evaluation

The second imperative for building evaluation is the need to study performance qualities in the occupied setting. Successes and failures in building environments are not the result of individual systems or components for the most part, but are the result of the effectiveness of those systems or components within their integrated setting (Rush 1986). Consequently, the building evaluation framework must demonstrate expertise in all of the building system areas, e.g., structure, enclosure, mechanical, lighting, interior, the generic choices, and their history of performance in the integrated setting. Trained in architecture, building science, and engineering, the authors of this chapter have created a checklist of technical attributes of building systems (TABS) that are critical to the measured and perceived thermal, acoustic, visual, spatial, and air quality of the integrated and occupied setting (see Sect. 3.4.3).

3.4 POE + Measurement Is Critical to Building Evaluation

User satisfaction studies reveal that there are significant gaps between the design intent and the performance of buildings and systems over both time and occupancy changes (Brager et al. 2000). An integrated approach to building evaluation can only identify possible cause and effect if subjective surveys are matched by field measurements (POE + M).

In partnership with the GSA, the CBPD at Carnegie Mellon University developed a three-prong approach (see Fig. 3.3) to post-occupancy evaluation plus measurements creating the National Environmental Assessment Toolkit (NEAT), in use for over 15 years. On-site measurements for IEQ conditions in occupant workstations are matched by comprehensive records of the attributes of building systems and the "right now" results of user satisfaction questionnaires. Through statistical analysis of study findings, the effects of building physical and environmental conditions on occupancy satisfaction can be determined, and recommendations can be developed to enhance environmental sustainability as well as contribute to occupant satisfaction, comfort, health, and performance.

3.4.1 User Satisfaction Surveys: Right-Now and Year-Round

To correlate occupant satisfaction and short term measured indoor environmental conditions at individual workstations, it is critical to have a shorter "right now"

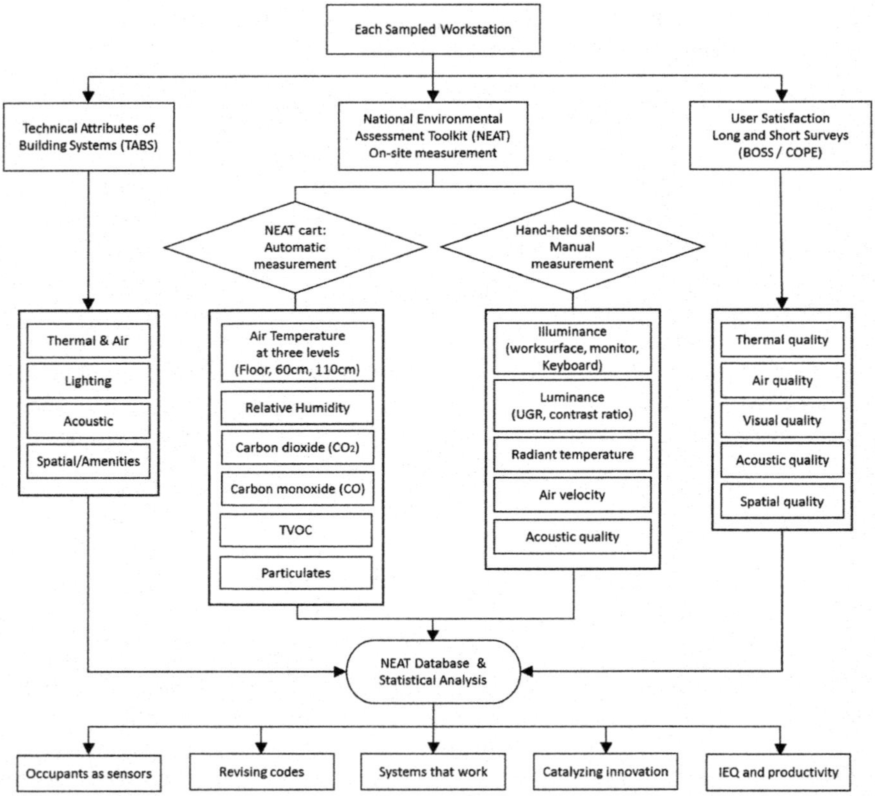

Fig. 3.3 Three-leg approach to building evaluation is key to occupants as sensors, revising codes, systems that work, catalyzing innovation, and linking IEQ and productivity. *Source* Authors

survey of user satisfaction with environmental quality and workplace settings. The NEAT field studies adopted a variation of the Cost-Effective Open-Plan Environments (COPE) short survey developed by Public Works Government Services Canada, for use in IEQ measurement and user satisfaction studies in various open-plan office environments (Veitch et al. 2007). The 25 questions in the COPE2 "right-now" survey assess user satisfaction with subset attributes of all performance variables to provide an overall satisfaction signature for the workplace studied (see Fig. 3.4) (Aziz et al. 2007; Park et al. 2013). Year-long Building Occupant Satisfaction Surveys (BOSS) were also developed at the CBPD to assess long-term satisfaction of employees with the quality of their work environment. Both right-now and long-term user satisfaction surveys have been deployed in over 75 public and private sector buildings in concert with IEQ measurements (Park 2015).

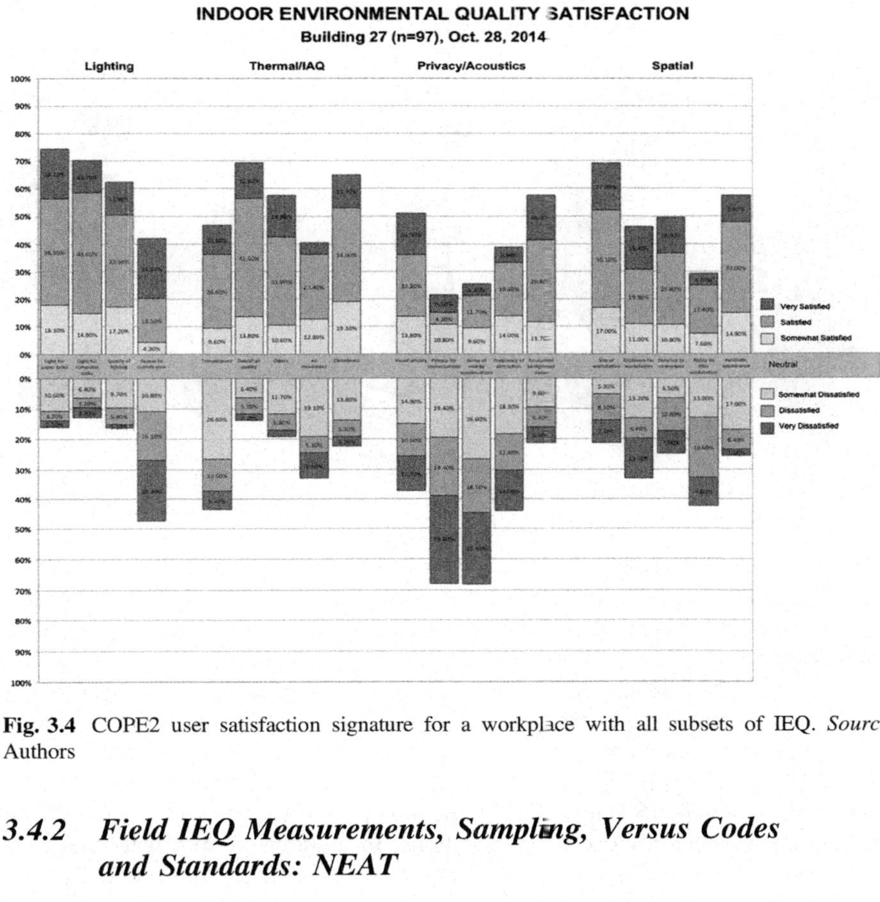

Fig. 3.4 COPE2 user satisfaction signature for a workplace with all subsets of IEQ. *Source* Authors

3.4.2 Field IEQ Measurements, Sampling, Versus Codes and Standards: NEAT

While user satisfaction surveys are growing in popularity, integrated field environmental measurement toolkits are less pervasive due to the cost and expertise required for their field deployment. First launched in 2000, Carnegie Mellon's portable suite of instruments on the NEAT cart has evolved over the years (see Fig. 3.5) and will continue to become more compact and robust as affordable sensor technology advances, and as field research reveals the attributes that truly need to be measured (CBPD 2009, 2013).

This cart was developed to ensure a simultaneous environmental assessment of the thermal, visual, acoustic, and air quality environment. Positioned in place of the occupant's chair at each sampled workstation, the cart collects temperature data at 1.1, 0.6 and 0.1 m from the floor, relative humidity, carbon dioxide (CO_2) and carbon monoxide (CO) concentrations, total particulates (PM2.5 and PM10), and volatile organic compounds (TVOCs) at 1.1 m, defined as the "breathing zone" (ASHRAE 1997). Hand-held instruments measure light levels on the monitor, work surface, and keyboard, as well as the horizontal and vertical radiant temperature

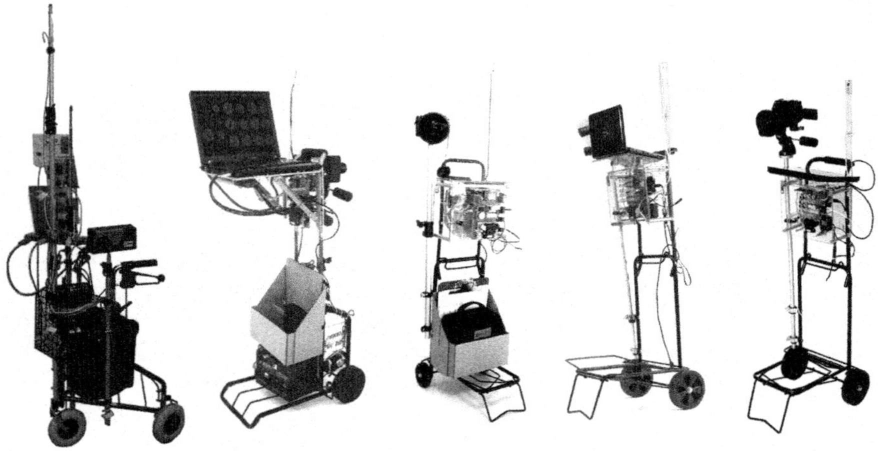

Fig. 3.5 NEAT Cart evolution. *Source* Authors

differences, and air velocity. A data logger connected to a tablet PC records data from the instruments for analysis, and a single-lens reflex (SLR) camera with a fisheye lens and Photolux software is used for brightness contrast and glare analyses.

While the physical measurements are recorded, the occupant is asked to sit nearby and complete the two-page 25-question COPE2 Questionnaire, to correlate satisfaction with conditions at the time of measurement. The sampling rates of spot measurements are typically 10–15% of the total number of office workstations on each floor, or at least 15 workstations if the workgroup is small, with a mix of open and closed, perimeter and core workstations. In addition to spot measurements, continuous measurements of air and thermal quality over a 24-h period are also taken. Since sampling may occur during cooling, heating, and swing seasons, the size of the multiple building database is critical for cross-sectional analyses against codes and standards. Code analyses are based on ASHRAE-55 and -62.1 (ASHRAE 2010a, b) as well as Environmental Protection Agency (EPA) guidelines for thermal and air quality assessment (EPA 2007), IESNA (IESNA 2011) for lighting quality assessment, and ASHRAE Handbook (2013) for workplace acoustic quality assessment.

3.4.3 Recording the Technical Attributes of Building Systems: TABS

The CBPD team developed expert walkthrough worksheets to ensure that comparable data is recorded for the attributes of building systems that affect thermal and air quality (mechanical, enclosure, interior), lighting and visual quality (enclosure, lighting and

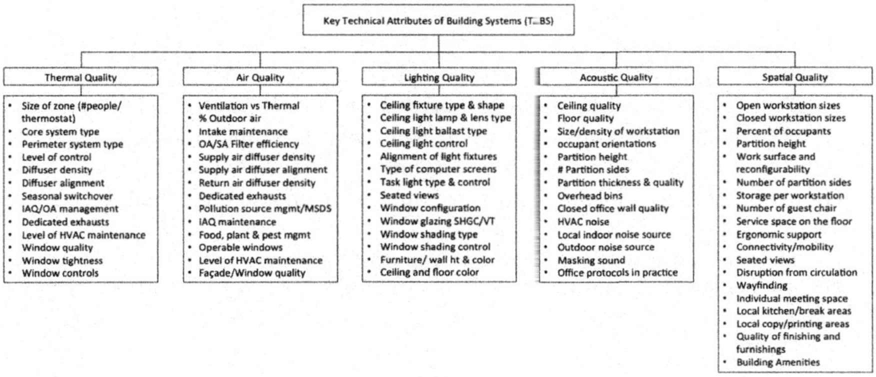

Fig. 3.6 Technical attributes of building systems are critical to measured and perceived thermal, acoustic, visual, spatial and air quality. *Source* Authors

interior), acoustic quality (mechanical, enclosure, interior) and spatial/ergonomic quality (individual and collaborative interior conditions as well as amenities).

Capturing over 57 attributes with subsets, illustrated in Fig. 3.6, the technical attributes of the building systems in the NEAT database support statistical analyses of the relationships between specific building design and engineering attributes, user satisfaction, and IEQ measured performance.

3.4.4 Putting It All Together: Databases and Statistical Analyses

The CBPD has a database of over 1600 workstations from more than 75 buildings across North America, and some in Europe, with statistically significant findings linking the measured environmental conditions with user-perceived quality of the indoor environment as well as the technical attributes of building systems.

A range of statistical methods and data mining algorithms is utilized to test research hypotheses formalized in POE studies. The adopted tools include Descriptive Statistics, Two-sample *t* tests, Analysis of Variance, Principal Component Analysis, and Pearson Correlations.

3.5 POE + M Results

There are at least five reasons why building owners, facility managers, and occupants should actively participate in field evaluation efforts that combine user satisfaction questionnaires with field IEQ measurements as well as 'as-built' records on the conditions of each building subsystem (COPE + NEAT + TABS):

- To recognize the interrelated nature of spatial, thermal, air, acoustic, and visual qualities—the human as an integrated sensor; and to promote occupants as sensors and controllers for environmental gains.
- To identify technologies and systems that work well and ensure investment where it matters: Linking satisfaction with environmental conditions and with the technical attributes of building systems.
- To revise codes and standards to reflect the occupied and integrated setting over time.
- To catalyze innovation for high-performance buildings.
- To prove that place impacts health and productivity: Linking occupant satisfaction, environmental conditions, and the technical attributes of building systems, to health, productivity and life cycle costs.

3.5.1 POE + M to Promote Occupants as Integrated Sensors and Controllers

CMU's NEAT studies clearly reveal that thermal comfort is affected by far more than air temperature, the only factor measured at the thermostat. The disconnect between thermal comfort standards that are written to ensure 80% occupant satisfaction, and the sober reality of less than 40% satisfaction with thermal comfort in many work environments, must be definitively established through POE + M. In Fig. 3.7, CMU's analysis of the NEAT database reveals 58% dissatisfaction with thermal comfort in the summer, and a statistically significant peak in summer satisfaction at 76.5 °F and in dissatisfaction at the colder 73.5 °F ($p < 0.05$, $n = 446$).

3.5.2 POE + M to Identify Technologies and Systems that Work

There is amazingly little POE work that carefully records the physical attributes of the building systems, even in before-and-after studies.

In a before-and-after NEAT study of the GSA Chicago Federal center, significant gains in user satisfaction with lighting may have been due to a number of critical changes: lower overall light levels, i.e., from 740 to 340 lx on the work surface; replacement of flush lenses with parabolic louvers to eliminate direct glare; and lower partitions to increase seated views from 40 to 80% of the occupants (Loftness et al. 2007). Indeed, lowering light levels below present standards, to as low as 250 lx, could universally improve satisfaction with 'the amount of light on

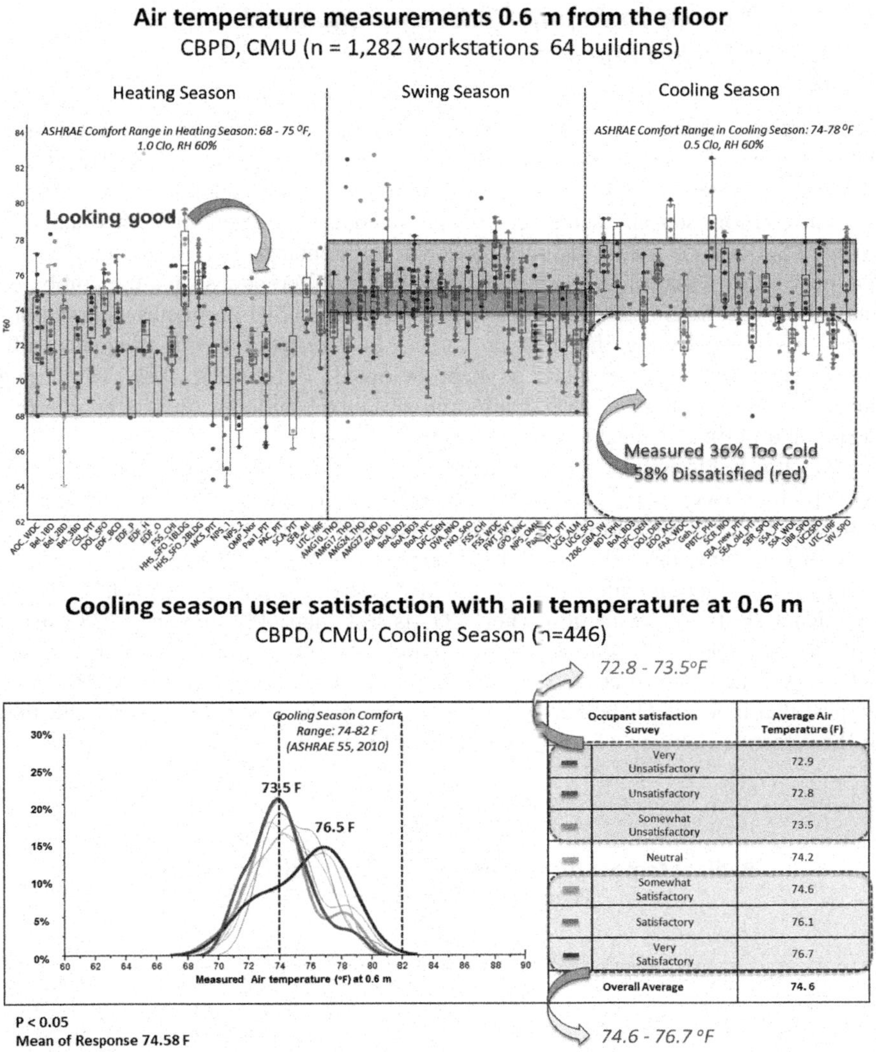

Fig. 3.7 Measured seasonal temperatures in workstations across the US correlated with satisfaction definitively showing that warmer temperatures are desired in summer. *Source* Authors

the desk'. These findings argue that lighting standards should be rewritten to shift away from combined task and ambient lighting at 500 lx designed to serve both computer and paper tasks, to significantly lower ambient levels to reflect predominant computer-based tasks, adding low wattage LED task lights as needed for paper-based tasks.

3.5.3 POE + M to Refine Codes and Standards for Performance

Deep analysis of the three-legged NEAT database has uncovered significant links between user satisfaction, environmental measurements, and the technical attributes of building systems (Park 2015). Three significant recommendations for refining thermal comfort and air quality codes and standards emerged from this analysis. While ASHRAE-55 standards allow an 18°F horizontal radiant asymmetry, anticipating windows that might be very cold or hot in different seasons, the combined analysis of user satisfaction and measured conditions in 391 perimeter workstations reveal a statistically significant improvement in 'satisfaction with thermal conditions' when radiant asymmetry is kept below 3 °F (see Fig. 3.8). Tightening thermal comfort standards would substantively move building owners to improve façade and perimeter mechanical system performance.

CO_2 thresholds are not yet code mandated, but guidelines suggest a threshold of 1000 ppm as acceptable for work environments, given outdoor CO_2 levels at 350–450 ppm. CO_2 is a measurable indicator of ventilation effectiveness, the delivery of outdoor breathing air to the nose of the occupant. CMU's NEAT database reveals occupant 'satisfaction with overall air quality' is strongly linked to CO_2 levels, with significant shifts to satisfaction when CO_2 is less than 600 ppm (n = 1282 in 64 buildings, $p < 0.05$). This confirms the findings of a number of both controlled and field experiments (Choi et al. 2012; Satish et al. 2012; Seppänen et al. 2004) that suggest that indoor CO_2 be kept within 200 ppm of outdoor levels as an indication of the effective delivery of breathing air (see Fig. 3.9).

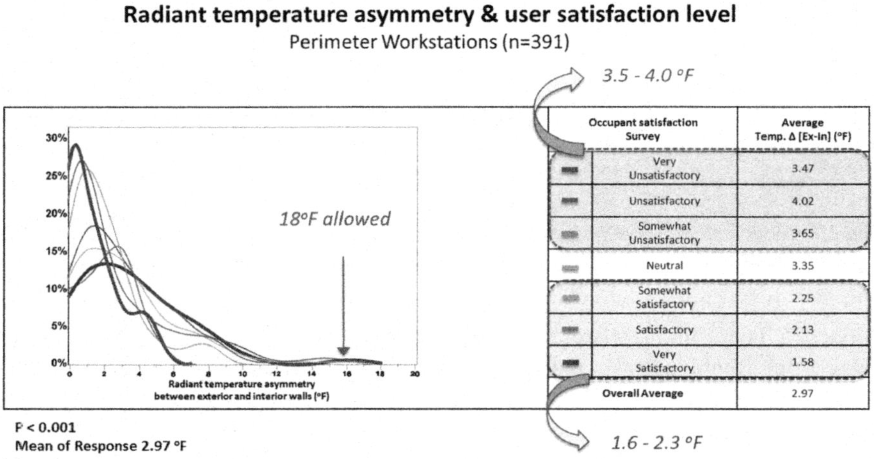

Fig. 3.8 Occupant comfort suggests limiting horizontal radiant asymmetry to 6 °F not 18 °F as written in today's codes. *Source* Authors

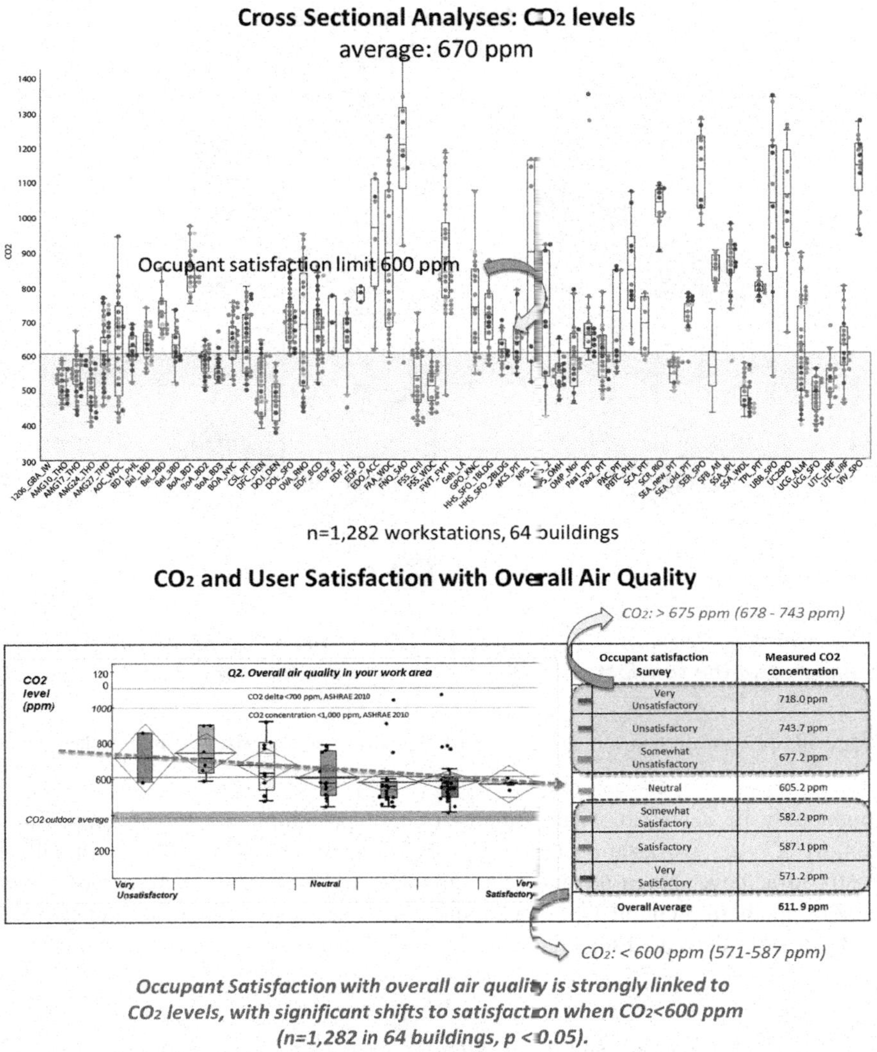

Fig. 3.9 CO_2 thresholds should be less than 200 ppm above outdoors to ensure satisfaction with IAQ. *Source* Authors

At present, there are no code mandates limiting the size of thermal zones, with value engineering often reducing the number of engineered zones before construction even begins. The TABS records of 75 buildings show open plan zone sizes ranging from 5 to 200 people sharing a thermostat, with terminal reheat disconnected in the 200. Closed office zone sizes range from 1 to 20 offices sharing a thermostat, often wrapping from east to north or south to west with significant variations in solar and wind loads. POE + M studies reveal that 80% satisfaction

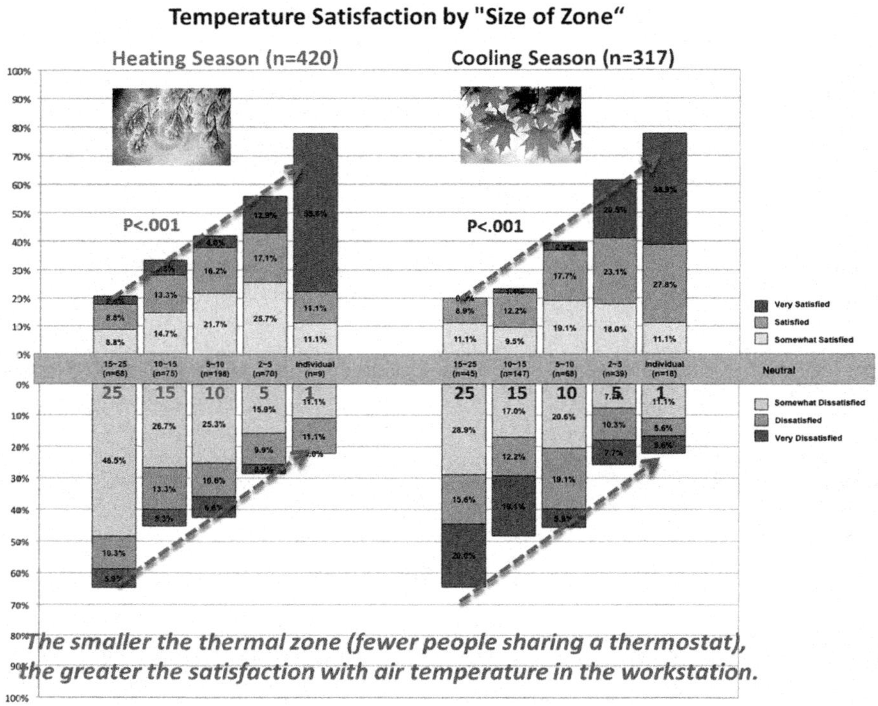

Fig. 3.10 80% satisfaction can only be achieved with smaller zone sizes. *Source* Authors

might only be achievable with 'micro-zoning,' providing a level of temperature control at *every* workstation (see Fig. 3.10). Today, this is no longer difficult or costly to achieve, given advances in variable refrigerant flow technologies, innovations in thermo-coupled perimeter units, and separate ventilation and thermal systems, e.g., dedicated outside air system, each supporting occupant control.

3.5.4 POE + M to Promote Innovation for Performance

There are significant disadvantages to the present U.S. tradition of combining thermal conditioning, i.e., heating and cooling, with ventilation, i.e., breathing air supply, in a single ducted system. The NEAT field studies of work environments illustrate that combined thermal and ventilation systems cannot consistently deliver good air quality or thermal comfort, especially in swing seasons or with changing occupant and equipment densities. Combining thermal and ventilation systems has also led engineers to pressurize buildings to ensure effective breathing air delivery, which in turn has eliminated the opportunity for operable windows, not just in high-rise buildings, but in low-rise offices, schools, community centers, and more.

POE + M reveals that there are significant advantages to separating heating and cooling from breathing air supply. The separation can support an energy effective increase in the levels of outdoor air supply to the occupants through effective ventilation delivery, economizer cycles and most significantly through natural ventilation. Moreover, a growing body of international studies has linked increases in outdoor air supply to both productivity and health gains in the workplace (Mendell et al. 2008).

3.5.5 POE + M to Prove that Place Impacts Health and Productivity

Proving that the quality of the built environment directly impacts health and productivity is extremely difficult, especially in field studies. While building clients support POE + M to gather key building attributes, measure environmental quality, and survey user satisfaction, the critical financial, health, or productivity data is rarely shared to prove that better buildings have better life cycle paybacks. The challenge is to link the technical attributes of building systems with occupant satisfaction and measured environmental conditions to actual health, productivity and life cycle costs of the organization.

Productivity data continues to be an elusive measure in the white-collar office. However, other financial data is more readily available, including absenteeism, turnover, customer base, material waste, facility management costs, and energy use, as shown for offices, schools, and hospitals in Fig. 3.11.

Offices	Schools	Hospitals
O&M, Energy & Water	O&M, Energy & Water	O&M, Energy & Water
Worker Health	Teacher Health Student Health	Patient Health/ Recovery Rates Patient Falls Staff Health
Attraction/ Retention Individual Productivity	Teacher Turnover Student Test Scores College Placement	Staff Turnover
Absenteeism/ Presenteeism	Absenteeism/ Presenteeism	Absenteeism/ Presenteeism
Organizational Productivity Market Share/ Customer Speed to Market	Drop-out Rates No Child Left Behind	Bed Vacancies Cost/Bed Profit/Bed
Waste Cost/Benefits	Waste Cost/Benefits	Waste Cost/Benefits
Litigation Insurance/ Tax SBS		Medication Errors

Fig. 3.11 Range of cost-benefit indices for POE in offices, schools and hospitals. *Source* Authors

3.6 The Future of POE + M with Measures of Productivity and Health

The value of instrumented POE + M is significant. It allows building occupants and managers to take back control of building systems; to identify technologies and systems that work; to prove that place impacts health and productivity; to ensure investment where it matters; to recognize the importance of behavior in environmental gains; and to catalyze innovation that can meet today's challenges.

Compared to past work environments, the design of a modern work environment must anticipate high levels of spatial and technological change by providing responsive thermal and air quality delivery systems, as well as flexible technology infrastructures. Building occupants will require indoor environmental conditions that can successfully support computer intensive activities, as well as paper-based tasks, while dealing with changing densities and configurations of workstations, plus changing levels of open planning. However, the current standards and guidelines for indoor environments were predominantly developed based on experiments involving human subjects in environmental chambers without consideration of these modern office variables (Jones 2002; Loftness et al. 2009; Van der Linden et al. 2006). This limitation may have resulted in higher levels of occupant dissatisfaction with IEQ, as well as unnecessary energy use.

At the same time, the creation of high-performance building facades, especially with the introduction of energy efficient windows and glazing technologies, has enabled construction of narrower building sections. It allows more occupants to be located at the perimeter of a building with access to daylight, enhanced views, and natural ventilation. This can contribute to increased thermal and visual comfort levels and a potential decrease in building energy consumption. Although numerous studies have underscored the benefits of access to nature for human physiological and emotional well-being (e.g., Heschong 2002), present thermal comfort standards may not be adequate to ensure appropriate design and engineering of building enclosures to achieve these benefits (de Dear and Brager 2002; Van der Linden et al. 2006). Modern workplaces would greatly benefit from field-informed POE + M with productivity and health tested modifications of buildings and indoor environmental conditions to increase user satisfaction, comfort, health, and work effectiveness.

Acknowledgements This National Environment Assessment Toolkit research and development was made possible by support from the U.S. General Services Administration. The authors would like to thank Kevin Kampschroer, Judy Heerwagen, Kevin Powell, and Michael Atkinson of the GSA, as well as a host of Ph.D. and Masters students of the Center for Building Performance and Diagnostics at Carnegie Mellon University, who supported data acquisition, processing, analysis and reporting.

References

ASHRAE. (1997). *Standard 129-1997: Measuring air change effectiveness*. Atlanta: American Society of Heating, Refrigerating and Air-Conditioning Engineers.

ASHRAE. (2010a). *Standard 62.1-2010: Ventilation for acceptable indoor air quality*. Atlanta: American Society of Heating, Refrigerating and Air-Conditioning Engineers.

ASHRAE. (2010b). *Standard 55-2010: Thermal environmental conditions for human occupancy*. Atlanta: American Society of Heating, Refrigerating and Air-Conditioning Engineers.

ASHRAE. (2013). *Handbook for workplace acoustic quality assessment*. Atlanta: American Society of Heating, Refrigerating and Air-Conditioning Engineers.

Aziz, A., Loftness, V., & Srivastava, V. (2007). Post-occupancy evaluation: Indoor environmental assessment toolkit—EnviroBot. In F. Shafii & M. Zahri Othman (Eds.). *Proceedings of conference on sustainable building Southeast Asia* (pp. 182–190). Johor, Malaysia: Institute Sultan Iskandar of Urban Habitat and Highrise, Level 4 Dewan Sultan Iskandar, University.

BRAB (Building Research Advisory Board). (1988). *Electronically enhanced office buildings*. Washington, DC: National Research Council, National Academy of Science.

Brager, G., Heerwagen, J., Bauman, F., Huizenga, C., Powell, K., Ruland, A., et al. (2000). *Team spaces and collaboration: Links to the physical environment*. Berkeley: Center for the Built Environment, University of California.

CBPD (Center for Building Performance and Diagnostics). (2009). *Energy savings and performance gains in general services administration buildings: Seven recommendations*. US General Services Administration.

CBPD (Center for Building Performance and Diagnostics) (2013). *National environmental assessment toolkit (NEAT) manual*. Carnegie Mellon University. http://neat-cbpd.arc.cmu.edu/neat/manual/NEAT-Manual.pdf. Accessed on 12 Jan 2017.

Choi, J.-H., Loftness, V., & Aziz, A. (2012). Post-occupancy evaluation of 20 office buildings as basis for future IEQ standards and guidelines. *Energy and Buildings, 46*(3), 167–175.

de Dear, R., & Brager, G. (2002). Thermal comfort in naturally ventilated buildings: Revisions to ASHRAE standard 55. *Energy and Buildings, 34*(6), 549–561.

EPA (Environmental Protection Agency). (2007). Testing for indoor air quality, section 01 81 09. US Environmental Protection Agency. https://www.epa.gov/sites/production/files/2014-03/documents/018109.pdf. Accessed 1April 2009.

Hartkopf, V., Loftness, V., & Mill, P. (1985). Integration for performance. In R. Rush (Ed.), *The building systems integration handbook* (pp. 231–316). New York: Wiley.

Hartkopf, V., Loftness, V., & Mill, P. (1986). The concept of total building performance and building diagnostics. In G. Davis (Ed.), *Building performance: Function, preservation, and rehabilitation, STP23009S* (pp. 5–22). West Conshohocken, PA: ASTM International.

Heschong, L. (2002). Daylighting and human performance. *ASHRAE Journal, 44*(6), 65.

IESNA (Illuminating Engineering Society of North America). (2011). *The IESNA lighting handbook* (10th ed.). New York, NY: Illuminating Engineering Society of North America.

Jones, B. (2002). Capabilities and limitations of thermal models for use in thermal comfort standards. *Energy and Buildings, 34*(6), 653–659.

Loftness, V., Aziz, A., Choi, J., Kampschroer, K., Powell K., Atkinson, M., et al. (2009). The value of post-occupancy evaluation for building occupants and facility managers. *Intelligent Buildings International, 1*(4), 249–268.

Loftness, V., Aziz, A., Hua, Y., Srivastava, V., & Yang, X. (2007). *GSA WP 20•20 Environmental Quality Report: Kluczynski Federal Building, Before- and After-renovation Comparison*. Chicago, IL: US General Services Administration.

Mendell, M., Lei-Gomez, Q., Mirer, A., Seppänen, O., & Brunner, G. (2008). Risk factors in heating, ventilating, and air-conditioning systems for occupant symptoms in US Office Buildings: The US EPA BASE Study. *Indoor Air, 18*(4), 301–316.

Park, J. (2015). *Are humans good sensors? Using Occupants as sensors for indoor environmental quality assessment and for developing thresholds that matter.* Ph.D. Thesis, Carnegie Mellon University, Pittsburgh, PA.

Park, J., Aziz, A., & Loftness, V. (2013). Post-occupancy evaluation for energy conservation, superior IEQ and increased occupant satisfaction. In *Proceedings of the IFMA's World workplace conference*, Philadelphia, PA. https://ifma.confex.com/ifma/ww2013/webprogram/Session3887.html. Accessed 12 Jan 2017.

Preiser, W. F. E., Rabinowitz, H. Z., & White, E. T. (1988). *Post-occupancy Evaluation.* New York: Van Nostrand Reinhold Company. (republished 2015, London: Routledge).

Rush, R. (1986). *The building systems integration handbook: American Institute of Architects.* New York, NY: Wiley.

Satish, U., Mendell, M. J., Shekhar, K., Hotchi, T., Sullivan, D., Streufert, S., et al. (2012). Is CO_2 an indoor pollutant? Direct effects of low-to-moderate CO_2 concentrations on human decision-making performance. *Environmental Health Perspectives, 120*(12), 1671.

Seppänen, O., Fisk, W., & Mendell, M. (2004). Association of ventilation rates and CO_2 concentrations with health and other responses in commercial and institutional buildings. *Indoor Air, 9*(4), 226–252.

Van der Linden, A., Boerstra, A. C., Raue, A. K., Kurvers, S. R., & De Dear, R. (2006). Adaptive temperature limits: A new guideline in the Netherlands: A new approach for the assessment of building performance with respect to thermal indoor climate. *Energy and Buildings, 38*(1), 8–17.

Veitch, J. A., Charles, K. E., Farley, K. M., & Newsham, G. R. (2007). A model of satisfaction with open-plan office conditions: COPE field findings. *Journal of Environmental Psychology, 27*(3), 177–189.

Author Biographies

Vivian Loftness FAIA is the Paul Mellon University Professor of Architecture, and served a decade as Head of the School of Architecture. She is a renowned researcher, author and educator focused on environmental design and sustainability, climate and regionalism in architecture, and the integration of advanced building systems for health and productivity. Vivian is a key member of Carnegie Mellon's leadership in sustainability and contributor to the ongoing development of the Intelligent Workplace—a living laboratory of commercial building innovations for performance—as well as the Adaptable Workplace Lab at the U.S. General Services Administration and the Laboratory for Cognition at Electricity de France. She has served on the Board of Directors of the US Green Building Council, the AIA Committee on the Environment, Turner Construction, and the World Business Council for Sustainable Development. She is senior editor of the Reference Encyclopedia Sustainable Built Environments by Springer Publishing, a member of twelve National Academy of Science committees and the Board on Infrastructure and the Constructed Environment, and has given four Congressional testimonies on sustainable design. Recognized as a LEED Fellow, a Senior Fellow of the Design Futures Council, and one of 13 Stars of Building Science by the Building Research Establishment at the Royal Academy of Engineering, Vivian received a National Educator Honor Award from the American Institute of Architecture Students and a "Sacred Tree" Award from the US Green Building Council. Vivian Loftness has a Bachelors of Science and a Masters of Architecture from MIT.

Volker Hartkopf Ph.D. is the Director of the Center for Building Performance and Diagnostics (CBPD) and a Professor of Architecture at Carnegie Mellon University. With the support of the National Science Foundation, he initiated the first inter-disciplinary graduate program merging architecture, engineering and planning to address the performance of the built environment—with hundreds of Masters and Ph.D. graduates working around the world. A frequent keynote speaker in

Europe, Asia, Australia and the Americas, Dr. Hartkopf has authored over 100 peer reviewed technical publications, books and book chapters on sustainable construction and high performance buildings. He started the Advanced Building Systems Integration Consortium (ABSIC), the first public/private research partnership in the building industry, designated as a National Science Foundation Industry/University Cooperative Research Program (NSF I/UCRC) and winning the Alex Schwarzkopf Award in 2013. Professor Hartkopf's leads advanced research and demonstrations of high performance work environments around the world, including: the Robert L. Preger Intelligent Workplace (IW) at Carnegie Mellon, awarded the AIA Honor Award and the Business Week Architectural Record Award; the LG Kangnam Tower in Seoul; the MOST Agenda 21 Headquarters and the Tsinghua University Low Energy Demo Building in Beijing; the EdF Laboratory for Design of Cognition in Paris; and the GSA Adaptable Workplace Lab in Washington, DC. Professor Hartkopf was appointed Chair of the United Nation's Environmental Programme Think Tank on Sustainable Buildings and Construction Initiative (SBCI) in 2007. Volker Hartkopf received his Diplom Ingenieur and PhD at the University of Stuttgart and a Master of Architecture at the University of Texas at Austin.

Azizan Aziz is Assistant Research Professor at the Center for Building Performance and Diagnostics in the School of Architecture at Carnegie Mellon University. He is a trained architect with extensive experience in sustainable design, the design of integrated systems for high performance buildings, post occupancy evaluation, building performance monitoring and energy data analytics. He is a central member of the team designing living laboratories for high performance workplaces, including the Intelligent Workplace at Carnegie Mellon University, the EdF Laboratory for Design of Cognition in Paris, and the GSA Adaptable Workplace Lab in Washington, DC. Azizan Aziz has led innovative energy data analytics projects with federal, corporate, campus and municipal leaders. The data analytics and his innovative controls research efforts have resulted in innovative technologies and interfaces to support building owners, facility managers and building occupants in reducing energy and resource consumption while improving indoor environmental quality in commercial buildings. In addition to his more than 20 years in research, Azizan teaches in the undergraduate and graduate programs for the School of Architecture, with an unwavering focus on design for a sustainable built environment. Azizan Aziz has a Bachelors of Architecture and a Masters in Building Performance and Diagnostics at Carnegie Mellon University.

Joon-Ho Choi Ph.D. is Assistant Professor of Building Science at the University of Southern California (USC) and directs the Human-Building Integration Research Group. Prior to taking the position, he was assistant professor of Architectural Engineering at Missouri University of Science and Technology. Dr. Choi's primary research interests are in the areas of advanced controls for human-building integration, sustainable building design and performance, comprehensive post-occupancy evaluation, and indoor environmental quality. He has published more than 40 research papers in international journals and peer-reviewed conference proceedings, receiving Best Papers Awards from the Architecture Institute of Korea in 2012 and the International Conference on Indoor Air Quality, Ventilation, and Energy Conservation in 2016. He also received a 2014–2015 New Investigator Award from the U.S. Architectural Research Centers Consortium. Dr. Choi's research focuses on the relationship between human factors, indoor environmental quality and advanced building systems and controls. As an interdisciplinary researcher, he is actively collaborating with faculty from multiple disciplines, including Medicine, Computer Science, and Electrical Engineering, to develop user-centered environmental systems and controls for enhancing occupant productivity and health, as well as environmental sustainability. Joon-Ho Choi earned his Bachelor's in B.S. and M.S. Architectural Engineering, Yonsei University, Seoul, Korea, his Master of Science in Architecture at Texas A&M University, and his Ph.D. in Building Performance and Diagnostics at Carnegie Mellon University.

Jihyun Park Ph.D. is the Research Associate in the Center for Building Performance and Diagnostics at Carnegie Mellon University. Prior to joining CMU, she served as a lecturer at Seoul National University of Science and Technology, Korea. Her research interests include indoor environmental quality, post occupancy evaluation, and integrated building energy assessment and optimization focused on occupant comfort, satisfaction and health. She has worked on IEQ assessment projects and big data analysis worldwide including Bank of America Tower, David Lawrence Convention Center, and Électricité de France. Dr. Park is currently working on the General Services Administration (GSA) Energy Savings Research and the Consortium for Building Energy Innovation project funded by U.S. Department of Energy (DOE) accelerating the adoption of advanced energy retrofits. In addition to her presence in academia, Dr. Park is also an architecture designer with professional experience in sustainability. She has worked on architectural design for over 40 projects including offices, schools, resorts, residential, medical facilities, and government buildings. She has LEED AP BD + C certification from the U.S. Green Building Council and is a Certified Passive House Consultant (CPHC) by Passive House Institute U.S. Jihyun Park received Bachelor of Architecture, Bachelor of Public Health, and M.S. in Architecture from Ewha Womans University, Seoul, Korea, and her Ph.D. in Building Performance and Diagnostics at Carnegie Mellon University.

Chapter 4
Informed Design: A Post-Occupancy Evaluation Toolkit for Courthouses

Erin Persky, Jay Farbstein and Melissa Farling

4.1 Introduction

Courthouses are a unique building type. Justice architects and planners are tasked with ensuring courthouses meet the functional and security needs of many different user groups, including the entity that owns the building, court staff, judges, litigants, attorneys, public visitors, jurors, and those in custody, among others—each with distinct concerns, requirements, and expectations for the building. Courthouse planning guidelines have become more comprehensive and available (Judicial Council of California/Administrative Office of the Courts 2011) and, at the same time, courts must adhere to strict building performance and efficiency standards (California Natural Resources Agency 2016). One way to discern whether or not courthouse features are responsive to these needs and requirements is to evaluate the courthouse's performance by conducting a post-occupancy evaluation (POE) during its occupancy. While a building performance evaluation (BPE) covers the entire life-cycle of a building, the POE focuses on the phase following building occupancy, as detailed in this chapter (Preiser and Schramm 1997; Preiser and Vischer 2005; Preiser et al. 2015).

In conjunction with the American Institute of Architects – Academy of Architecture for Justice (AIA-AAJ), and a multidisciplinary advisory committee, the authors have developed a Post-Occupancy Evaluation (POE) Toolkit for

E. Persky (✉)
3511 State Street, San Diego, CA 92103, USA
e-mail: erinpersky@gmail.com

J. Farbstein
1500 Rustic Lane, Pacific Palisades, CA 90272, USA
e-mail: jayfarbstein@gmail.com

M. Farling
HDR, 3200 East Camelback Road, Suite 350, Phoenix, AZ 85018, USA
e-mail: Melissa.Farling@hdrinc.com

W.F.E. Preiser et al. (eds.), *Building Performance Evaluation*,
DOI 10.1007/978-3-319-56862-1_4

Courthouses. This chapter introduces the Courthouse POE Toolkit as a strategy for the dissemination and administration of an integrated system for evaluating courthouse performance.

4.2 POE Toolkit Project Parameters

During the POE planning process, it may be the case that stakeholders express a range of interests about the features upon which they would like to focus the investigation. The court architect may wish to demonstrate that her client is satisfied with the design and find out how well design features, materials, or systems are working; court managers may be looking for ways to improve the efficiency of operations, customer satisfaction or the levels of maintenance that are required; or the building owners may wish to develop design guidelines for future courthouses in their jurisdictions. Each of these perspectives requires different types of data and information gathering techniques. The Toolkit is intended to be flexible enough to respond to all these potential uses within a standardized framework.

Furthermore, the "toolkit" concept offers two primary applications: first, to assess the performance of a particular court building and to provide feedback to its owners, occupants and/or designers—this is the most common purpose of a POE. Second, the standardization of instruments provided in the Toolkit affords the opportunity to aggregate and compare findings from multiple POEs, allowing general conclusions to be drawn about what works and what does not work, and to tie outcomes, e.g., ratings, to specific design features. This level of analysis provides opportunities to develop and catalog evidence-based findings that would provide a valuable resource for planning future courthouses and for developing performance-based design guidelines.

This chapter describes key features of the Toolkit, including:

- A discussion of the purpose and applications of the Toolkit;
- A description of each instrument and its contribution toward measuring design performance;
- Suggestions about who should be involved in the process. Successful operation of a courthouse requires collaboration among many agencies, and the input of these groups is essential;
- Instrument and fieldwork methodologies;
- Discussion of how POEs can aid in the successful development and application of evidence-based design principles to courthouses.

The POE Toolkit is part of a broad effort by the AIA to disseminate knowledge pertaining to best practices in justice facility design. As such, users will be asked to share their data and findings in order to be able to access the instruments and instructions. The results will contribute to a database of information about courthouse design that will inform substantial improvement in the field and allow researchers to examine the relationships among physical variables of courthouse design and their outcomes.

4.3 The Toolkit

The Courthouse POE Toolkit is configured to offer building evaluations of a broad range of scopes and depth of analysis. The Toolkit consists of:

- Guidance and forms for planning the POE, for example, as to who should be involved, roles, timing, scheduling, and the like;
- Information-gathering instruments;
- On-site fieldwork recommendations;
- Suggestions on data analysis methods and how to present the report.

In determining whether to conduct a POE and how to proceed with it, it is essential to be explicit about the goals and types of information desired, as well as available resources. A relatively brief POE, with a tour of the building and interviews with the court and building managers, may suffice. On the other hand, much more detail may be desired or even required, especially if a set of comparative POEs is being considered.

The Toolkit provides the opportunity to gather and analyze information about several aspects of a building's performance. Examples of areas investigated during a courthouse POE include, but are not limited to:

- Functional area operations; for example, the usefulness of clerk service windows, effectiveness of maps and signage, or the efficiency of security screening areas.
- User and occupant satisfaction; including workstation comfort, access to natural light, adequacy of support spaces, etc.
- Maintenance requirements and technical performance; such as frequency of equipment or repairs, condition of materials, or performance of mechanical/electrical/plumbing systems.
- Safety and security; assessed through evaluations of central holding facilities, the performance of security systems, and user satisfaction, etc.
- Energy and environmental sustainability; for example, water and energy usage, utility costs, material selection, and waste management protocols.

Generally, less resource- and labor-intensive POEs would gather fewer types of information, namely perhaps only quantitative data, whereas more intensive POEs might gather several types, i.e., both quantitative and qualitative. For any level POE, however, it is highly recommended that multiple methods of data collection be utilized since it is always valuable to look at an issue from multiple perspectives.

4.3.1 Data-Gathering Forms

The Courthouse POE Toolkit includes the following data-gathering instruments:

- Plan Review Form: aids in documentation of the physical characteristics of and functions within the courthouse building and its site. This form should be completed in advance of the site visit, with any missing or ambiguous items completed on site (see Fig. 4.1).
- Building Conditions Survey and Interview: evaluate the condition and performance of many features of the materials and systems of the courthouse. The survey (see Fig. 4.2) is completed while on the facility tour with its facilities and operations managers and other knowledgeable parties (see Sect. 4.4 "Who should be involved in the POE?"). A more detailed semi-structured interview (see Fig. 4.3) is also provided to be completed with the facilities manager about courthouse features as a follow-up to issues that may have arisen during the building conditions survey tour.
- Court Employee Survey: assesses the degree to which the design of the courthouse building supports the work-related tasks carried out by courthouse staff. Topics include the courthouse site, building access, staff areas and workspaces, and courtroom functionality (see Fig. 4.4).
- Visitor Survey: assesses the degree to which the design of the courthouse supports the functions people visit the courthouse to accomplish. Topics include the courthouse site, building access, wayfinding, safety, circulation, publicly accessible functional areas, the courtroom, and designated jury spaces (see Fig. 4.5).

Courtroom (duplicate and complete this form for **each type** of courtroom)

Main type of proceedings (check all that apply):

☐ civil ☐ criminal ☐ arraignment ☐ traffic ☐ family ☐ drug ☐ juvenile delinquency
☐ juvenile dependency ☐ other ______________________

Number of courtrooms of this type: ________

Courtroom area: ____________ net square feet

Courtroom dimensions: ____ feet wide x ___ feet long x ____ feet high

Well/Litigation Area

Bench arrangement:
☐ center
☐ corner
☐ "re-centered" (not in geometric center of courtroom but aligned with entry door)

Fig. 4.1 Plan review form sample. *Source* Authors

	Very Good	Good	Neutral	Poor	Very Poor	Not Applicable	Comments
Main Entry & Lobby							
Signage/directories	☐	☐	☐	☐	☐	☐	
Kiosks	☐	☐	☐	☐	☐	☐	
Seating	☐	☐	☐	☐	☐	☐	
Other ___________	☐	☐	☐	☐	☐	☐	
Courtroom(s) (typical)							
AV systems	☐	☐	☐	☐	☐	☐	
Projectors/screens	☐	☐	☐	☐	☐	☐	
Access controls	☐	☐	☐	☐	☐	☐	

Fig. 4.2 Building conditions survey form sample. *Source* Authors

2. Are there any features of this facility's design, including systems, that require excessive maintenance or that are showing excessive wear? If so, please list them and describe the issues or problems.

3. Is there any defective construction work that requires repair? If yes, please explain.

4. Are there any areas or systems that require little maintenance or that are "wearing well"? If so, what are they?`

Fig. 4.3 Building conditions interview form sample. *Source* Authors

IF YOU WORK IN A COURTROOM...

If so, please check the box below that best describes the type and size of courtroom you usually work in:

- ☐ Hearing room with limited spectator seating and no jury
- ☐ Non-jury courtroom
- ☐ Jury courtroom

The size of the **well/litigation area** is adequate to efficiently conduct courtroom proceedings.

Strongly Agree	Agree	Neutral	Disagree	Strongly Disagree
☐	☐	☐	☐	☐

Fig. 4.4 Court employee survey form sample. *Source* Authors

If you spend time in a courtroom, please answer the following questions. **If you did not spend time in a courtroom,** please check this box [ ] and skip to the last question.

How satisfied were you with the **waiting spaces outside the courtroom?**

Very Satisfied	Satisfied	Neutral	Dissatisfied	Very Dissatisfied	Did Not Use
☐	☐	☐	☐	☐	☐

How satisfied were you with **your ability to see all participants in the courtroom?**

Very Satisfied	Satisfied	Neutral	Dissatisfied	Very Dissatisfied
☐	☐	☐	☐	☐

Fig. 4.5 Court visitor survey form sample. *Source* Authors

Detailed instructions for the administration of these instruments are included with each.

Opportunities to collect varying amounts of data are built into each instrument. For example, the Building Conditions Survey supports quantitative data collection by means of a checklist for rating the performance of many courthouse building and site features. Additionally, the survey offers space for commentary on each area in case elaboration is required. The Building Conditions Interview, the complement to the survey, supports entirely qualitative data collection via an in-depth, open-ended set of questions to allow for more thorough evaluation.

Furthermore, redundancy is built into the forms to obtain multiple perspectives on courthouse functions. Information can be cross-referenced against other instrument data to develop a more robust understanding of areas of inquiry that architects and other POE users can use to identify and solve existing problems.

4.3.2 Supplemental Studies

Though the Toolkit offers instruments to conduct POEs of considerable depth, it is possible that even greater depth may be achieved using instruments beyond those provided in the Toolkit. Examples of in-depth assessments not provided in the Toolkit include:

- Energy and environmental performance (see below for a list of references).
- Measurements of ambient conditions, such as acoustics and illumination.
- Comprehensive building condition assessments.

The following energy, environmental, and sustainability evaluation references may be useful:

- Leadership in Energy and Environmental Design (LEED)™ Project Design Checklists;
- AAJ Sustainable Justice Committee's "Sustainable Justice 2030: Green Guide to Justice" and "Sustainable Justice Guidelines";
- The Living Future Institute's "Living Building Challenge";
- General Service Administration (GSA) "High Performance and Sustainable Buildings Guidance", "GSA Sustainability Matters", and "GSA Sustainable Facilities Tool".

4.4 Who Should be Involved in the POE?

An evaluation team must be assembled for the POE, and its composition will depend on the purpose and depth of the evaluation. These decisions about scope will suggest who should conduct the POE: can the design team, the building occupants, or the owners conduct the POE? Or should it include representatives of all of them and perhaps an outside professional or academic who specializes in building evaluation? If surveys are to be completed by court staff and/or visitors, or if the Building Conditions Interview is going to be conducted with the facilities manager, the latter is highly recommended. If environmental and ambient conditions are to be measured, expert assistance is critical. The project delivery method might also influence who should be involved in the subsequent POE.

Regardless of the intended POE scope, a successful evaluation must involve input from at least the following representatives:

- The design team;
- Those in court management who know and understand the operations of the courthouse under investigation, e.g., court operations and facilities managers;
- The owner agency;
- Someone with knowledge of and experience in conducting POEs. Such an expert may be contracted to conduct the POE, but in any case, will engage with representatives of the other groups.

4.5 The POE Process

The Toolkit provides preparation, on-site, and post-site visit guidance for evaluators to optimize their use of the POE instruments. Below are excerpts of instructions provided for these portions of the POE process.

4.5.1 Preparation

As the evaluation team is assembled, a member of the team must contact the facility to state their intent to conduct a POE. At this time the details of the POE should be explained, including:

- The staff members who should be present for tours and interviews;
- Types of data collection that will be carried out including methodology and requirements to carry out each one;
- Areas of the facility that must be accessed;
- Length of time the team will be on site, with proposed days and times;
- Specific requests. For example, copies of plans or other documents, permission to photograph.

Contact should be made well in advance of the visit so that the client can assemble the appropriate staff members to participate in the site tour, alert court staff and other parties of the POE and of their potential involvement, and receive necessary security clearances for the evaluation team.

After all of the relevant court staff is assembled, a site tour, interviews, and Court Visitor Survey recruitment times should be scheduled. The POE on-site activities should be scheduled according to optimal times for the completion of each form: for example, the Visitor Survey should be scheduled according to peak times of visitor volume in order to gain the most possible survey responses. The Building Conditions Tour should be organized for a time at which all courthouse areas can be accessed without interruption, which might be outside of business hours. The tour should also be done early in the visit to allow the team to gain a general understanding of the layout of the facility and observe some of the operations, which will benefit subsequent data collection. The Building Conditions Interview should be scheduled during a time at which the Facilities Manager and others can participate without interruption and at a time that will not conflict with visitor recruitment. The POE schedule should be sent to participating staff approximately one to two weeks prior to the beginning of the evaluation.

4.5.2 On-Site

The evaluation team should arrive early to review the strategy for the day, confirm responsibilities, and discuss last-minute items related to the site visit. While on the facility tour, list follow-up questions that can be asked during the interview. Take as many photographs as possible, if permitted to do so. Permission should be arranged in advance of the site visit. If building drawings, construction documents, or other building information was provided, bring these on the tour. Look for relevant details or background information that had not been provided and note accordingly.

Unanticipated events are common and a strategy for handling them should be considered beforehand. For example, the visitor survey recruiter may experience a low acceptance rate, or disruptions to the court schedule may impact visitor survey recruitment or significantly skew the types of visitors at the courthouse during the POE. If a staff member is suddenly unavailable for an interview, prepare to reschedule the interview for as early as possible after the site visit.

4.5.3 *After the Site-Visit*

A debriefing session with the evaluation team is highly recommended and should occur as soon as possible after the site visit - if necessary, by conference call - to discuss:

- General thoughts about the success of the POE in terms of process and meeting its objectives;
- Review of impressions and findings about the building's quality and performance that should be captured for the report;
- Additional information that may require follow-up;
- Methodological concerns that could impact the data;
- The delegation of next steps, including responsibility for conducting analyses and drafting sections of the report.

4.6 Pilot Application of the POE Toolkit

In December 2015, a pilot POE was conducted on a large courthouse in North America to test the methodologies of each Toolkit instrument. The POE evaluated several areas of building performance, including site conditions and access, building systems, furnishings and amenities, functional area components (e.g., security screening, courtrooms, clerks service areas), circulation systems, and other topics, listed above. Staff and visitor satisfaction were assessed on a number of topics including courthouse building and site, functional areas utilized while in the building (e.g., staff workstations, publicly-accessible areas), and courtrooms. In addition, a number of qualitative features of the courthouse were assessed, including the degree to which the courthouse conveys a sense of openness, transparency, dignity, justice, and fairness.

The goals of this pilot study were to validate each of the POE Toolkit instruments, to assess technical building performance, to gauge courthouse staff and visitor satisfaction with the courthouse building and site, and to learn from the experience about what would be most helpful to include in the Toolkit instructions.

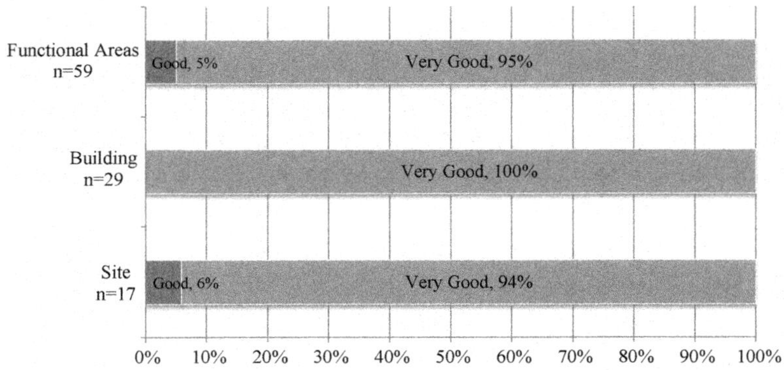

Fig. 4.6 Building conditions overall results by area. *Source* Authors

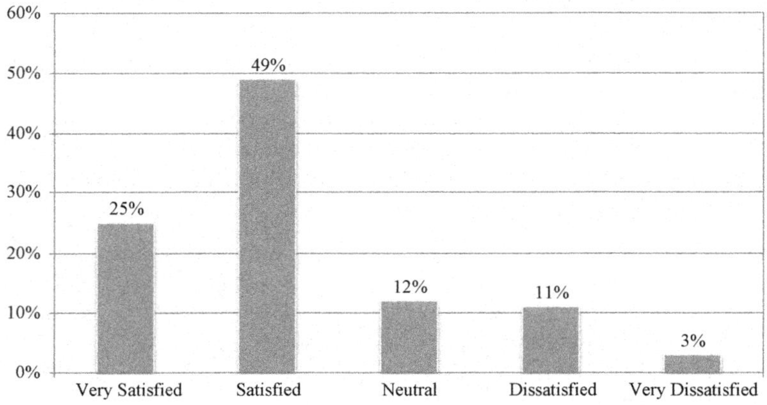

Fig. 4.7 Courthouse employee general workspace satisfaction. *Source* Authors

The courthouse scored highly in almost all areas, with facilities reporting over 95% of building performance items as "very good" or "good" (see Fig. 4.6).

Courthouse employees also rated the courthouse highly, including public areas, staff-only areas, and workstations (see Fig. 4.7).

Courthouse visitors were largely satisfied with the courthouse, particularly with regard to maintenance, cleanliness, and safety (see Fig. 4.8). One area requiring significant attention was the need for additional parking accommodations.

This pilot study garnered confidence in the results yielded by the data obtained with the Toolkit.

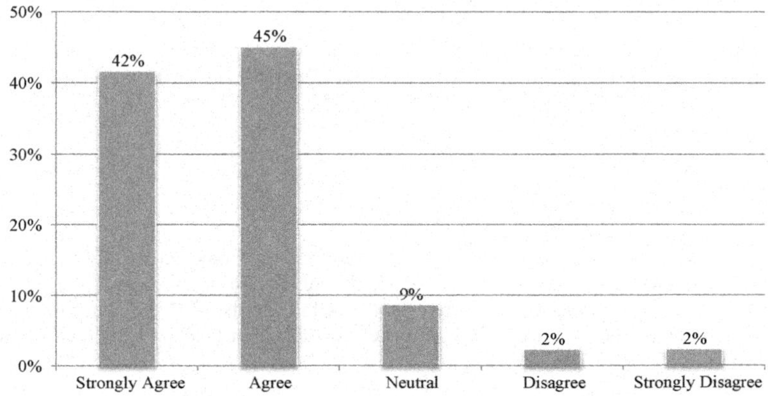

Fig. 4.8 Courthouse visitor feelings of safety and security *Source* Authors

4.7 Conclusion—Opportunities to Contribute to Best Practices

While a single POE can provide valuable feedback to the clients, users, and designers of that particular building, as a "case study" it may be of little interest or value in terms of drawing general conclusions or results that can be applied to future designs. On the other hand, when many POEs are performed, the opportunity to draw general conclusions about lessons learned is greatly enhanced. Conducting regular POEs provides the opportunity for continuous improvement in building standards, procedures, designs, and operations.

There is little precedent for a systemized catalog of evidence-based design strategies specific to building type (Pati 2005). The Toolkit structure, as a function of the use of standardized instruments, supports consistency in the types of data collected and the means by which collection is completed. The database of results affords direct comparison of the data collected on multiple types of courthouse-specific design strategies to determine which are most effective. This data aggregation is particularly effective for building types that contain repetitive building programs, as is the case with courthouses As application of best practices is crucial for the evolution of justice facility design, and access to evidence-based design strategies is of the utmost importance for ensuring that design decisions reflect best practices, this database will simplify access to the latest evidence-based design data pertaining to courthouses. The data collected can be used to contribute to the body of knowledge pertaining to courthouse design. Research studies can be conducted utilizing the data collected via the toolkits, resulting in substantial time and resource savings. This data can also contribute to the development and refinement of courthouse design guidelines.

The benefits of POE as part of the overall BPE process cannot be overstated. The goals of this Courthouse POE Toolkit are to streamline the process of conducting building evaluations, to increase the frequency with which POEs are conducted, and to encourage courthouse research endeavors, in order to develop a resource for the support of widespread application of best practices to courthouse design.

References

California Natural Resources Agency. (2016). California environmental quality act. Retrieved from http://leginfo.legislature.ca.gov/faces/codes.xhtml. Accessed on 26April 2016.

Judicial Council of California/Administrative Office of the Courts. (2011). California trial court facility standards. Retrieved from www.courts.ca.gov/documents/ctcfs2011.pdf. Accessed on 5 July 2016.

Pati, D. (2005). *Maximizing the benefits of courthouse POEs in design decision support and academic inquiry through a unified conceptual model*. Doctoral Dissertation, Georgia Institute of Technology Retrieved from Georgia Institute of Technology SMARTech Database. hdl.handle.net/1853/6843. Accessed on 1 May 2016.

Preiser, W. F. E., & Schramm, U. (1997). Building performance evaluation. In D. Watson, M. Crosbie, & J. H. Callender (Eds.), *Time-saver standards for architectural design data* (7th ed., pp. 231–238). New York: McGraw-Hill.

Preiser, W. F. E., & Vischer, J. C. (Eds.) (2005). *Assessing building performance*. Amsterdam: Elsevier (republished in 2015, London: Routledge).

Preiser, W. F. E., Davis, A. T., Salama, A. M., & Hardy, A. E. (Eds.). (2015). *Architecture beyond criticism: Expert judgment and performance evaluation*. London: Routledge.

Author Biographies

Erin Persky, Associate AIA, CCHP is a San Diego-based planner specializing in justice facilities, and has over 10 years of experience conducting justice-related research. Her research career began in the field of psychology, where she investigated mental illness in the probation and inmate populations of several counties in southern California. She later moved onto civic and courts research, studying ways in which specific site and building features encourage or hinder the relationship between government and constituents. Her research currently focuses on Post-Occupancy Evaluation, and Erin is now working with the Academy of Architecture for Justice (AAJ) Research Committee on the development of a "Toolkit" of building evaluation and user satisfaction instruments for architects or owners to use to evaluate their courthouses.

Erin has presented her work to the American Institute of Architects (AIA), National Association for Court Management (NACM), and the National Commission on Correctional Healthcare (NCCHC), and has been featured in *Architect Magazine* for her research. She works closely with the AIA, having served on several committees and is serving a five-year term on the AAJ Leadership Group. Erin holds a Bachelor's Degree in Psychology and Social Behavior, *summa cum laude*, and a Master's Degree in Political Science, both from the University of California, Irvine. She also holds a Master's Degree in Architecture, *magna cum laude*, from the NewSchool of Architecture and Design. Erin is an Associate Member of the AIA, a Certified Correctional Health Professional with the NCCHC, and a bronze member of the San Diego Regional Chamber of Commerce.

Jay Farbstein, PhD, FAIA President of Jay Farbstein & Associates, Inc., Jay has more than 30 years of professional experience and is nationally recognized for his contributions in the field of facility planning, programming, and post occupancy evaluation—on which he has spoken and published widely and for which he has received many awards. Mr. Farbstein has led or participated in numerous research projects for clients including the National Institute of Corrections, US Postal Service, the US Department of Labor, The US Department of State, the World Bank, and the Bruner Foundation.

Jay has published widely on facility programming and evaluation, including *People in Places* (Prentice Hall), and articles in the AIA's Architects Handbook on Facility Programming, as well as in Wolfgang Preiser's books *Facility Programming; Programming the Built Environment, The Professional Practice of Programming*, and, as well as in many other journal's and edited volumes. He has been quoted in the New York Times, Time magazine, and many design journals, and has delivered papers to many professional and academic groups, including the AIA, Human Factors Society, Environmental Design Research Association, ASTM, and at symposia in the US, Canada, England, France, Spain, Germany, and Japan. Recently, Jay led a study of the application of neuroscience to the evaluation of correctional environments.

Mr. Farbstein earned an M. Arch from Harvard University and a PhD from the University of London. He is a Fellow of the American Institute of Architects where he co-chairs the Academy of Architecture for Justice's research program. He was chair of the Environmental Design Research Association, which awarded him its lifetime career achievement award and the CORE Certificate of Research Excellence, including a special Award of Merit for one of his research projects (done with Melissa Farling and others).

Melissa Farling, FAIA, LEED AP is managing principal of HDR in Phoenix, Arizona. She has a long track record of leadership in the profession and communities where she works. She is well-known for her research into the impacts of architecture on people, believing this knowledge is essential to enable the creation of sustainable and appropriate environments.

Ms. Farling is a member of the Leadership Group for AIA's Academy of Architecture for Justice (AAJ), served as an AAJ Research and Technology Committee co-chair from 2006–2015, is an active member on the AAJ Sustainable Justice Committee, and is on the Advisory Council for the Academy of Neuroscience for Architecture. Melissa was one of the principal investigators on a National Institute of Corrections funded study to examine impacts of views of nature on stress in a jail intake area. This study received the inaugural CORE Certificate of Research Excellence as well as a special Award of Merit from the Environmental Design Research Association. Her experience has focused on criminal justice facilities and public projects, and has led post-occupancy evaluations on criminal justice, educational, and behavioral health facilities. Melissa gives frequent presentations on evidence-based design applications and is a contributing author to the *AIA AAJ Sustainable 2030: Green Guide to Justice*, *Arizona School Design Primer: The Basic Elements of School Design* by Marlene Imirzian, and contributed the chapter, "From Intuition to Immersion: Architecture and Neuroscience" in Mind in Architecture: Neuroscience, Embodiment, and the of Design (MIT Press).

Melissa is a registered architect in Arizona holding a Bachelor's degree in Architecture from the University of North Carolina at Charlotte and Bachelor of Architecture and Master of Architecture from the University of Arizona.

Chapter 5
Synthetic Populations of Building Office Occupants and Behaviors

Jennifer A. Senick, Clinton J. Andrews, Handi Chandra Putra, Ioanna Tsoulou and MaryAnn Sorensen Allacci

5.1 Introduction

The goal of this chapter is to convey a novel approach to overcoming the limitations of case study research of building occupant behavior in workplace settings by pooling samples and creating a synthetic population of building occupants and behaviors. Synthetic populations can be used by researchers and designers of buildings to develop more accurate models of performance and behavior (Andrews et al. 2016). In the example presented here, three disparate field studies of workplace settings are combined into a larger database that is enhanced through the generation of a statistically similar synthetic data set.

5.2 Building Occupant Behavior and Synthetic Databases

Behavioral researchers know that occupants influence energy use and other aspects of building performance by their heterogeneous choices over building environmental conditions and adaptive behaviors (Bordass et al. 2001; Hewitt et al. 2015). Computer simulation models are increasingly used to assist in the building design process, but few incorporate the influence of occupant behavior (Andrews et al. 2016; Hong et al. 2015). Similarly, designers struggle to optimize comfort, satisfaction, health, and productivity of commercial office building occupants (Leaman and Bordass 1999; Heerwagen 2000), without knowing attendant parameters a priori, e.g., before the building or its occupants exist.

J.A. Senick (✉) · C.J. Andrews · H.C. Putra · I. Tsoulou · M.S. Allacci
Rutgers University, New Brunswick, NJ 08901, USA
e-mail: jsenick@rci.rutgers.edu

W.F.E. Preiser et al. (eds.), *Building Performance Evaluation*,
DOI 10.1007/978-3-319-56862-1_5

The synthetic population approach offers advantages over current methods for incorporating quantitative insights about occupant behavior into the design process: it preserves co-varying and interdependent relationships among behavioral and environmental variables and does not risk compromising confidential data. Additionally, it makes insights more transferable across behavior settings because data on contextual variables accompany the behavioral data. This allows hypothesis testing regarding the generalizability, or context-dependence, of behavioral patterns.

While procedures for developing synthetic populations have not previously been applied to the study of building occupant behavior, they have been well represented in other fields such as transportation planning and public health. An important resource in this area is FRED (Framework for Reconstructing Epidemic Dynamics), an open-access census-based synthetic population used in agent-based epidemic modeling (Grefenstette et al. 2013). Another large synthesized database created by RTI International assigns US population in 50 states and the District of Columbia to schools, workplaces, military bases, dormitories, prisons, and nursing homes and other contextual settings (Wheaton et al. 2009).

In generating synthesized populations, there are four commonly utilized approaches: deterministic reweighting, conditional probability, Monte Carlo simulation, and combinatorial optimization, e.g., simulated annealing. Each of these has advantages and disadvantages, depending on such factors as amount and quality of data, and computing resources available (Harland et al. 2012; Wheaton et al. 2009). This implementation using R, an open source platform for statistical computing, relies on classification and regression trees (CART) to generate synthetic populations (Nowok et al. 2015).

5.3 Creating a Synthetic Population of Building Occupants

This section presents a 10-step process of assembling three diverse occupant behavior datasets to serve as the basis for a synthetic population of building occupants.

- Establishment of a theoretical/methodological rationale based on variables of interest.
- Identification of studies that include these variables and measures.
- Location of data and coding for each study.
- Identification of common variables of interest.
- Establishment of an equivalence basis for converting the coding schemes to a common one.
- Pooling of the discrete datasets into one database.
- Analysis of key multivariate relationships in the underlying datasets.

- Reproduction of these investigations utilizing the aggregated database.
- Creation of synthesized database.
- Validation of synthetic database.

Rationale for Combining Datasets (Step 1):
The rationale for combining datasets is to gain greater predictive power in understanding occupant behaviors, while establishing the foundation for a synthetically enhanced database as a robust and reliable representation of expected building occupant behavior. For instance, in Andrews et al. (2016), it is demonstrated that better information about adaptive behaviors greatly improves energy model accuracy.

Identification of Existing Datasets (Steps 2–3):
The individual datasets drawn upon here are comprised of workplace-based, cross-sectional, longitudinal, time-series POE (post-occupancy evaluation) studies, and a large-scale ASHRAE RP-884 dataset. They are variable in terms of occupant age, sex, daily average outdoor temperature, relative humidity, and indoor air temperature. However, all sets show some significantly consistent patterns, especially regarding variables of interest, e.g., availability of portable space heaters, availability of portable fan, and their frequency of use. Links to all three datasets are provided in the bibliography.

- Cross-Sectional Dataset

The cross-sectional dataset contains variables on thermal and lighting comfort and satisfaction as drawn from 6 separate POEs conducted by Rutgers Center for Green Building researchers (2016) in 16 low- to mid-rise commercial office buildings between 2009 and 2014. Two of the studies include time-series data on occupants' responses to building conditions during actual and simulated demand response events, i.e., load shedding experiments; the majority of the data is cross-sectional. A total of 954 occupant records were collected.

- Longitudinal Dataset

This dataset consists of 24 occupant records of thermal comfort of responses to questions about thermal comfort and related behavioral adaptations from a longitudinal case study of a single office building. Twice-daily online surveys were administered for two-week periods in four seasons of one year for 2012 and 2013 and were accompanied by more frequent observations through datalogger measurements of indoor and outdoor temperatures and other environmental factors (Langevin 2015; Langevin et al. 2015).

- ASHRAE-RP-884 Dataset

Data on thermal comfort were collected from multiple projects and various researchers of 160 buildings worldwide to assemble this mix of cross-sectional and longitudinal data recorded during 1982–1997. All projects utilized a standard template that organized records in the following groups of variables: basic identifiers, thermal comfort questionnaire, indoor climate physical observations, calculated indices, personal environmental control, and outdoor meteorological observations. The resulting 20,215 occupant records additionally incorporated quality control measures throughout an adaptive modeling method (de Dear et al. 1997; de Dear and Brager 1998).

Establishing the Basis for the Pooled Database (Steps 4–5):
The individual datasets, although diverse, share a large number of variables in common. In some cases, the common variables of interest were coded differently. Such differences need to be addressed to create the pooled database. There were also a number of fields that existed in only some of the sets, but could be easily reproduced. Lastly, there were a number of fields that were relevant to the research scope but did not have values in all sets.

The next step is to create a standard template for the pooled database that takes into account similarities and differences among the datasets. The resulting template includes all common fields from the 3 sets, the number of fields that could be easily reproduced, as well as those that were incomplete, but still relevant to the scope of the research project and deemed valuable for future use.

In order to comply with the final database template fields, units and coding, the underlying datasets require modification. Illustratively, to the cross-sectional dataset 7 fields were added and sources, units and coding were standardized relating to such items as meteorological data, indoor environmental observations, lighting, and thermal adjustment access and use. Similar modifications were made also to the longitudinal and ASHRAE RP-884 datasets.

Creation of the Final Database (Steps 6–8):
To locate common explanatory variables to predict occupant behaviors, logistic regressions were run against the three data sets and the combined data set. Table 5.1 shows that with significantly more observations than the other two data sets, ASHRAE RP-884 performs the best in predicting a variable of interest—use of portable fan, while cross-sectional data performs the worst. Regression results for the pooled data set show that mean indoor air temperature, occupant's age, and occupant's gender significantly predict the use of portable fan at the 0.001 level. Outdoor air temperature explains the use of portable fan at the 0.01 level of significance.

The final database acquires characteristics of the three underlying datasets in proportion to the number of observations in each. As expected, the larger ASHRAE RP-884 dataset is most influential in the combined database, while the cross-sectional dataset, which is the smallest, is least influential. Compared to the others, ASHRAE RP-884 has a higher mean age, more naturally ventilated buildings, and more measurements from warmer climates.

Generating and Validating Synthetic Population (Steps 9 and 10)
Generating a synthetic version of data fits the original dataset to the assumed distribution and obtains its parameters estimates. For detailed instruction on how to generate the synthetic population using R, interested readers may consult the package documentation (Nowok et al. 2015).

Table 5.1 Regression results comparing the three data sets and synthetic database *Source* Authors 2016

Selection	Longitudinal		Cross-sectional		RP-884		Combined	
Data set	Observed	Synthetic	Observed	Synthetic	Observed	Synthetic	Observed	Synthetic
Dep. Var.	Hportfan							
Expl. Var.	Coeff (SD)							
Dayav_ta	−0.014* (0.006)	0.001 (0.006)	−0.03 (0.05)	0.004 (0.041)	0.032*** (0.004)	0.059*** (0.004)	0.01** (0.003)	0.03*** (0.003)
TA_M	0.182*** (0.04)	0.131*** (0.039)	0.43* (0.24)	0.05 (0.175)	0.421*** (0.011)	0.336*** (0.009)	0.419*** (0.010)	0.342*** (0.009)
Age	0.450*** (0.05)	0.427*** (0.046)	−0.24* (0.137)	−0.129 (0.169)	−0.283*** (0.023)	−0.246*** (0.023)	−0.067*** (0.018)	−0.04* (0.018)
Sex	0.405** (0.41)	0.266* (0.124)	−0.82* (0.367)	−0.535 (0.432)	0.185** (0.054)	0.296*** (0.063)	0.281*** (0.047)	0.366*** (0.05)
_cons	−7.55*** (0.96)	−6.12*** (0.915)	−8.12 (5.84)	−1.18 (4.54)	−11.2*** (0.263)	9.88*** (0.246)	11.29*** (0.249)	9.904*** (0.232)
Number of obs	2023	2071	147	128	10,496	10,289	12,666	12,488
LR $chi^2(5)$	193.92	167.67	13.18	2.67	4975.87	4488.54	4719.94	4204.32
Prob > chi^2	0.000***	0.000***	0.01*	0.6152	0.000***	0.000***	0.000***	0.000***
Pseudo R^2	0.0785	0.0642	0.0674	0.0191	0.354	0.3241	0.2813	0.2524

Asterisks show which variables are significant at the 0.1 level (*), 0.01 level (**), and 0.001 level (***)

Table 5.2 Descriptive statistics of pooled observed data and synthesized data. *Source* Authors 2016

Variable	Source	Mean	SD	Minimum	Maximum	N
Age	Observed	2.72	1.42	1	7	21,990
	Synthetic	2.73	1.42	1	7	22,004
Sex	Observed	1.51	0.5	1	2	27,500
	Synthetic	1.51	0.5	1	2	27,465
Dayav_ta	Observed	18.5	9.92	−24.9	35	28,299
	Synthetic	18.48	9.89	−24.9	35	28,262
TA_M	Observed	22.22	7.91	0	42.67	27,814
	Synthetic	22.21	7.92	0	42.5	27,804
Hportfan	Observed	0.37	0.48	0	1	16,780
	Synthetic	0.38	0.48	0	1	16,551

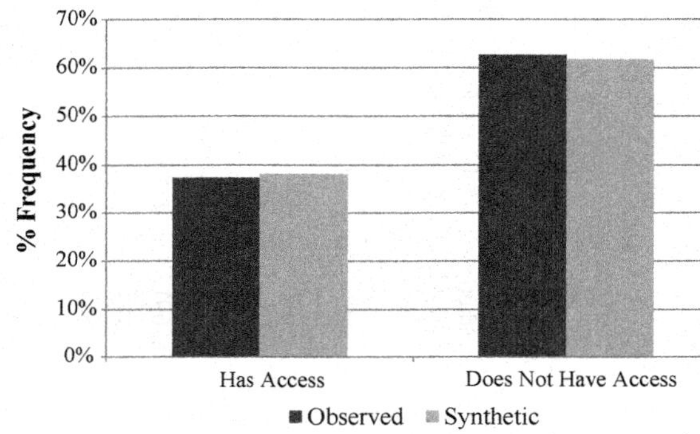

Fig. 5.1 Relative frequency distribution of use of portable fan (hportfan) for observed and synthetic data. *Source* Authors, 2016

After running the synthesis process, the resulting data needs to be validated with the original observed data by comparing statistical inferences. In this example, an ordered logistic regression is estimated where the dependent variable is the use of portable fan. For explanatory variables, the model uses outdoor air temperature, indoor air temperature, sex, and age. Descriptive statistics compare actual data and synthesized data. Results from the descriptive statistics support the conclusion that both data have similar measures of central tendency and similar measures of spread (see Table 5.2).

Another way to compare both data sets is by evaluating the relative frequency distributions of key variables. In this example, the frequency distributions of the use of portable fan in the observed data set and the synthetic one are compared (see Fig. 5.1).

Alternatively, the synthesis process can be validated by comparing goodness-of-fit for both the observed and synthesized data (see Fig. 5.2). Estimates of the 95% confidence interval and Z statistics from a logistic regression of the use of portable fan for both data were relatively similar. These findings show that in both data sets, mean indoor air temperature and gender are significant at 0.001 level. Both data sets also show that outdoor temperature is significant at the 0.001 level for the cohort of buildings with HVAC.

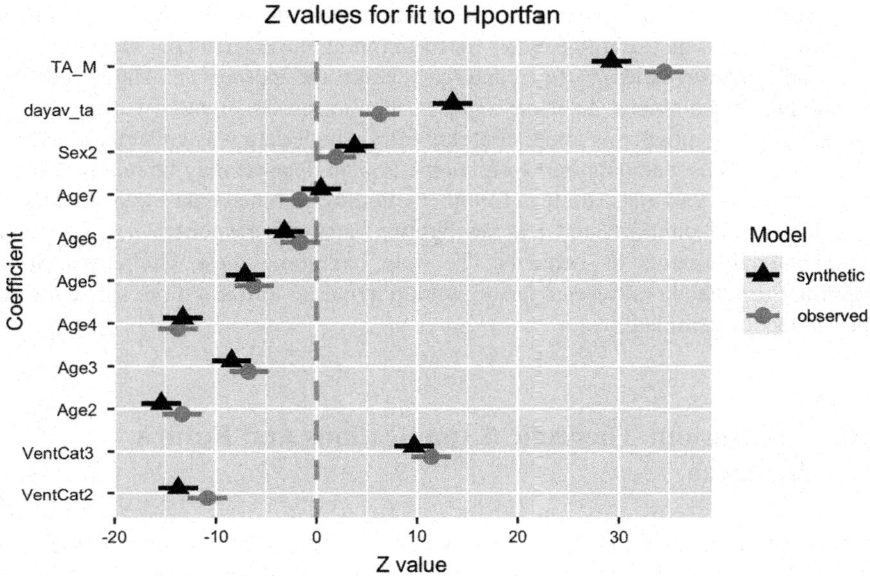

Fig. 5.2 Estimates for Z statistics from a logistic regression of the use of portable fan for both observed and synthetic data. *Source* Authors, 2016

5.4 Uses of Synthesized Data

The creation of a synthetic population of building occupants and their behaviors enables researchers and designers of buildings to use these synthetic data in Building Energy Modeling (BEM), Agent Based Modeling (ABM), and other co-simulation methods thereby extending the framework previously advanced by Andrews et al. (2011), and Andrews et al. (2012). Synthesized data is additionally a promising basis for supplementing the practice of generative POE (Wener et al. 2016), allowing for grounded-hypotheses derived from case study data to be evaluated using the larger synthetic database.

5.5 Methodological Implications and Limitations

This chapter demonstrates that creating a synthetic set of generic building occupants is feasible for leveraging smaller POE-based studies of building occupant behavior. Guidance for assembling disparate databases is provided, resulting in an adequate foundation of building occupant characteristics and behaviors from which to generate a representative population using the R statistical software package. The synthetic database preserves confidentiality while capturing building occupants' interactions and covariation in commercial buildings. The application of long-standing synthetic population techniques to the study of building occupant behavior presents an exciting frontier for overcoming the current constraint of many case studies. Nevertheless, a few cautionary notes are appropriate. The process of assembling the combined database requires attention to detail and a willingness to make subjective judgments about contexts from which data was collected, missing data fields, and needed data transformations. Pooled datasets may be relatively rich in data for some variables and lean for others. In the combined dataset, there is more data on thermal comfort, and less on lighting, adaptive responses and social or organizational aspects of behavior. To help overcome these challenges, it is important for behavioral researchers to work together to utilize a more standardized data collection template.

5.6 Conclusion: Theoretical Implications and Future Research

In future and on-going work, the authors extend the synthetic database approach beyond standard hypotheses about building occupant behavior to a series of analyses that delve deeper into the roles of organizational determinants of behavior and contexts, and seek to integrate energy modeling and agent based modeling simulations. These analyses intend to draw on data derived from POEs as well as from large scale representative databases such as the US Department's Commercial Building Energy Consumption (CBECS) database and the American Community Survey of the US Census Bureau. Other opportunities for enhancing studies of building occupant behavior with synthetic data reside in 3-D imaging techniques for more detailed characterizations of building contexts.

Acknowledgements This research was supported by the Consortium for Building Energy Innovation, sponsored by the US Department of Energy Award Number DE-EE0004261.

References

Andrews, C. J., Yi, D., Krogmann, U., Senick, J. A., & Wener, R. E. (2011). Designing buildings for real occupants: An agent-based approach. *IEEE Transactions on Systems, Man, and Cybernetics-Part A: Systems and Humans, 41*(6), 1077–1091.

Andrews, C. J., Senick, J. A., & Wener, R. E. (2012). Incorporating occupant perceptions and behavior into BIM. In S. Mallory-Hill, W. Preiser, & C. Watson (Eds.), *Enhancing building performance* (pp. 234–246). Oxford, UK: Blackwell Publishing.

Andrews, C. J., Allacci, M., Senick, J. A., Putra, H. C., & Tsoulou, I. (2016). *Using synthetic population data for prospective modeling of occupant behavior during design.* Accepted for publication in Energy and Buildings special issue on occupancy behavior.

Bordass, B., Leaman, A., & Ruyssevelt, P. (2001). Assessing building performance in use 5: Conclusions and implications. *Building Research & Information, 29*(2), 144–157.

de Dear, R. J., Brager, G. S. & Cooper, D. (1997). ASHRAE RP-884 dataset link: http://sydney.edu.au/architecture/staff/homepage/richard_de_dear/ashrae_rp-884.shtml. Accessed 04 May 2016.

de Dear, R. J., & Brager, G. S. (1998). Towards an adaptive model of thermal comfort and preference. *ASHRAE Transactions, 104*(1), 145–167.

Grefenstette, J. J., Brown, S. T., Rosenfeld, R., DePasse, J., Stone, N. T. B., Cooley, P. C., et al. (2013). FRED (a framework for reconstructing epidemic dynamics): An open-source software system for modeling infectious diseases and control strategies using census-based populations. *BioMed Central Ltd.* doi:10.1186/1471-2458-13-940.

Harland, K., Heppendstall, A., Smith, D., & Birkin, M. (2012). Creating realistic synthetic populations at varying spatial scales: A comparative critique of population synthesis techniques. *Journal of Artificial Societies and Social Simulation 15*(1):1. http://jasss.soc.surrey.ac.uk/15/1/1.html. Accessed 04 May 2016.

Heerwagen, J. (2000). Green buildings, organizational success and occupant productivity. *Building Research & Information, 28*(5), 353–367.

Hewitt, E., Andrews, C., Senick, J., Wener, R., Krogmann, U., & Sorensen Allacci, M. (2015). Distinguishing between green building occupants' reasoned and unplanned behaviors. *Building Research & Information.* doi:10.1080/09613218.2015.1015854.

Hong, T., D'Oca, S., Turner, W. J. N., & Taylor-Lange, S. C. (2015). An ontology to represent energy-related occupant behavior in buildings, Part I: Introduction to the DNAS framework. *Building and Environment, 92*(2015), 764–777. doi:10.1016/j.buildenv.2015.02.019.

Langevin, J. (2015). Longitudinal dataset. http://en.openei.org/datasets/dataset/one-year-behavior-environment-data-for-medium-office. Accessed 04 May 2016.

Langevin, J., Wen, J., & Gurian, P. L. (2015). Simulating the human-building interaction: Development and validation of an agent-based model of office occupant behaviors. *Building and Environment, 88,* 27–45. doi:10.1016/j.buildenv.2014.11.037.

Leaman, A., & Bordass, B. (1999). Productivity in Buildings: the "Killer" variables. *Building Research & Information, 27*(1), 4–20.

Nowok, B., Raab, G. M., & Dibben, C. (2015). *Synthpop: Bespoke creation of synthetic data in R. Package.* vignette http://cran.r-project.org/web/packages/synthpop/vignettes/synthpop.pdf. Accessed 22 Nov 2015.

Rutgers Center for Green Building. (2016). Cross-sectional dataset (2009–2014). http://en.openei.org/datasets/dataset/ob-commercial-building. Accessed 04 May 2016.

Wener, R., McCunn, L., & Senick, J. (2016). Did that plan work? Post occupancy evaluation. In R. Gifford (Ed.), *Research methods for environmental psychology* (pp. 249–270). Hoboken, NJ: Wiley-Blackwell.

Wheaton, W. D., Cajka, J. C., Chasteen, B. M., Wagener, D. K., Cooley, P. C., Ganapathi, L., et al. (2009). Synthesized population databases: A US geospatial database for agent-based models. *Methods Report.* doi:10.3768/rtipress.2009.mr.0010.0905.

Author Biographies

Jennifer A. Senick Rutgers Center for Green Building and Edward J. Bloustein School of Planning and Public Policy at Rutgers University.

Senick holds a Ph.D. in Planning and Public Policy from Rutgers, an M.A. in Political Science from UCLA, and an A.B. in Government from Bowdoin College. She is Executive Director of the Rutgers Center for Green Building at the Edward J. Bloustein School of Urban Planning and Policy Development, Rutgers University and is well published on topics in green building and sustainable development, environmental behavior, evaluation methods. Current appointments include as Board Member and Treasurer of the Environmental Design Research Association (EDRA) and co-chair of the EDRA Sustainable Planning Design and Behavior Network. Senick has worked in numerous private sector positions in areas of emerging technology and has founded two successful start-up companies in energy planning and environmental assessments.

Clinton J. Andrews Rutgers Center for Green Building and Edward J. Bloustein School of Planning and Public Policy at Rutgers University.

Andrews, Ph.D. PE AICP, was educated at Brown and MIT as an engineer and planner. He is Director of the Center for Green Building, Professor of Urban Planning and Policy Development, and Associate Dean for Planning and New Initiatives at the Edward J. Bloustein School, Rutgers University, New Brunswick, New Jersey. His research addresses behavioral, policy and planning questions related to energy use in the built environment. He has worked in the private sector as a design engineer and technology assessor, helped launch an energy planning project at MIT, and helped to found a science policy program at Princeton. At Rutgers, he has launched initiatives in energy planning and green building. He has published over 100 scholarly and popular articles and his books include *Humble Analysis: The Practice of Joint fact-Finding*, *Regulating Regional Power Systems*, and *Industrial Ecology and Global Change*. He is the new editor (with Frank Popper) of the *Journal of Planning Education and Research*.

Handi Chandra Putra Rutgers Center for Green Building and Edward J. Bloustein School of Planning and Public Policy at Rutgers University.

Chandra Putra holds a Master of City and Regional Planning (MCRP) from Rutgers University. He is a Research Assistant at the Rutgers Center for Green Building and a doctoral student at the Edward J. Bloustein School of Planning and Public Policy at Rutgers University. His dissertation investigates the effects of coastal flooding on housing market dynamics by using spatial econometrics and agent-based modeling (ABM). He has served as a reviewer to the Journal of Artificial Societies and Artificial Social Simulation.

Ioanna Tsoulou Rutgers Center for Green Building and Edward J. Bloustein School of Planning and Public Policy at Rutgers University.

Tsoulou received a Master of City and Regional Planning (MCRP) from the Edward J. Bloustein School of Planning and Public Policy, Rutgers University, NJ with concentration in Environmental Planning and Design and a Master of Engineering from the Department of Civil Engineering, University of Patras, Greece with Concentration in Transportation and Environmental Engineering. She is a Research Assistant at the Rutgers Center for Green Building and a doctoral student at the Edward J. Bloustein School of Planning and Public Policy at Rutgers University. Her research interests lay in the areas of Sustainability and Environmental Land Use Planning, Intelligent Transportation Systems, Green Buildings, Neighborhood Planning, Urban Design and Smart Growth.

MaryAnn Sorensen Allacci Rutgers Center for Green Building and Edward J. Bloustein School of Planning and Public Policy at Rutgers University.

Sorensen Allacci, has a Ph.D. from CUNY's Environmental Psychology Program. She has been the Research Project Coordinator for the Rutgers Center for Green Building and specializes in community based collaborative research and engagement with organizations and diverse groups including low income, ethnic minority communities, and individuals with disabilities. Her particular focus is environmental health and environmental justice concerns in housing and community environments, and she teaches these and other aspects of to interior designers. Dr. Sorensen Allacci is also president of the board for Projects for Environmental Health, Knowledge, and Action, Inc., a non-profit organization.

Chapter 6
From Pre- to Post-occupancy Evaluations: Acceptance of Intelligent Building Technologies

Ulrich Schramm, Sybille Reichart and Dominic Becking

6.1 Introduction

This chapter is based on several studies that indicate an increasing use of building automation systems in new buildings currently and in the near future, while other sources—at the same time—list a few factors that continue to limit the acceptance of smart home technology by occupants (Hille 2013; Spath 2012). Therefore the focus of this study is on building users' needs, worries, and concerns throughout the lifecycle of a building. What are the barriers to acceptance? Which measures will increase well-being and acceptance?

In order to find answers to these questions, a team of professors, research assistants, and master's students from the disciplines of architecture, psychology, and computer science at the Bielefeld University of Applied Sciences started to implement the multi-disciplinary research project 'Well-Being and Acceptance in the Intelligent Building', funded through the State Ministry of Innovation, Science and Research, located in the state of North Rhine-Westphalia, Germany. Using the construction of the demonstration and research facility 'Intelligent Building Technologies' on the Minden Campus as a major case study (Schramm in press), pre-occupancy evaluations in the form of surveys, behavioral studies, and usability tests were carried out with students. In the near future, results from these

U. Schramm (✉) · D. Becking
Bielefeld University of Applied Sciences, Artilleriestrasse 9,
32427 Minden, Germany
e-mail: ulrich.schramm@fh-bielefeld.de

D. Becking
e-mail: dominic.becking@fh-bielefeld.de

S. Reichart
Bielefeld University of Applied Sciences, Interaktion 1, 33619
Bielefeld, Germany
e-mail: sybille.reichart@fh-bielefeld.de

W.F.E. Preiser et al. (eds.), *Building Performance Evaluation*,
DOI 10.1007/978-3-319-56862-1_6

pre-occupancy evaluations will be complemented by investigative post-occupancy evaluations (POE). The relevant POE techniques were developed and successfully tested using another intelligent university building on the Bielefeld Campus. With more POEs applied to various settings within the new Campus building in Minden, further insights into building users' well-being and acceptance of the intelligent building as well as short, medium, and long term perspectives on how to improve the situation will be generated.

6.2 Pre-occupancy Evaluations

6.2.1 Building Users' Expectations, Concerns, and Attitudes Regarding Intelligent Buildings

Perceived usefulness and perceived ease of use are the two most important criteria for the acceptance of a technological innovation such as intelligent building technologies (Davis 1989). However, perceived pressure from technical configurations, the so-called techno-stress (Tarafdar et al. 2011) causes rejection. Future occupants will start to develop an attitude towards the new building and its building technology from the announcement of the new construction or modification of an existing building. This attitude also includes an estimation of anticipated benefits and anticipated stress.

Another important element for user acceptance of intelligent building technologies is the perceived influence on the design and the technology of a building. A lack of influence and understanding can result in decreasing user acceptance and increasing user resistance towards building technologies. This behavior is the result of 'reactance', a reaction towards a constraint of possibilities to act (Dickenberger et al. 2002) and may lead to disuse of a building technology system, and in worst cases even to its destruction.

Students at the Minden Campus were given a questionnaire and asked about their expectations and concerns, as well as perceived participation possibilities. 344 students participated in this survey, which took place a year before the actual occupancy of the new building. The students expressed high expectations regarding the range of functions of building technologies and only minor concerns regarding malfunctions and their usefulness. On the contrary, they more often expressed fears about personal data security and expected deficits in upholding data security in an intelligent building (Schimweg 2014). A study about students' needs regarding participation in planning and implementation showed a significant interest in the architecture and civil engineering disciplines, and a lower need for participation in the other fields of study. Simultaneously, 98% of students reported that they actually did not participate in the new Campus building project at all – neither in programming, in design development, or in any construction site visits (Szymura 2014).

6.2.2 Usability-Testing of Room Control Unit

Besides the perceived usefulness, the perceived ease of use determines the acceptance of a technological innovation (Davis 1989). Therefore, in general, usability tests are performed with prospective users testing a technical product with a human-machine-interface in a properly equipped laboratory. Here, the test participants perform representative tasks in order to test a product's ease of use. The objective is to deduce references for potential improvement of man-machine-interaction (Sarodnick and Brau 2011). Therefore, focusing on pre-occupancy evaluations, the need became obvious to test the usability of intelligent building technologies before moving in.

Both offices and seminar rooms of the new Campus building in Minden were planned to be furnished with a room control unit. It is a multifunctional device with at least ten room control functions regarding heating and air conditioning, as well as lighting and blinds. The device consists of an LCD-display and eight control keys.

Eleven people performed the usability test of the room control unit. The participants were asked to open and close the blinds, switch on the lights, and activate a special lighting scenario ("presentation"). As the test was carried out at the research laboratory of Business Psychology, the effect of each test was indicated by LED-lamps mounted on a wooden panel. For each task of the test, the numbers of activated control keys, time to perform each task as well as type and number of handling errors were recorded. The results indicate that some of the participants needed up to ten times as many key strokes than others to perform their tasks (see Fig. 6.1). Based on the operating errors, important recommendations to rearrange the key assignments have been developed (Maahs 2015).

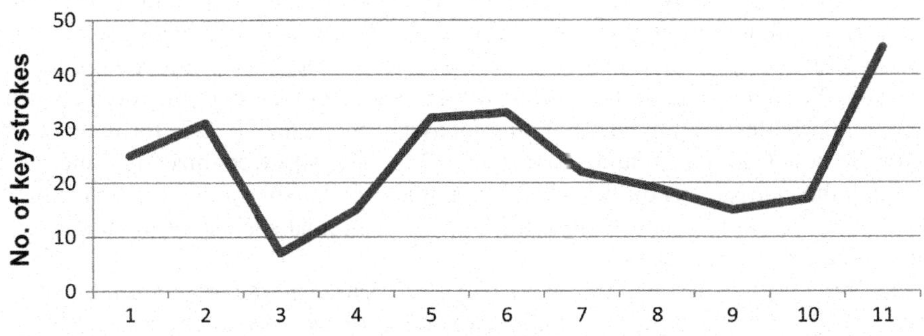

Fig. 6.1 Number of key strokes per test person to lower the blinds for sun protection. *Source* Ulrich Schramm based on Maahs (2015, p. 30)

6.2.3 Proto-Type Testing of Micro-Curricula Applications via Smart Phones

New and unusual operating concepts can prove to be an obstacle for interacting with a technical device, especially if the device is something as complex as a building with modern building technologies. The more complex the technology becomes the bigger is the risk for unwanted and unforeseen side effect such as 'reactance', as mentioned before. Thus, Human-Building Interaction is becoming a part of the scientific field of Human-Computer Interaction. But one has to bear in mind that the 'traditional' interaction between occupant and building—as opposed to computers—is deeply influenced by affordances (Gibson 1979; Chemeno 2001). However, modern human-building interaction devices lack these affordances and offer different and unknown means of operating features of the building. Therefore, micro-curricula modules are proposed—short situation and user awareness teaching units—to enhance user experience, acceptance, and well-being of occupants in intelligent buildings, especially where user interaction is not influenced by affordances. The goal is to offer these teaching units to all users and visitors of non-residential buildings who are not familiar with the specific human-building interaction devices. To achieve this, the user and situation awareness is essential because aspects of a user model, e.g., language, access rights, and earlier experiences, need to be reflected in the formulation and the presentation of the micro-curricula modules. Situation and user awareness can be achieved if the means of presentation are at the same time the means of detection of the situation, as well as location and identification of the user. Thus, the decision to use the users' own mobile devices for this purpose came naturally.

In the context of the new Campus building in Minden, a mobile app was developed which presents micro-curricula in augmented reality (Becking et al. 2015). Augmented reality for mobile learning is a widely discussed subject. Olsson et al. (2013), for example, did research on user expectations and found that information should be context-sensitive, intuitive, lively, and entertaining as well as meaningful and efficient. To address these findings the app as a prototype is used to measure the effects of context-sensitive micro-curricula on users of the room control unit, a poorly designed human-building interface (Siemens UP 227) for interaction with the new Campus building (see Sect. 6.2.2). The app recognizes the interface through the camera lens of the users' smartphones and displays the micro-curricula modules needed to explain steps to achieve a certain goal such as lifting the blinds of a window (see Fig. 6.2).

Usability tests have been run in a lab environment using methods from business psychology. Test participants were confronted with the user interface and had to solve tasks with the help of the micro-curricula. To quantify results, the key strokes needed to achieve a goal were counted and compared with the minimal number of key strokes. Interviews and surveys were done as well to measure acceptance of the app and the underlying degree of assistance to the users. Most of the test participants found the app helpful and would prefer micro-curricula in augmented reality over printed or pdf instruction manuals (Maldener 2015).

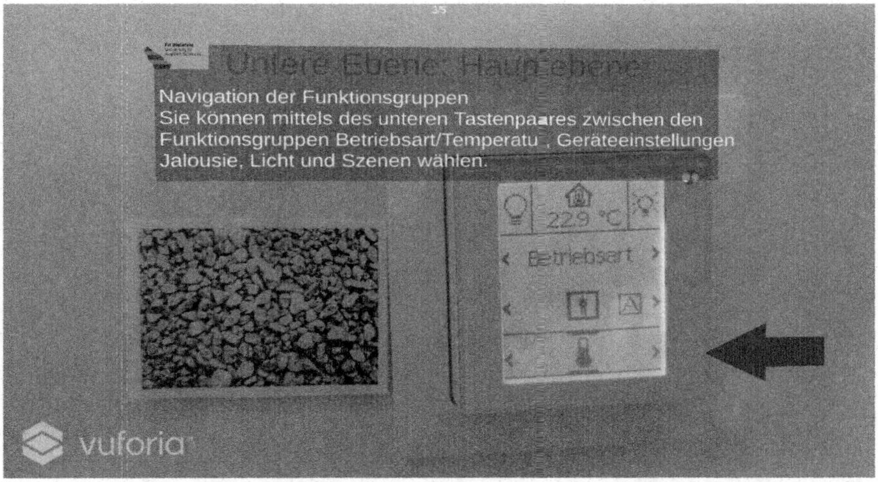

Fig. 6.2 App screenshot (with provisional photo marker to facilitate orientation). *Source* Dominic Becking based on Becking et al. (2015, p. 175)

6.3 Post-occupancy Evaluations

In the context of the new Campus building in Minden, the results of these pre-occupancy evaluations will be complemented by investigative POEs. The relevant techniques were developed and tested using another intelligent University building on the Bielefeld Campus.

6.3.1 POE of the Cognitive Interaction Technology Center of Excellence (CITEC), Bielefeld Campus

- The building and its users:
  The CITEC-building was opened in 2013. The 230 building users conduct basic research in the field of future technical systems, focusing on the interaction of man and machine. The cube-like building is five stories high (see Fig. 6.3), with a basement for technical equipment and parking; the first floor has lab core facilities and a conference room for 200 persons; the upper three floors offer office space, including three green interior courtyards. In total, almost 8000 sqm (86,000 sq ft) net usable space is available.
- Conducting the POE:
  An investigative POE was done by Köhn in his Master thesis (Köhn 2014), which was honored with the grand prize of the German Facility Management Association (GEFMA) in 2015. The aim was to determine the building's

Fig. 6.3 The CITEC building on the Bielefeld Campus with automatic blinds for sun protection lowered. *Source* Ulrich Schramm

strengths and weaknesses in general, and the factors influencing building users' acceptance of intelligent building technologies in particular. Three tools for data collection were used: walk-through evaluations; survey with an online-questionnaire covering 170 aspects, answered completely by 60% of the users; and, finally, interviews with field related experts.

- Selected findings:
  Building users were asked about the usefulness of certain technologies in general and their personal satisfaction with the technology in the CITEC-building in particular. For example, 60% of the users do not see the motion sensor that regulates artificial light to be useful in their office, and even 70% are not satisfied with the sensors as light is turning off while they are working (see Fig. 6.4). More than 80% of the employees are not satisfied with the automatic sun protection in their office as the system is very sensitive to wind movement: the blinds are moving up exposing the monitor to bright sunshine.

  It turns out that the indoor climate and thus comfort of building users has a great impact on the acceptance of the building. In this context, on the one hand, personal control of indoor climate factors is considered to be important by 90%

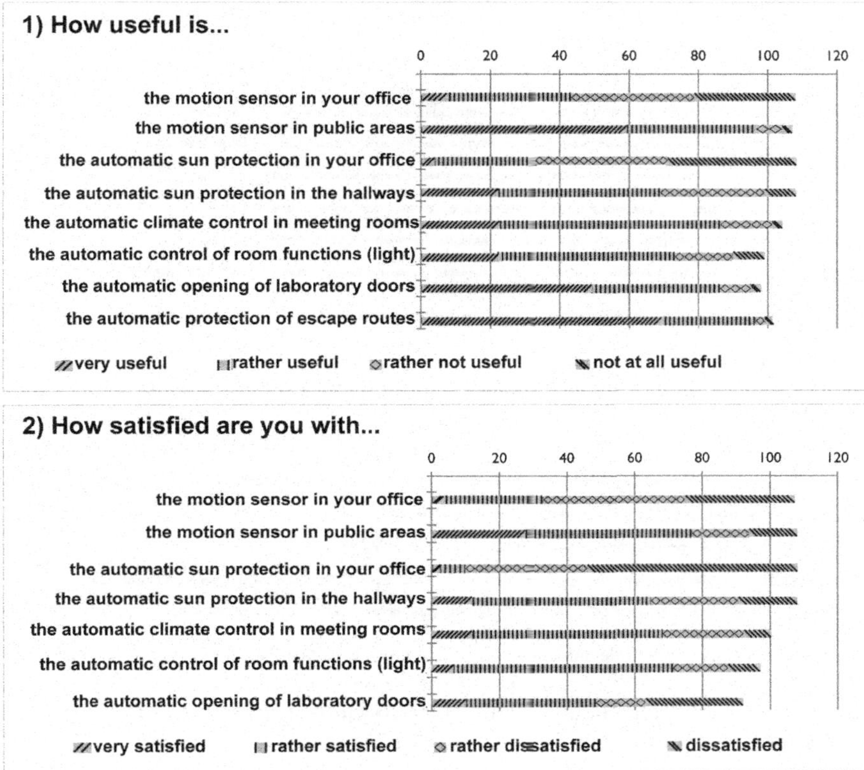

Fig. 6.4 Usefulness (*1*) and satisfaction (*2*) with intelligent building technologies. *Source* Ulrich Schramm based on Köhn (2014, p. 103)

of the building users (see Fig. 6.5 above). On the other hand, and in reality, only 50% say that they have influence on indoor climate factors like room temperature or sun protection (see Fig. 6.5 middle, and Fig. 6.3). In general, the study shows that people have little knowledge about the building technologies installed in the facility: 70% consider the information level to be poor (see Fig. 6.5 below). Moreover, 72% say they would have loved to participate in the development of measures related to intelligent building technologies. In reality, 95% of the occupants did not participate in the planning of the building's architecture, and 98% were not involved regarding intelligent building technologies (Köhn 2014).

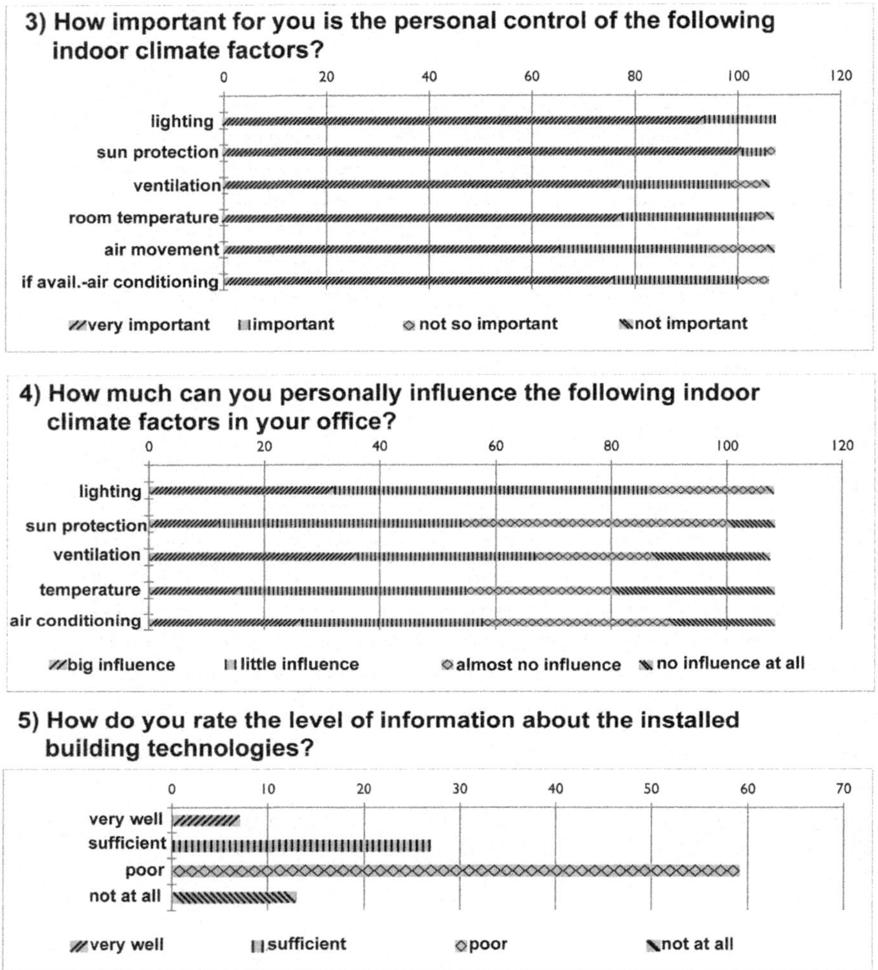

Fig. 6.5 Importance of personal control (*3*), level of influence (*4*) on indoor climate factors, and level of information (*5*) about installed intelligent building technologies. *Source* Ulrich Schramm based on Köhn (2014, pp. 107 and 112)

6.3.2 POE of the New Intelligent Campus Building, Minden Campus

- The building and its users:
  With the expansion of the Minden Campus an entirely new five-story building with more than 3000 sqm (32,000 sq ft) net usable space was planned and opened in Fall 2015: simply said, the basement is used for technical equipment,

the first floor for the cafeteria with 230 seats, the second floor for the library with an atrium (see Fig. 6.6), and the third and fourth floor for seminar rooms, labs, and offices. The energy-efficient building uses intelligent technologies like

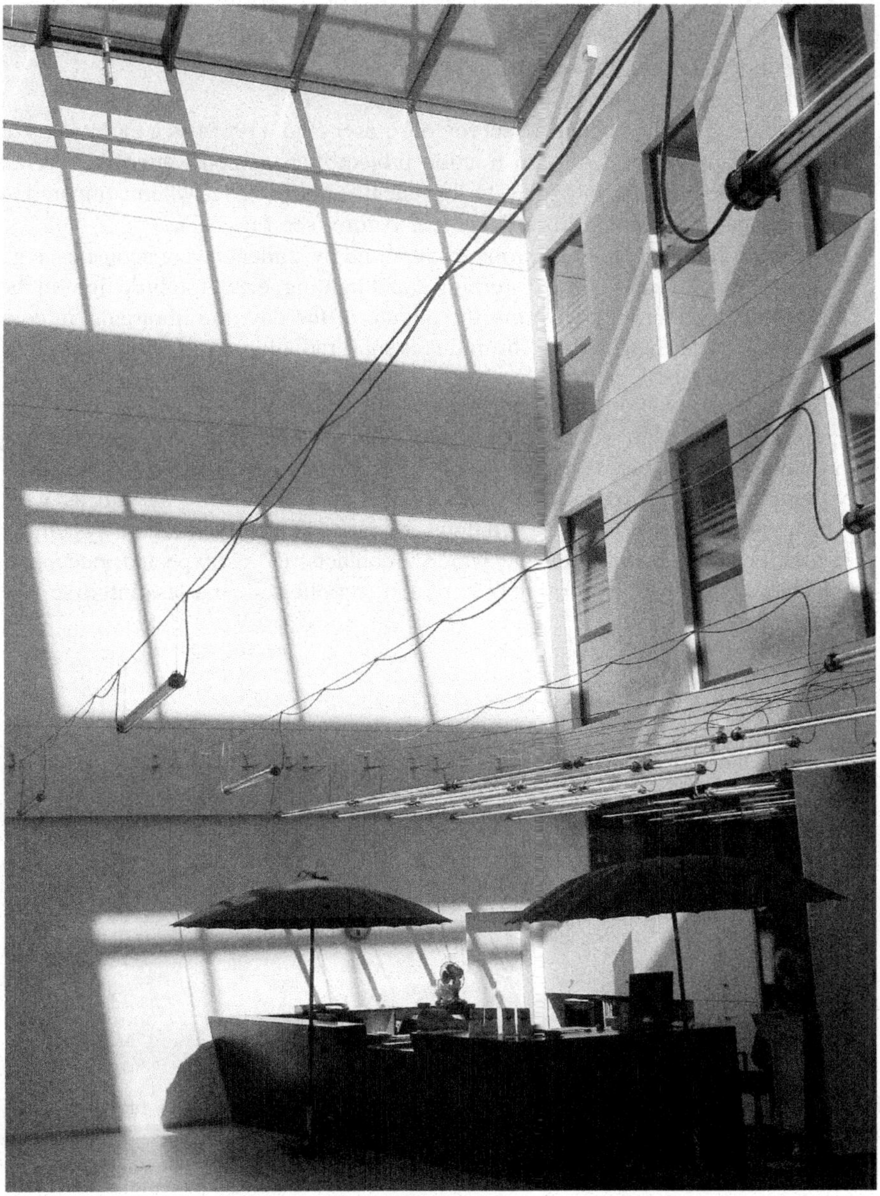

Fig. 6.6 Atrium in the library with improvised sunshades for the librarians. *Source* Ulrich Schramm

geothermal energy, photovoltaics, and building automation systems. Given these technologies, it is considered to be a demonstration and research facility.

- Conducting the POE:
After a settling-in period of an entire year, the building's performance will be objectively assessed by investigative POEs based on Köhn's proven instruments (see Sect. 6.3.1). Nevertheless, some indicative POEs and related psychological studies on well-being have already been carried out.
- Selected findings - Library:
Using the established occupant survey of Preiser et al. (1988) as a tool, teams of students from the field of 'construction project management' rated the overall quality of the new library 'good' to 'excellent'—much better when compared to the evaluation of the old library the year before (see Fig. 6.7).
Right now, the most obvious problems reported by students were acoustics, e.g., "sound is perceived as being disturbing", and lighting, e.g., "too little light in the periodicals reading area". During the middle of the day, the librarians, in contrast, suffer at their workplace from direct solar radiation through the glass roof of the atrium (see Fig. 6.6).
- Selected findings—Offices:
Empirical studies of students from the field of 'business psychology' show problems with intelligent technologies: (1) occupants are disturbed by noise, i.e., volume and pitch, generated by the lighting and ventilation system as well. Thus, individuals open their windows to turn off the ventilation system as regulation depends on door and window contacts; (2) as expected, occupants experience the room control units to be too complicated, and less intuitive (see Sect. 6.2.2).

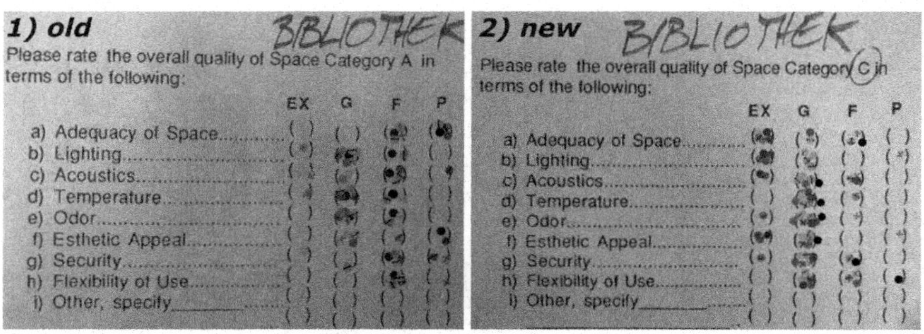

Fig. 6.7 Overall quality of the old library in 2015 (*1*) and the quality of the new one in 2016 (*2*). *Source* Ulrich Schramm

6.3.3 Recommended Actions

- Short-term actions:
  Electronic ballasts were replaced to operate the fluorescent lamps in the offices with less noise. A horizontal shading system in the atrium is in the planning stage in order to make the sun shade redundant in the near future. Furthermore, additional absorbing elements, also in the planning stage, will reduce noise in the library. However, most significant at this point—building users have to be provided with additional information: (1) about the idea of an intelligent building in general, including the related benefits for its users, e.g., energy efficiency, and comfort; (2) about the usability of the installed technologies in particular, including the individual possibilities and restrictions, in order to control the personal environment, i.e., room control unit, motion sensor.
- Medium-term actions:
  With the increasing enrollment of students and the dynamic expansion of focal areas of research, the new Campus building in Minden—just one year after its opening—is already too small to meet the current need for space in a satisfactory manner, since, as Bain puts it, "direct space implications are that academic environments must encourage, more than ever, interaction, interdisciplinary exchanges and informal, serendipitous encounters" (see Sect. 21.2). Such POE findings relating to students' activities or professors' research processes will be directly applicable to the programming and design phases of any new constructions, or further adaptations of the existing buildings on the Minden Campus. Findings like occupants' need to control their indoor environment require types of regulation that are easy to understand and simple to operate.
- Long-term actions:
  With the investigative or diagnostic POEs and further psychological studies to be done after the settling-in period of an entire year, more in-depth information is expected to support the development of design guidelines and criteria for future campus buildings and similar intelligent facilities as well.

6.4 Conclusions

With the pre- and post-occupancy evaluations that were carried out to date, a couple of conclusions became obvious: (1) future building users have high expectations towards the functioning of intelligent building technologies and doubts about data security as well; (2) uncontrollable intelligent technologies like blinds for sun protection and motion sensors for lighting have the potential to decrease well-being; (3) lack of control of indoor climate, e.g., complex room control units, and lack of information about installed technologies are barriers for acceptance, and may even grow into a psychological 'reactance' of the building occupants.

Therefore, additional information for building users is needed, a demand Vischer is postulating as well regarding occupants' psychological comfort in general (see Sect. 10.6). On the one hand, information helps occupants to use technology in terms of occupancy efficiently, as the usability testing of smart phone application has shown, and for less digitally versed people, a printed user manual has been developed recently by a Master's student (Plettenberg 2016). On the other hand, beyond information, user involvement and feedback is required throughout the entire building delivery process. Therefore, participatory planning with recurrent building performance evaluation (BPE)—from programming to occupancy—turns out to be a key concept. This is true for complex facilities in general and for intelligent buildings in particular.

References

Becking, D., Budke, J., & Reichart, S. (2015). Microcurricula in augmentierter Realität – Ein Prototyp zur Hilfe bei der Interaktion mit intelligenten Gebäuden. In D. Cunningham, P. Hofstedt, K. Meer, & I. Schmitt (Eds.), *INFORMATIK 2015—Lecture Notes in Informatics (LNI)* (pp. 171–180). Bonn: Gesellschaft für Informatik.

Chemeno, A. (2001). An outline of a theory of affordances. *Ecological Psychology, 15*(2), 181–195.

Davis, F. (1989). Perceived usefulness, perceived ease of use, and user acceptance of information technology. *MIS Quarterly, 13*(3), 319–340.

Dickenberger, D., Gniech, G., & Grabitz, H. J. (2002). Die Theorie der psychologischen Reaktanz. In D. Frey & M. Irle (Eds.), *Theorien der Sozialpsychologie* (pp. 243–273). Seattle: Hans Huber.

Gibson, J. J. (1979). *The ecological approach to visual perception*. Boston: Houghton Mifflin.

Hille, N. (2013). Auftragschance Automatisierung. *Deutsches Architektenblatt, 4*, 52–53.

Köhn, P. (2014). *Zur Akzeptanz intelligenter Campusbauten am Beispiel des Forschungsgebäudes CITEC in Bielefeld*. Unpublished Master Thesis, Fachhochschule Bielefeld, Germany.

Maahs, M. (2015). *Akzeptanz von moderner Gebäudetechnik: Eine empirische Untersuchung der Bedienbarkeit des Raumbediengerätes UP227 von Siemens*. Unpublished Bachelor Thesis, Fachhochschule Bielefeld, Germany.

Maldener, S. (2015). *Usability Testing – Augmented Reality Smartphone Applikation zur Erklärung der Gerätebedienung intelligenter Gebäudetechnik*. Unpublished Bachelor Thesis, Fachhochschule Bielefeld, Germany.

Olsson, T., Lagerstam, E., Kärkkäinen, T., & Väänänen-Vainio-Mattila, K. (2013). Expected user experience of mobile augmented reality services: A user study in the context of shopping centres. *Personal and Ubiquitous Computing, 17*(2), 287–304.

Plettenberg, S. (2016). *Nutzerhandbuch als Voraussetzung für eine nachhaltige Nutzungsphase von Gebäuden am Beispiel des intelligenten Neubaus am Campus Minden*. Unpublished Master Thesis, Fachhochschule Bielefeld, Germany.

Preiser, W., Rabinowitz, H., & White, E. (1988). *Post-occupany evaluation*. New York: Van Nostrand Reinhold (republished in 2015, London: Routledge).

Sarodnick, F., & Brau, H. (2011). *Methoden der Usability Evaluation. Wissenschaftliche Grundlagen und praktische Anwendungen*. Bern: Hans Huber.

Schimweg, C. M. (2014). *Vermeidung von Technostress Eine empirische Untersuchung des Einflusses von Vorwissen und Technikaffinität auf Erwartungen und Befürchtungen im Kontext intelligenter Gebäude*. Unpublished Bachelor Thesis, Fachhochschule Bielefeld, Germany.
Schramm, U. (in press). An adaptive evolution from Prussian artillery barracks to intelligent campus buildings. In W. Preiser, A. Hardy, & J. Wilhelm (Eds.), *Adaptive architecture: Changing parameters and practice*. London: Routledge.
Spath, D. (Ed.). (2012). *Arbeitswelten 4.0: Wie wir morgen arbeiten und leben*. Stuttgart: Fraunhofer.
Szymura, M. (2014). *Smart Campus Minden – Pre-occupancy evaluation von Neubau und Gebäudetechnik*. Unpublished student research project, Fachhochschule Bielefeld, Germany.
Tarafdar, M., Tu, Q., Ragu-Nathan, T. S., & Ragu-Nathan, B. S. (2011). Crossing to the dark side: Examining creators, outcomes, and inhibitors of technostress. *Communications of the ACM, 54*(9), 113–120.

Author Biographies

Ulrich Schramm is Professor in the Department of Architecture and Civil Engineering at the Bielefeld University of Applied Sciences, Minden Campus, in Germany. His appointment includes teaching and research responsibilities in the field of facility programming, building performance evaluation and building technologies. He is a trained architect and an experienced facility programmer. He received his doctorate in Architecture from the University of Stuttgart and a post-doctoral fellowship from the German Research Foundation (Deutsche Forschungsgemeinschaft, DFG) for his stay at the University of Cincinnati as Visiting Professor of Architecture. Results of his studies within the International Building Performance Evaluation (IBPE) consortium and the campus-based research project Intelligent Building Technologies (InteG-F) have been presented at conferences of the Environmental Design Research Association (EDRA) since 1995 and published in several articles and book chapters. Some recent chapters he has written were published in *Enhancing Building Performance* (Wiley-Blackwell, 2012) and *Architecture beyond Criticism: Expert Judgment and Performance Evaluation* (Routledge, 2015). He also co-authored the first book on Facility Programming in Germany *Nutzerorientierte Bedarfsplanung: Prozessqualität für nachhaltige Gebäude* (Springer, 2011). Currently, he is serving at the German Institute for Standardization (Deutsches Institut für Normung, DIN) in Berlin as expert and co-chair of the revised edition of the German building standard DIN 18205 *Brief for Building Design*.

Sybille Reichart is Professor of Business Psychology in the Department of Business and Health at the Bielefeld University of Applied Sciences in Germany. Her main interest in teaching and research is the alignment of modern technology and the requirements of people who use innovative technical solutions. Sybille Reichart started her career at the Technical University of Munich (TUM), where she worked on a research project on software-ergonomic issues. When she worked for the Corporate Technology department at the Siemens AG in Munich, she designed concepts for user participation in the development of innovations in medical engineering. In 2002, she received her doctorate in Psychology from the University of Munich (LMU). The topic of her dissertation was an empirical investigation into the success of different participatory approaches. In 2009, she was appointed Professor at the Bielefeld University of Applied Sciences in Germany. She is currently a member of the research group Intelligent Building Technologies (InteG-F). Here, she investigates the influence of smart buildings on the well-being and the acceptance of users.

Dominic Becking is Professor of Databases and Information Systems in Computer Science at the Bielefeld University of Applied Sciences, Minden Campus in Germany. His research interests (apart from core database technology, data science and big data) are in the field of computer science in and for intelligent buildings especially smart office buildings. After getting a MSc equivalent in Physics and a MEd equivalent in Chemistry he worked as a research assistant to the chair for Databases and Information Systems at the University of Hagen (Prof. Schlageter) where he received a doctorate degree in Computer Science. He authored an extensive computer based training manual on SQL, published in the field of E-Learning and M-Learning and was an invited speaker at the Joint Workshop on Cognition and Learning through Media-Communication for Advanced E-Learning (JWCL), Tokyo 2005. He worked as a senior consultant at Sybase (later Sybase/SAP) before joining Bielefeld University of Applied Sciences in 2011. First research results on acceptance and well-being in automated buildings have been presented and published in the Proceedings of the INFORMATIK2015. Future research projects will cover the correlation (and prediction) of user behaviour, energy efficiency and the influence of aspects of the social environment.

Chapter 7
Evaluating the Built Environment from the Users' Perspective: Implications of Attitudinal Models of Satisfaction

Guido Francescato, Sue Weidemann and James R. Anderson

7.1 Introduction

This chapter is intended as a contribution to the theory of building performance evaluation (BPE) and as an update of a prior chapter on this subject (Francescato et al. 1989). It focuses on the concept of users' satisfaction and relies primarily on research conducted by the authors and others on specific environmental types: multifamily housing and work environments. But, the perspective presented here is equally applicable to other types of environments. In the first section of the chapter, users' satisfaction is defined as an evaluation criterion. The reasons for its utilization are outlined. In the second section, the utility of models is discussed. The third section summarizes the evolution of a number of models of satisfaction. A model of satisfaction based on attitude theory is described in some detail. Finally, a number of implications of using such a model are discussed.

G. Francescato (✉)
144 Bungalow Avenue, San Rafael, CA 94901, USA
e-mail: guidof@umd.edu

S. Weidemann
169 Huntley Road, Buffalo, NY 14215, USA
e-mail: sueweide@buffalo.edu

J.R. Anderson
1107 W. Charles, Champaign, IL 61821, USA
e-mail: jim.r.anderson@att.net

W.F.E. Preiser et al. (eds.), *Building Performance Evaluation*,
DOI 10.1007/978-3-319-56862-1_7

7.2 Satisfaction as an Evaluation Criterion

To evaluate is to assess performance with respect to a criterion. Assessment, in turn, requires that performance be measured, but there is always the problem of how to deal with multiple aspects of performance, some of which will be relevant to the intended assessment and some of which will not. This is the problem of choosing adequate operational definitions. Post-occupancy evaluation (POE) stressed the value of assessing environments in use, rather than prior to, or independent from, occupancy. An emphasis on use alone, however, is insufficient to determine the relative importance of the aspects of performance to be measured. The resolution of this problem requires a criterion that expresses the value attributed to the whole, and to which the values attributed to each aspect of that whole may be related. The degree to which the multiple users of a building feel satisfied with their experience of that environment is one such a criterion.

The choice of users' satisfaction as an evaluation criterion has been frequently discussed in the literature and a number of conceptual definitions have been offered, e.g., Czepiel and Rosenberg (1977) for the generic construct of customers' satisfaction; Canter and Rees (1982) for the more specific case of environmental satisfaction; Francescato et al. (1987) for that of residential satisfaction. But the application of satisfaction to environmental assessments hinges primarily on Canter's early observation that "the environment is used and perceived, rather than simply looked at" (Canter 1983, p. 662) and on the assumption that many problems in the built environment are in fact the result of neglecting the users' point of view.

In operational terms, satisfaction may be defined as an index of a number of items from users' self-reports, rather than a single one, for two reasons. The first is technical: the reliability of the criterion can be increased by using an index rather than the single question "How satisfied are you with [living, working, studying, etc.] here?" (Francescato et al. 1987; Kim 1997; Weidemann et al. 2003). The second is conceptual, reflecting the hypothesis that satisfaction is a multifaceted construct in which cognitive, affective, and conative variables coexist.

7.3 Utility of Models

Moore et al. (1985) suggest a scheme in which theoretical orientations, frameworks, and models are seen as leading to the formulation of explanatory theories. They stress the need for explicitness and for models and theories that can be tested. In the particular case of environmental evaluation, it is possible to develop models that can make explicit the theoretical orientations and assumptions that underlie a research approach. This explicitness is essential in the process of interpretation to which the results of any study must be submitted. Interpretation confers meaning to data, so that they can make a contribution to knowledge and serve as a guide for action. But interpretation is impossible in the absence of some kind of conceptual formulation.

Models can also illuminate the potential linkage to work done in other fields, either directly, by describing congruence with existing models in different areas of study, or indirectly, by making clear what is the domain of concern. For example, the model presented in this chapter identifies a linkage to attitude theory and shows how that theory allows a representation of the relationships among environment, satisfaction, and behavior.

Finally, models provide a structured means with which research utilizing different approaches or focusing on specific concepts, sets of factors, or groups of variables can be classified. Classification of research is not only desirable to gain a clearer understanding of the strengths and weaknesses of each study, but also to identify areas of potential overlap.

7.4 Evolution of Users' Satisfaction Models

In the original version of this chapter, the authors discussed in some detail the evolution of models of residential satisfaction in multifamily housing (Francescato et al. 1989). That evolution included both conceptual and empirical work, and resulted in an attempt to reconcile somewhat different views: one that conceived of satisfaction as a global affective appraisal (Zajonc 1980; Weidemann and Anderson 1985), another that viewed it as a cognitive event (Mandler 1982, 1984), and a third that doubted whether cognition and affect were defined in precise enough terms to decide the issue (Russel and Snodgrass 1987).

To arrive at a more inclusive definition, and one applicable as well to environments other than the residential, the authors invoked a classical construct in social psychology that defines global evaluations of psychological objects as 'attitudes.' In reviewing the development of this construct, Ajzen and Fishbein (1980) mention that by the end of the 1950s attitudes were already considered by most psychologists "as complex systems comprising the person's beliefs about the object, [their] feelings toward the object, and [their] action tendencies with respect to the object" (p. 19), that is, systems made up of cognitions, affects, and conations.

Since the 1950s, a number of studies had focused on attitude and its components. Rosenberg (1956) proposed an expectancy-value model of attitude in which evaluations were seen as strongly dependent upon people's expectations or beliefs that the evaluated object furthered or hindered the attainment of their goals. Triandis (1964) and Fishbein (1964) showed that affective and conative variables are highly interrelated in evaluative processes involving attitudes. Campbell (1947), Bettelheim and Janowitz (1950), and Ostrom (1969) provided further evidence that measures of cognition, affect, and conation all contribute to explain attitudes.

Ajzen and Fishbein (1980) developed their Theory of Reasoned Action, in which they suggest that the link between attitudes and behavior is mediated by intentionality with respect to a specific behavior. They also postulated an important distinction: that between attitudes toward “targets,” e.g., people, institutions, objects; and attitudes toward behavior. Further, they determined that while attitudes toward targets only indirectly affect behavior, attitudes toward behavior, when combined with measures of subjective norms and behavioral intentions, do indeed predict behavior.

Within this line of thinking, users’ satisfaction can be conceived of as a complex, multidimensional, global appraisal combining cognitive, affective, and conative facets, thus fulfilling the criteria for defining it as an attitude. This is the point of departure for the construction of the comprehensive model of satisfaction presented in the next section. It is worth noting that this view is reflected operationally in the elaboration of satisfaction indices composed of items that span the entire spectrum of cognition, affect, and conation.

7.5 An Attitudinal Model of Satisfaction

Central to the model proposed here are two considerations suggested by Ajzen and Fishbein (1980). The first restricts the term “attitude” to a person’s evaluation of any psychological object and distinguishes between beliefs, attitudes, intentions, and behavior. The second distinguishes between prediction and understanding. In this sense, variables that have the power of strengthening prediction are considered direct “determinants,” i.e., predictors of either satisfaction or behavior. Their relationships to the criterion are defined as “stable”. On the other hand, variables that do not improve the accuracy of prediction of the criterion are considered “external.” External variables are still worth measuring and including in models, because they still exhibit some relation to the criterion, albeit often an indirect one. Consequently, they are useful when attempting to gain a better understanding of the phenomenon under study, but their effect on the criterion may be indirect and have volatile, or low predictive strength.

With the distinction between external and predictor variables clearly established, it is possible to construct a conceptual model of users’ satisfaction based on attitude theory (see Fig. 7.1). In such a model, the external variables consist of the objective characteristics of the physical, social, and organizational environments; the demographic variables of the users; and the personal characteristics of the respondents. The predictor variables entail cognitive aspects, i.e., beliefs; affective variables, i.e., emotions; and conative aspects, i.e., behavioral intentions. All these variables are then operationally defined to measure each of these aspects.

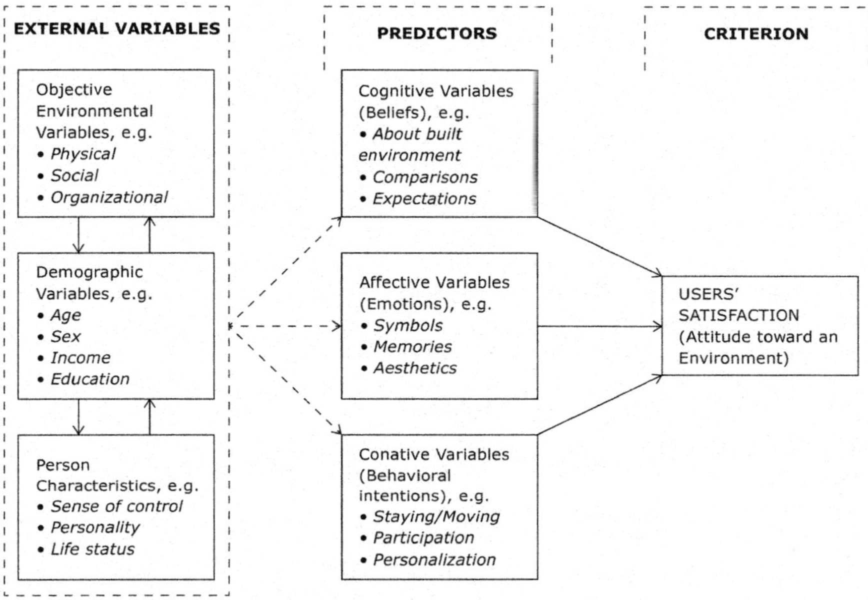

Fig. 7.1 An attitudinal model of users' satisfaction. *Source* Authors

7.6 Implications of Attitudinal Models of Satisfaction

A first implication of attitudinal models of satisfaction regards the universe of variables that are to be included in a specific performance assessment. Is the evaluation aimed merely at identifying factors that will accurately predict a criterion, i.e., users' satisfaction, for the limited purpose of intervening in the environment or its management in order to maximize that criterion? Or, is it aimed at explaining the relationships and mechanisms underlying the system being studied, i.e., producing generalizable knowledge applicable to a variety of different circumstances? The former may be viewed essentially as a POE process; the latter as a much more complex process of "feedforward" that aims to yield knowledge in the form of performance criteria, databases, standards, and other information that can be used, as BPE promises, to "improve the quality of decisions made at every phase of the building life cycle, i.e., from strategic planning to programming, design and construction, all the way to facility management and adaptive reuse" (Preiser and Vischer 2005, p. 8).

Inspection of the pattern of relationships displayed in the model begins to suggest explanations for certain research findings that have been occasionally puzzling. For example, there has been a long tradition of belief among planners and designers that improving certain objective physical environmental conditions, such as structural soundness, sanitation standards, and thermal comfort, would lead to more satisfying environments. Experts discussing "quality" tend to emphasize such objective criteria, e.g., CIB (1988). But research has often shown weak or indirect correlations between improvements of this kind and satisfaction, e.g. U.N. Economic Commission for Europe (1973) and Francescato et al. (1979).

Conversely, objectively less desirable conditions were found to correlate with high levels of satisfaction under certain conditions. In an example, G. I. Bill students housed in temporary barracks while attending college after their return from World War II were highly satisfied with their residential environment in spite of its poor objective quality (Schorr 1963). In another study, residents living in multi-family high-rise developments, but aspiring to move to single-family detached homes, were more satisfied the more they believed they would be able to move to the environment of their choice in the future (Michelson 1977).

The model proposed here accounts for these findings by clearly identifying the mediating role of beliefs, emotions, and behavioral intentions between objective variables and satisfaction. In sum, the model makes it clear that objective environmental characteristics are not, in and of themselves, strong predictors of satisfaction.

If users' satisfaction is a central criterion in the evaluation of environmental performance, what is the relationship that links satisfaction to behavior? In other words, can interventions in an environment result in changes not only in the users' satisfaction with that environment, but also in changes–one would hope for the better–in the personal or social behavior of environmental users? For instance, Newman (1973, 1976) devoted a great deal of attention to design features that might deter crime. Anderson et al. (1994) and Weidemann et al. (2003) focused on the impact of design features on the satisfaction and productivity of office workers in Army bases.

Before that question is addressed within the framework of the model of satisfaction discussed here, it is necessary to recall Ajzen and Fishbein's (1980) distinction between attitudes toward a target and attitudes toward a behavior. Users' satisfaction is to be viewed as an attitude toward a target, that is toward a system composed of people, institutions, and physical objects, not as an attitude toward a behavior. When that distinction is kept in mind, it is possible to arrive at the comprehensive model of the satisfaction-behavior relationship shown in Fig. 7.2.

Structurally, this comprehensive model is similar to that proposed by Ajzen and Fishbein (1980) and further refined by Fishbein and Ajzen (2015). But there is one important change with respect to the model of satisfaction displayed in Fig. 7.1: users' satisfaction, as an attitude toward a target, can no longer be included among the predictors, because it is only the attitude toward the behavior that will predict a specific behavior.

When viewed as aspects of behavior rather than aspects of satisfaction, environmental settings, demographic and person characteristics, beliefs, emotions, behavioral intentions toward the target, and users' satisfaction must all be considered as external variables. As such, their relationship to a specific behavior is not "stable," that is, it has low and volatile predictive strength. A second implication of an attitudinal model of satisfaction then, is not that there are no relationships

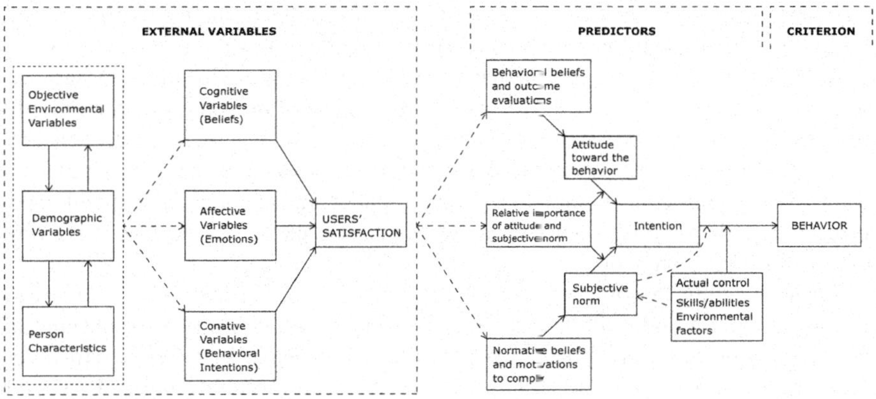

Fig. 7.2 Comprehensive model of the satisfaction-behavior relationship. *Source* Authors

between environment, satisfaction, and behavior but rather that such relationships are indirect and mediated by predictors of specific personal and social behavior. And even then, as Fishbein and Ajzen make clear in their more recent version of the "Reasoned Action" model (2015), the variables that predict behavior are mediated by people's actual control over performing the behavior. The complex web of relationships of the model of Fig. 7.2 clearly dispels any notion of a direct relationship between satisfaction and behavior and especially of the naive assumption that changes in the physical or social environment may predictably result in specific behavior.

A third implication of the comprehensive model of the satisfaction behavior relationship is that it identifies those elements, such as beliefs, emotions, and behavioral intentions that are more likely to change over time than other aspects such as objective building characteristics. This suggests that monitoring those variable elements over time, rather than simply measuring them at one point in time, should be a preferred mode of satisfaction evaluations.

7.7 Conclusion

Satisfaction can be a powerful construct with which to perform assessments sensitive to the users' point of view. It has been used in many fields, e.g., information technology (Wixom and Todd 2006). But it does not follow that it should constitute the only basis for evaluation. In other words, satisfaction is a necessary but not sufficient criterion. Ideally, environmental evaluations should also include other criteria, such as economic, ecological, technological, and functional soundness. This is, of course, nothing but common sense. But it is worth noting that one cannot achieve comprehensive BPEs without also taking into account important aspects of the total environmental performance other than users' satisfaction.

Environments are systems with multiple stakeholders, hence with multiple objectives (Francescato et al. 1987). Therefore, they must be evaluated using multiple criteria. In assessing whether or not a particular environment is satisfactory, one must ask: satisfactory for whom? In the case of housing, it is not only the inhabitants who must be satisfied; but the planners and designers, with their professional ideologies and interests; the developers, who must be able to make a profit; the governmental bodies that regulate and perhaps provide financial assistance, with their political ideologies and constraints; and so on. And so it is in the case of other environments.

When engaging in comprehensive BPE, one must still define a set of relevant evaluation criteria and, within that set, a ranking or weighing of the contribution that each single criterion should make to a comprehensive assessment. However, defining and ranking a set of criteria cannot be done through empirical research alone. It is still an activity that also requires judgment in balancing the often contradictory objectives of the various stakeholders against each other.

References

Ajzen, I., & Fishbein, M. (1980). *Understanding attitudes and predicting behavior*. Englewood Cliffs, NJ: Prentice Hall.

Anderson, J. R., Weidemann, S., Heinen, D. C., Adeoye, B., & Beazly, S. J. (1994). *Evaluating office environments: A case study* (USACERL Technical Report FF-94/22). Champaign, IL: U.S. Army Construction Engineering Research Laboratory.

Bettelheim, G., & Janowitz, M. (1950). *Dynamics of prejudice: A psychological and sociological study of veterans*. Englewood Cliffs, NJ: Harper and Row, Publishers.

Campbell, D. T. (1947). *The generality of a social attitude*. Unpublished doctoral dissertation. University of California, Berkeley.

Canter, D. (1983). The purposive evaluation of places: A facet approach. *Environment and Behavior, 15,* 659.

Canter, D., & Rees, K. (1982). A multivariate model of housing satisfaction. *International Review of Applied Psychology, 31,* 185.

CIB, International Council for Building Research Studies and Documentation. (1988). *CIB 89: Quality for building users throughout the world, Call for papers*. Paris: Centre Scientifique et Technique du Bâtiment.

Czepiel, J. A., & Rosenberg, L. J. (1977). The study of consumer satisfaction: Addressing the so what question. In H. K. Hunt (Ed.), *Conceptualization and measurement of consumer satisfaction and dissatisfaction* (pp. 92–119). Cambridge, MA: Marketing Science Institute.

Fishbein, M. (1964). The relationship of the behavioral differential to other attitude instruments. *American Psychology, 19,* 40.

Fishbein, M., & Ajzen, I. (2015). *Predicting and changing behavior: The reasoned action approach*. New York: Routledge.

Francescato, G., Weidemann, S., Anderson, J. R., & Chenoweth, R. (1979). *Residents' satisfaction in HUD-assisted housing: Design and management factors*. Washington: U. S. Department of Housing and Urban Development.

Francescato, G., Weidemann, S., & Anderson, J. R. (1987). Residential satisfaction: Its uses and limitations in housing research. In W. van Vliet, W. Michelson, & D. Popenoe (Eds.), *Housing and neighborhoods: Theoretical and empirical contributions* (pp. 43–58). New York: Greenwood Press.

Francescato, G., Weidemann, S., & Anderson, J. R. (1989). Evaluating the built environment from the users' point of view. In W. F. E. Preiser (Ed.), *Building evaluation* (pp. 181–198). New York: Plenum Publishing Corporation.

Kim, S-H. (1997). *Modeling resident satisfaction: Comparison of the Francescato and Fishbein-Ajzen TRA models*. Doctoral dissertation, University of Illinois at Urbana-Champaign.

Mandler, G. (1982). The structure of value: Accounting for taste. In M. S. Clark & S. T. Fisk (Eds.), *Affect and cognition* (pp. 55–78). Hillsdale, NJ: Erlbaum.

Mandler, G. (1984). *Mind and body: Psychology of emotion and stress*. New York: Norton.

Michelson, W. (1977). *Environmental choice, human behavior, and residential satisfaction*. New York: Oxford University Press.

Moore, G. T., Tuttle, D. P., & Howell, S. C. (1985). *Environmental design research directions: Process and prospects*. New York: Praeger.

Newman, O. (1973). *Architectural design for crime prevention*. Washington: U. S. Government Printing Office.

Newman, O. (1976). *Design guidelines for creating defensible space*. Washington: U.S. Government Printing Office.

Ostrom, T. M. (1969). The relationship between the affective, behavioral, and cognitive components of attitude. *Journal of Experimental Social Psychology, 5,* 12.

Preiser, W. F. E., & Vischer, J. C. (2005). The evolution of building performance evaluation: An introduction. In W. F. E. Preiser, & J. C. Vischer, (Eds.), *Assessing building performance* (pp. 3–14). Oxford: Elsevier Butterworth-Heinemann (republished in 2015, London: Routledge).

Rosenberg, M. J. (1956). Cognitive structure and attitudinal affect. *Journal of Abnormal and Social Psychology, 53,* 367.

Russel, J. A., & Snodgrass, J. (1987). Emotion and environment. In D. Stokols & I. Altman (Eds.), *Handbook of environmental psychology* (pp. 245–281). New York: Wiley.

Schorr, A. (1963). *Slums and social insecurity*. Washington: U.S. Department of Health, Education, and Welfare, Social Security Administration.

Triandis, H. C. (1964). Exploratory factor analysis of the behavioral components of attitude. *Journal of Abnormal and Social Psychology, 68,* 420.

U.N. Economic Commission for Europe. (1973). *Housing requirements and demand: Current methods of assessment and problems of estimation*. Geneva: United Nations.

Weidemann, S., & Anderson, J. R. (1985). A conceptual framework for residential satisfaction. In I. Altman & C. Werner (Eds.), *Home environments* (pp. 153–185). New York: Plenum Publishing Corporation.

Weidemann, S., Jenicek, E. M., Lister, D. B., & Zimmer, C. (2003). *Assessing the relationship between worker productivity and the indoor environment* (No. ERDC/CERL TR-03-7). Champaign, IL: U.S. Army Construction Engineering Research Laboratory.

Wixom, B., & Todd, P. (2006). A theoretical integration of user satisfaction and technology acceptance. *Information Systems Research, 16,* 1.

Zajonc, R. B. (1980). Feeling and thinking: Preferences need no inferences. *American Psychology, 35,* 151.

Author Biographies

Guido Francescato is Professor Emeritus of Architecture at the University of Maryland. He has designed buildings and urban interventions in North America, Argentina, Italy, and Ethiopia. His most recent design work includes the Mathews-Potter House and Studio in Ranchitos de Galisteo, NM, built in 2003 and the Tibbits House in Bethesda, MD, completed in 2009.

His publications include *Residents' Satisfaction in HUD-Assisted Housing: Design and Management Factors* (with Sue Weidemann and James R. Anderson; Washington: U.S. Department of Housing and Urban Development, 1979), *Residential Environments: Choice, Satisfaction and Behavior*, (with Juan Ignacio Aragonés and Tommy Gärling; Westport, Connecticut–London: Bergin & Garvey, 2002), chapters in edited books and encyclopedias, and journal articles. He has presented numerous papers at professional conferences.

His work on residents' satisfaction in publicly assisted housing (with Sue Weidemann and James R. Anderson) received a 1983 Award for Exemplary Research from the National Endowment for the Arts and a 1980 Progressive Architecture Research Award. He has been a consultant with a number of private and public organizations, including the World Bank and the U. S. Department of Housing and Urban Development. He is a Fellow of the Society for Human Ecology, has been a submission reviewer for the journal *Environment and Behavior*, and a proposal reviewer for the Social Sciences and Humanities Research Council of Canada.

Dr. Sue Weidemann is an environmental psychologist. Through her research, teaching, and consulting, she has studied the relationships between people and the places/spaces they use. She taught research-based design decision-making for 25 years at the University of Illinois, in Urbana-Champaign. While there, she authored or co-authored more than 40 articles, book chapters, and reports, most of them with Guido Francescato and James R. Anderson. Their team projects received eight national awards or recognition for applied research (via Progressive Architecture, ASLA, and the NEA).

Since joining BOSTI Associates in 1994 as Director of Research, her work has focused on the workplace, quantitatively measuring the effects of workplace design upon important business outcomes such as job satisfaction, team performance, and individual performance of office workers. The goal of this work has been to identify those qualities of the workplace that have the strongest effects on desired outcomes, enabling the prioritization of aspects that would lead to increased productivity and job satisfaction. Through collaborative work with designers, clients, and consultants, these aspects were incorporated into the design and planning of high-performance workplaces.

In 2011, she returned to the academic realm, at the University at Buffalo, NY, as a Visiting Professor in the Inclusive Design Research Group in the Department of Architecture. Weidemann also serves as a Senior Research Consultant to the Center for Inclusive Design and Environmental Access.

In 2014, she received the Environmental Design Research Association Career Award.

James R. Anderson is retired from the University of Illinois where he held positions as Associate Dean, Chair of the Building Research Council, Professor of Architecture, and Professor of Landscape Architecture.

Along with his colleagues, he has conducted a long-term program of research that seeks to explain satisfaction with an environment in terms of the physical, social and organizational characteristics of that environment. Settings for this past research have included multifamily housing, housing for the elderly, housing for disabled adults, correctional facilities, military housing, offices, and central business districts. He was the Principal Investigator for a study examining the ability of residents of section 8 housing to assess physical characteristics of their dwelling in terms of meeting HUD's Housing Quality Standards (HQS), and a subsequent study comparing data collected from tenants with data collected by on-site inspectors.

Professor Anderson has had extensive experience with implementation and evaluation of federal housing programs. He directed several contracts with the Department of Housing and Urban Development's Office of Native American Programs (ONAP). First he and his colleagues administered the Indian Housing Block Grant (IHBG) program formula. This work involved maintaining a database on over 575 tribes and using that data for calculating the annual IHBG allocations to each tribe. Subsequently he directed the Indian Housing Operating Cost Study ONAP.

Professor Anderson's teaching included seminars focused upon methods of research in designed environments and his design studios focused upon the application of research information and the inclusion of social and cultural issues in design.

Chapter 8
Towards a Hospital Activation Process Model

Wolfgang F.E. Preiser, John P. Petronis, John W. Petronis and Lexi Petronis

8.1 Introduction and Background

This chapter appeared in the original 1989 publication of *Building Evaluation*. It has since been updated for clarity, and to incorporate current findings and lessons learned in the succeeding time period.

As preparations are made for the occupancy of renovated or new facilities, there are many complex processes and interweaving coordinations that are necessary for consideration. This is particularly true in the case of hospitals and other healthcare facilities, which must remain in operation for patients even as critical equipment and structures are placed during facility transition. 'Towards A Hospital Activation Process Model" gives overviews as to the history and importance of activation, along with key findings of an evaluation performed by Architectural Research Consultants, Inc. The chapter concludes with a discussion as to how an activation process model may be built upon and used successfully for other purposes.

Wolfgang F.E. Preiser passed away in August 2016 during the final phase of chapter editing.

W.F.E. Preiser
Scottsdale, USA

J.P. Petronis (✉) · J.W. Petronis · L. Petronis
Architectural Research Consultants, 220 Gold Avenue Southwest,
Albuquerque, NM 87102, USA
e-mail: jppetronis@arcplanning.com

J.W. Petronis
e-mail: jwpetronis@arcplanning.com

L. Petronis
e-mail: lpetronis@arcplanning.com

W.F.E. Preiser et al. (eds.), *Building Performance Evaluation*,
DOI 10.1007/978-3-319-56862-1_8

Activation is a process of preparing people and a facility for occupancy and operation. It is complex and little understood, particularly for hospitals, where the health, safety, and patient care is the highest priority, and where systems must be operational and fault-free on move day to ensure a cost-efficient, smooth transition, especially when multi-phased construction and moves are involved. Activation includes selecting equipment and supplies, hiring and training staff, planning for the move, and building readiness and operation, all with the intent of relocating to an upgraded facility that promotes patient healing. The challenge was to clarify activation, to develop an Activation Process Model (APM), and to demonstrate that activation focuses on and expands the scope of the Post Occupancy Evaluation (POE) process (Preiser et al. 1988) (Fig. 8.1).

The research focused on in-depth study of a large, recently-activated facility. After gathering data and analyzing documents, ARC, Inc. conducted workshops with project and move coordinators from four similar facilities throughout the United States, two of which had recently activated their buildings, while the other two will activate within the next few years. This innovative process model views activation from the management perspective. With further development, it will evolve into a more detailed model that can be adapted and applied to the activation of any facility, but particularly medical institutions with patients' care of utmost importance.

This chapter describes key findings and recommendations that resulted from evaluating the activation of a major facility in Albuquerque, New Mexico. The study's findings, issues and recommendations were presented at a workshop to relevant representatives.

The evaluation's principal purposes were to:

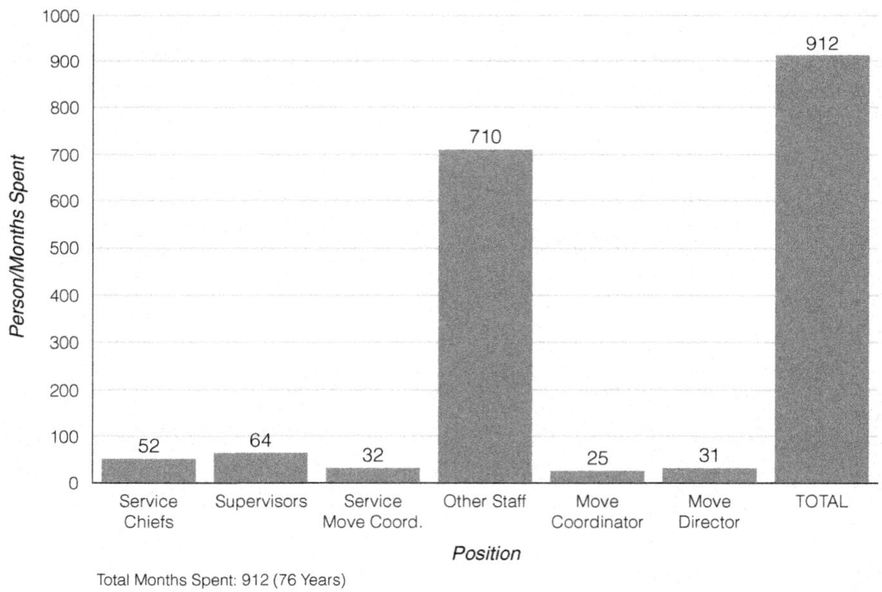

Fig. 8.1 Activation spans aspects of the building delivery process. *Source* ARC, Inc.

- Define activation.
- Develop a structured APM for use and development at other facilities.
- Identify key activation issues that affect facility staff, users, and patients.
- Develop and test an expanded scope of POE.

8.2 Methodology

ARC, Inc. used various methods to evaluate the facility's activation experience: interviews, analysis of activation literature and facility-specific documents, administration of an extensive questionnaire to departments, and workshops to clarify information.

8.2.1 Interviews and Survey Questionnaire

ARC, Inc. met with members of the facility's staff who had been a part of activation; they helped in understanding and refining the APM.

ARC, Inc. prepared and distributed a survey questionnaire to all departments-approximately 90% of which completed it. Departments provided opinions on adequacy of planning input, quality of training, and guidance materials, and the success of the move itself.

8.2.2 Document Analysis

Little has been written on activation and it is not readily available in current literature. The field, as one for formal study, is rather new, although there has since been documentation of related practices, such as the U.S. Army Corps of Engineers' Initial Outfitting & Transitioning, a POE of existing medical facilities in Korea.

The organization's headquarters provided useful documents and policies, which included four years of agendas from the facility's activation steering committee. These helped clarify the timing and activities that took place during activation. A formal move and transition plan explained each department's scheduling to carry out the changeover to the new building. Documents such as "showstopper" lists, critical item lists, "to do" lists, and completion lists indicated the type and complexity of facility-related items.

8.2.3 Findings

Replacing an aging 250,000-square-foot facility with a new 550,000-square-foot facility provided a unique opportunity to examine the activation process. The data gathering quantified significant information on complexities involved.

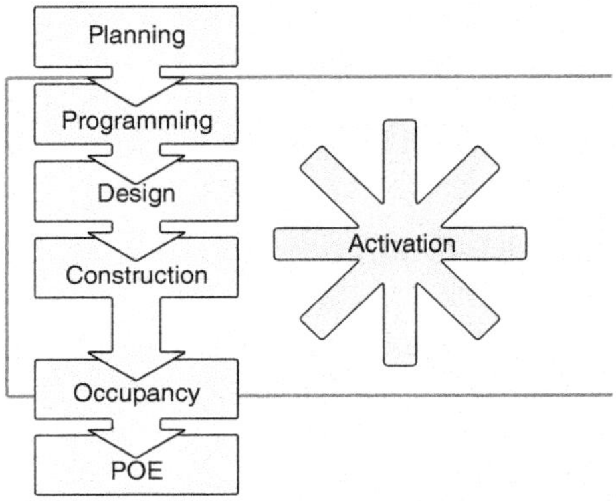

Fig. 8.2 Total time spent on move activities. *Source* ARC, Inc.

Time Spent on Activation

Questionnaire results showed that activation is time intensive. Facility department chiefs, supervisors, and staff reported time spent on activation as 76 person/years. If equated to dollars, that totals approximately $3 million (based on an average cost of $40,000/person/year, including salary and benefits). The amount represents approximately 3% of the total project cost, i.e., construction and equipment expenditures. Building Management (228 persons/months), Move Director's Staff (144 persons/months), and Engineering Maintenance/Repair (214 persons/months) reported the most time spent on activation. 57 of 76 person/years (75%) were spent on lengthy pre-start-up planning and tasks. 7.5 person/years were spent on the start-up phase and 10.6 person/years were expended during the operation phase. 59 of 76 person/years (78%) were by staff members, with the remaining 22% (or about 17 person/years) spent by department heads, supervisors, and Move Coordinators (Fig. 8.2).

Training and Orientation

Departments with high amounts of training were those with heavy responsibilities for facility operation, or departments with large staffs and new high-tech equipment. Departments rated training and orientation quality from providers on a scale from 1 (poor) to 5 (excellent). The quality of equipment training rated average or below: 1.76 for contractor-provided training and 2.5 for vendor training. Training provided by the facility's organization was rated at 2.21, while in-house orientation rated 3.03.

Activation Planning and Execution

Departments rated the quality of the move, again from 1 to 5. The quality of overall activation planning for the facility earned a 3.25 rating. Specific department moves rated their own move plans at 3.16, and the entire move execution at 3.38.

Temporary Staff

Departments hired a total of 60–65 temporary staff, with Engineering Maintenance and Repair, Building Management, and the Move Director's Staff requiring the bulk of temporary labor. Departments most needed additional staff during the people/facility readiness phase (six months before Move Day). Total temporary staff costs totaled $400,000 (Building Management Department); $372,000 (Engineering Maintenance and Repair); and ($200,000) Move Director's Staff.

8.2.4 *Workshops*

Personnel from the activated facility and headquarters held three workshops to present the Activation Process Model for critique. The major and final workshop was held in Albuquerque to a broader audience.

8.3 The Facility Activation Process Model

The Activation Process Model is a management tool outlining the logic and sequence of decisions, tasks, durations, and dependencies. The future challenge is testing this model at other facilities, refining it to a more generalized model. The model has five phases (Fig. 8.3).

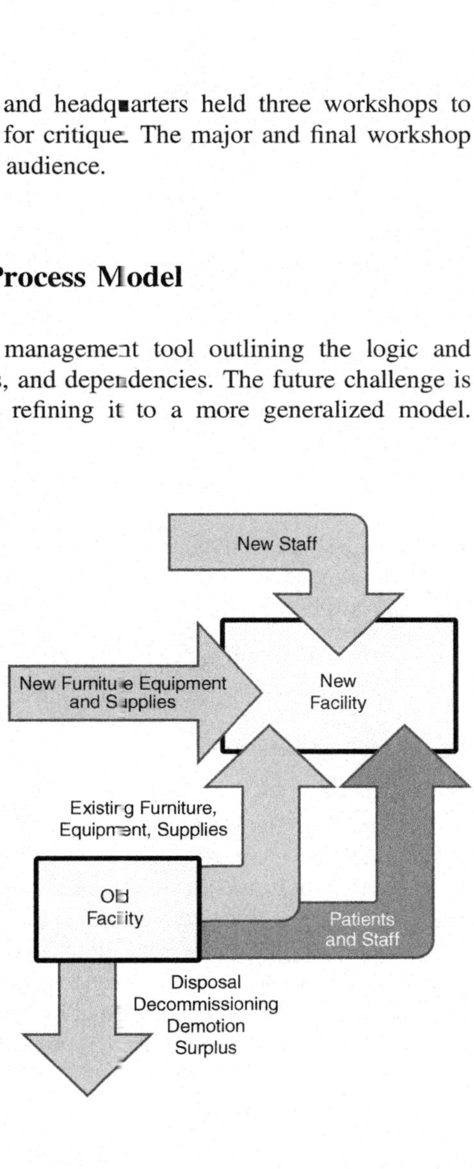

Fig. 8.3 Activation is complicated when a new building replaces an old one. *Source* ARC, Inc.

8.4 Phase I: Initiation

8.4.1 *Purpose*

The Initiation Phase identifies and requests activation funding, and plans and coordinates the specific funding and ordering of equipment and furnishings.

8.4.2 *Timing and Resources*

The activation process starts during programming and preliminary design, possibly more than 5 years before Move Day, depending upon facility size and complexity, and ends 12–18 months before Move Day. The facility's top management makes preliminary funding decisions and long-range staff allocation. An Activation Steering Committee may be readied. Facility and headquarters staff responsible for the construction project carry out the initiation phase.

8.4.3 *Major Tasks*

- Budgeting for Activation: Estimating and considering initial activation and impact funds (such as commercial movers and visits to other facilities) is necessary to the process.
  - Project staff and special personnel
  - Warehouse costs
  - Overtime
  - Contract personnel
  - Training costs
- Working with the Design Team to Determine Basic Equipment/Furniture Needs and Preliminary Specifications: A significant lead time is needed between design and activation to minimize potential problems.
- Equipment and Furniture Identification and Ordering: This process includes policies and guidelines related to equipment and furniture requirements for major and minor construction and renovation projects.
- Receiving and Warehousing Equipment and Furniture: Equipment and furniture starts arriving any time from this point to Patient Move Day. It is important to anticipate the major impact that activation planning will have on the Supply Department, including:
  - Budgeting for use of temporary storage space, on- or off-site.
  - Organizing the warehouse for efficiently transferring stored items.

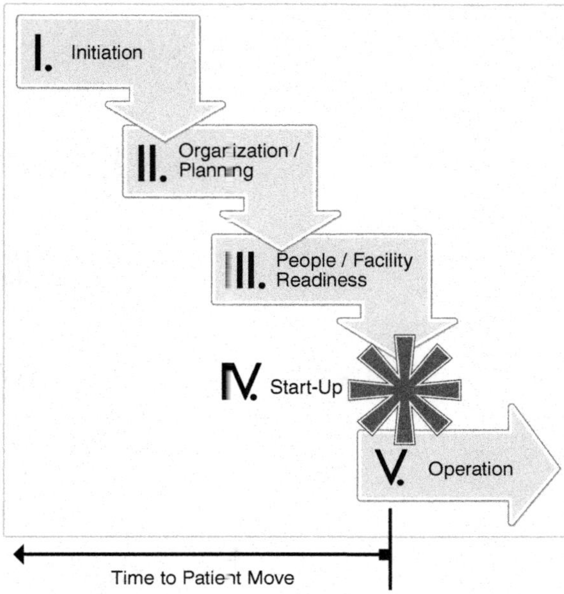

Fig. 8.4 Activation Process Model (APM). *Source* ARC, Inc.

  - Allowing adequate lead time for equipment orders, especially complex or one-of-a-kind items.

- Impact Funds Identification and Requests (Fig. 8.4).

8.5 Phase II: Organization/Planning

8.5.1 Purpose

This phase mobilizes facility management for organizing people, equipment, and resources that are necessary to create detailed procedures and actions before the move takes place (Fig. 8.5).

8.5.2 Timing and Resources

The phase starts 12–18 months before Patient Move Day (it can last up to 9 months) and ends 3–6 months before Move Day. It's important to not start this phase too early, as not to risk premature dissipation of enthusiasm for the activation process.

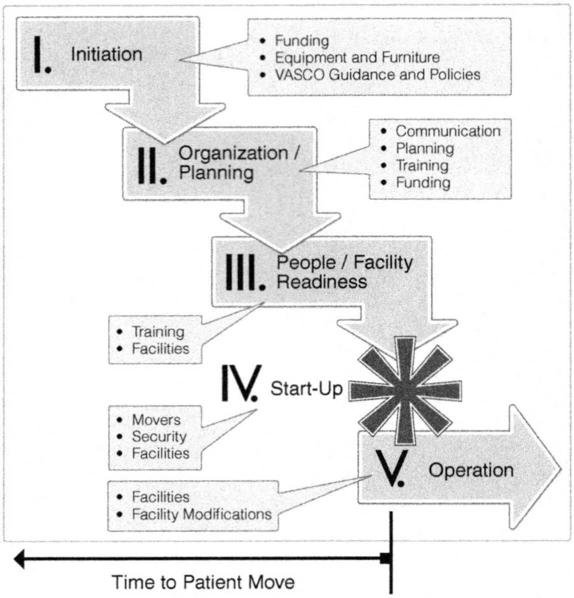

Fig. 8.5 The five phases of APM. *Source* ARC, Inc.

Resources include:

- Facility Top Management
- Move Steering Committee
- Move Director and Dedicated Staff
- Construction Project Manager and Staff Departments
- Headquarters

8.5.3 Major Tasks

- Appoint Activation Steering Committee.
  This committee represents the interests of top facility management and all major departments. It clearly defines activation roles and authority.
- Organize Move Committee.
  This committee implements activation and oversees the entire process. It monitors the moving of people and of furniture, equipment and supplies. It's vital that a professionally qualified coordinator oversees each function, supervised by the Move Director. The committee communicates frequently and effectively with everyone involved (patients, facility staff, headquarters

personnel, the public) to create "esprit de corps." Professional staff involvement is particularly important.

- Determine Move Approach.
  The Move Committee drives the overall move approach. The Albuquerque facility delegated move responsibilities to individual departments, and hired commercial movers for the majority of equipment and furniture. Other move approach elements include:
  - Identifying move priorities.
  - Moving those with preparatory or security functions at the earliest date.
  - Non-essential activities.
  - Essential activities.
  - Compressing move into a few weeks to minimize disruption.
  - Lowering the workload in critical areas to the maximum practical extent.
  - Keeping life support room and other necessary activities functional at all times.
  - Retaining adequate telecommunications.
- Delegate Responsibilities.
  Department chiefs are responsible for their own department's moves.
  - Prepare and Disseminate Move Plan Guidance.
  - The highest quality performance for department moves requires comprehensive guidance. Department plans include:
  - Describe department operations:

    How it will operate (before, during, after move).
    Key relationships to other departments.
    Backup requirements (duplicate supplies, additional personnel).
    Required staffing levels and overtime.
    Outside resources required.
    Needed training programs.
    Vacation rescheduling.

- How space will be used, including location of furniture and equipment.
- Supplies (amounts, placement).
- Equipment (required testing, training and maintenance).
- Personnel roles.
- Telephones, paging systems, computers and mail requirements.
- Transportation of people/equipment.
- Security (egress/ingress, type of locking doors).
- Parking and location directions.
- Work/activity/material flows.
- Prepare and Present Department Move Plans.
- Approve (Revise) Move Plans.
  - Provided by the Move Committee with facility management reviews.

8.6 Phase III: People/Facility Readiness

8.6.1 Purpose

This phase prepares the building and its occupants for the move. Equipment and furnishings are moved, checked and calibrated. Orientation to the building and training for equipment operation and maintenance are conducted.

8.6.2 Timing and Resources

The preparation process begins 3–6 months prior to the move.

Resources include:

- Move Director
- Resident Engineer
- Education Department
- Engineering Department
- Supply Department
- Building Management
- Equipment Vendors/Contractors
- Equipment Vendors/Contractors

8.6.3 Major Tasks

- Identify and Prepare Orientation and Training Materials.
  - A lead training group (the Education Department is a logical choice) implements a Training Master Plan, which establishes specifications and expectations for facility training and orientation. The plan budgets resources, including potential out-of-town equipment sessions, and includes directives for contractor and vendor training.
- Departments Conduct Tours and Prepare "To Do" Lists.
  - As soon as possible, individual departments are given ample time to visit the facility during construction, accompanied by Engineering Staff, and generate completion "to do" lists.
- Initiate Orientation and Training.
  - Soon after beneficial occupancy in the facility, orientation and training is underway. Use of "mock-up" clinical ward areas is especially helpful.

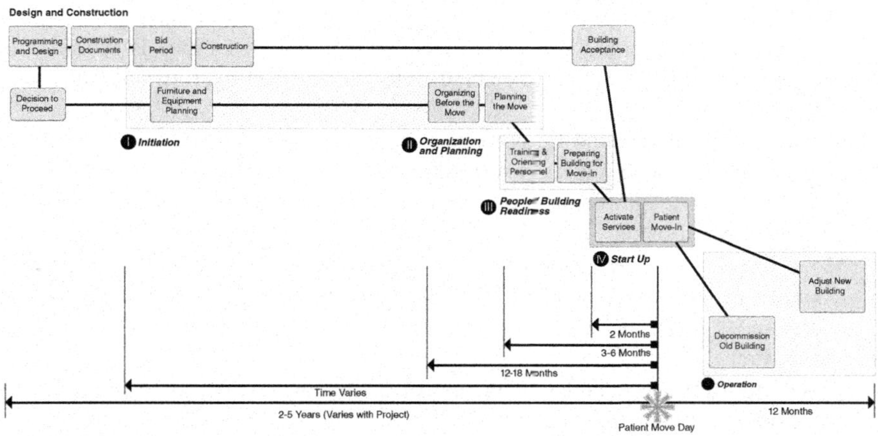

Fig. 8.6 Design and construction. *Source* ARC, Inc.

8.7 Phase IV: Start-up

8.7.1 Purpose

This phase assures a smooth move-in and transition, causing minimum disruption to facility operations. Move activities are phased according to priority, with "public" visitors no longer onsite (Fig. 8.6).

8.7.2 Timing and Resources

The start-up phase begins up to 4 months before the day of the move.

Resources include:

- Top Management
- Move Director and Staff
- Departments
- Commercial Movers
- Temporary/Auxiliary Department Providers

8.7.3 Major Tasks

- Estimate Move Requirements.
  The Move Committee and departments gauge the amount of equipment, furniture, supplies and personal belongings to be moved and time needed to do so.

- Identify and Hire Mover.
  A reputable commercial mover with experience and resources is required to perform the move during the selected time period. Facility staff can help with portions of the move.
- Move New Furniture and Equipment.
  New furniture and equipment should be moved into the facility as soon as possible. Certain equipment may require special contracts; commercial movers may need supervision.
- Move Departments by Priority.
  Departments move as established in the Move Plan, along with all equipment and supplies to be reused.
- Move Staff.
  This is the most significant milestone of activation. While some preparations occur up to 4 weeks before the move, the move itself occurs in the shortest time safely feasible. Delays can be costly, with additional expenses for movers and temporary staff.
- Preparations for the move include:
  - Reducing the facility population census by:
  - Temporarily discontinuing non-essential activities.
  - Limiting public visits.
  - Ensuring that signage is installed in time for department moves.
  - Providing adequate security at old and new facilities.
  - Assuring that the telephone system, intercom and computers are working.
  - Selecting staff for every activity and providing proper orientation.
  - Preparing staff:

    Obtaining staff consent for pictures and interviews.
    Identifying records and personal belongings.
    Strict organization and discipline are needed, requiring designating dispatchers and receivers for each ward. Various forms, checklists and logistical procedures aid in the effort.

8.8 Phase V: Operation

8.8.1 Purpose

This phase completes all move activities and the remaining "to do" list items, remaining punch list items dating from building acceptance, and any other minor adjustments needed as departments resume operation. The old facility is deactivated and essential adjustments are made to the new facility.

8.8.2 Duration and Resources

Operation begins the day of the move and lasts up to one year afterward.

Resources include:

- Project Director and Staff
- Move Director and Staff
- Supply Department
- Building Management
- All Affected Departments
- Engineering Department

8.8.3 Major Tasks

During facility operation, significant items requiring adjustment are identified. In some organizations, modifications requiring major construction have a moratorium of one year from building acceptance before funding. Tasks include:

- Secure old facility to prevent loss of equipment and furniture.
- Remove excess equipment and furniture.
- Decommission and demolish old facility.
- Identify and budget one-year modifications.

8.9 Conclusions

8.9.1 The Evaluation of the Facility Activation Successfully Demonstrated the POE Approach

The authors explored POE as it pertains to moving to a new facility in order to promote patient healing. In the study, the Chief Engineer and members of his staff were key in recognizing the need to evaluate and to document the experience, and to develop an APM that other facilities could use and develop.

8.9.2 Evaluation of the Facility Activation Process

For this study, POE included the assessment of activation, and also provided insight into establishing a process model for activating other facilities.

8.9.3 A More Focused and Proactive POE Approach Evaluated Facility Activations

By focusing on the process of activation and extending it by sharing and clarifying procedures, information, issues, and lessons learned, the sponsoring organization expects to guide new facilities through the complexities of activation, saving time, effort, and money.

8.9.4 The Evaluation Achieved Its Purposes

The evaluation of the activation process used in the startup of a new building succeeded in meeting the principal objectives of the study:

- An APM is now developed for immediate use, and for further development at other facilities.
- Key activation issues are identified and documented for immediate use.
- Specific task-oriented recommendations are available to organize and manage activation activities at new facilities.
- Major components of the current approach to conducting POE studies are further validated.

As a direct result of this study, many of the recommended improvements continue to be made in programming, designing, construction and project management practices and the activation of new facilities. Attention is now focused directly on the activation process and on addressing key issues that often determine the effectiveness, success, and acceptability of new facilities.

Acknowledgements The authors thank Mr. Ron Richter, Chief, Engineering Service and Mr. Tom Casper, Project Coordinator, VAMC Albuquerque, as well as all participating staff of the organization for their excellent collaboration in this project. Furthermore, we thank Mr. Irwin Axelrod and Mr. Dick Kelly of the Veterans Administration Central Office and Regional Office, respectively, for their foresight and guidance in this project.

References

Preiser, W., Rabinowitz, H., & White, E. (1988). *Post-occupany evaluation*. New York: Van Nostrand Reinhold (republished 2015, Routledge: London).

Author Biographies

Wolfgang F.E. Preiser held a Ph.D. in Man-Environment Relations from Penn State, and several architecture degrees from Virginia Tech, Karlsruhe Tech (Germany) and Vienna Tech (Austria). He had over 40 years of experience in teaching, research and consulting in the evaluation and programming of environments, including health care facilities, public housing, public libraries, cross-cultural and universal design, as well as design research in general. He published 20 books and over 130 chapters, monographs, and articles. Most recent books include: *Architecture Beyond Criticism: Expert Judgment and Performance Evaluation* (Routledge, 2015); *Enhancing Building Performance* (Wiley, 2012); *Universal Design Handbook* (McGraw-Hill, 2010); *Designing for Designers: Lessons Learned From Schools of Architecture* (Fairchild, 2007). Preiser lectured worldwide at 69 venues and conferences in the United States and Canada, as well as 86 overseas. He served on national committees with the American Institute of Architects, the Building Research Board of the National Academy of Sciences, and the National Institute for Disability Rehabilitation Research. Preiser received many awards, including: 2 Progressive Architecture Awards; 2 Professional Fellowships from the National Endowment for the Arts; The Career Award from the Environmental Design Research Association (EDRA), 2 EDRA Achievement Awards, as well as other awards while at the University of Cincinnati.

John Petronis, AICP, AIA founded Architectural Research Consultants, Incorporated (ARC) in 1976. ARC is the largest and longest-established planning firm in the state. ARC specializes in facility programming and master planning, and has successfully completed and implemented hundreds of projects. Mr. Petronis holds a Bachelor of Arts degree from Gettysburg College, and Master's degrees in Architecture and Business Administration from the University of New Mexico. He is both a licensed architect and a certified planner. Mr. Petronis' work has been featured in a number of facility planning and programming text books.

John W. Petronis is a graphic designer in Albuquerque, New Mexico. Mr. Petronis has worked with ARC since 1999 to develop illustrations, page layouts, and user interfaces designs. His background includes experience in digital animation, video editing for broadcast television, and information technology management services. He also employs rendering and visualization techniques to analyze potential facility locations and architectural changes, helping architects and planners understand and conceptualize an area's needs. He strives to distill visual information to its essence, maintaining a graphic aesthetic that is clean, concise, and compelling.

Lexi Petronis received a B.A. in creative writing and music performance from the University of New Mexico, and an M.S. in magazine journalism from the Columbia Graduate School of Journalism. Since 1998, Ms. Petronis has written for national publications, worked as a copywriter for corporations and nonprofit organizations, and provided content for consumer web sites and blogs. In addition to serving as the editor-in-chief of Albuquerque's most prominent lifestyle magazine, she has authored *47 Things You Can Do for the Environment* (Zest Books, 2012), and edited the award-winning books *Our Favorite Recipes* (Rio Grande Books, 2008) and *One Nation, One Year* (Rio Grande Books, 2010).

Part III
Advances in Evaluation Knowledge

Preamble

Martin Hodulak, Munich, Germany

Part III of this book is bridging Part II 'Frontiers of Building Evaluation' and Part IV 'Advances in Evaluation Methods'. It contributes a broad and diverse scope of topics on current evaluation knowledge by authors from Canada, Denmark, Germany, the USA, and New Zealand. The range of the authors' individual backgrounds, regarding their nationality, culture, area of expertise, profession and work experience provides a rich diversity of perspectives and insights.

All evaluation knowledge in this section is originating and closely related to the authors' academia or practice experiences. One part of the contributions focus on description and discussion of POE or BPE frameworks in use, while others place experiences, derived from their case studies in practice into the center of their considerations. The case studies include commercial, governmental as well as higher education projects. This further enriches the range and diversity of insights.

In Chap. 9, which is based on a case study, the author highlights the impact of a new work environment on employees' behavior and their work styles within a tradition steeped company. To ensure a work environment which would fully reflect the company's goals, culture and work styles, a Bavarian brewery commissioned a programming phase prior to the office design and further conducted a POE. In this chapter, the author focuses on the methodologies and findings.

In Chap. 10, Jacqueline Visher reflects on the evolution of Post-Occupancy Evaluation (POE) towards Building-In-Use Assessment (BIU) and its role for diagnosing building performance as well as for creating new work settings. She provides an overview on ideas, concepts and observations based on feedback gathered from office building occupants in various countries. She highlights the shift from assessing the user's satisfaction towards a better understanding of how effectively the physical environment supports work.

Linda Nubani investigates in her chapter the effectiveness of the Space Syntax methodology to support client needs in the course of office design. She discusses and compares the visual properties of six different offices in Dubai regarding their terms of intelligibility and expected level of face-to-face communication. Based on the insights, she then compares her examples against mainstream workplace concepts, using Space Syntax techniques.

Mille Sylvest takes a broad approach to building evaluation by further expanding the areas which are traditionally considered as necessary and important for evaluations. Instead of primarily evaluating technical issues, space requirements and user satisfaction, she extends and focuses her assessment on social behavior patterns among users and the opportunities of social interactions, collaborations or informal meetings within built environments.

Duncan Joiner describes the process of establishing and maintaining POE practice in New Zealand government agencies. POE was initiated in the Ministry of Works and Development with the aims to contribute to improvement of existing buildings and to create policy to future design. Even though POE proved partially successful the challenge to create a sustaining and continuous POE demand remained. The author describes the strategies, methods and outcomes of this process.

Thierry Rosenheck provides insights on POE methodology used within an agency of the United States Government for evaluating international governmental facilities of a repetitive building type. In the course of a detailed step-by-step description of the standardized evaluation process, the author shares experiences from practice and discusses approaches for best-practice and for further future improvements, such as a better understanding of feedforward or more effective reporting methods.

Greg Barker illustrates the use of building evaluation to assess and adapt technical requirements used by the U.S. Army in the design of its headquarter facilities. In his chapter he focuses on methods to determine the gross building area based on net space requirements of each building's design program. He acknowledges that further steps such as performance-based evaluations will be necessary to validate the value of setting space requirements based on the empirical evaluation of existing buildings.

Chapter 9
Merging Tradition and Innovation: Programming New Workplaces for Tradition-Steeped Companies

Martin Hodulak

9.1 Introduction

New open space office concepts are gaining popularity among larger national and international corporations in Germany. However, within smaller and more traditional companies of the German "Mittelstand" (mid-sized companies), established cellular office concepts prevail. German medium-sized companies, key to the country's economic success, are often highly innovative in terms of their products, production technologies, and services. At the same time, they are rooted in local context and regional cultural values. These have a strong impact on the company's culture, their staff's expectations, as well as on workplace design.

The Regensburg-based "Brauerei Bischofshof" is a traditional Bavarian brewery, established in 1649, owned by the Church and represented by the Bishop of Regensburg. After the restoration of production facilities in 2006, the management decided to renovate the administration facilities as well. In order to ensure functionality, efficiency and, most of all, acceptance by the owner, staff, and clients, the author was commissioned to develop a functional program in 2008. In further steps, he was involved in the development and implementation of new office concepts in 2011 and 2012.

Based on the case study of the "Brauerei Bischofshof", this chapter is about the challenges and solutions when merging values and expectations of tradition-steeped companies with new and progressive workplace concepts.

M. Hodulak (✉)
Daiserstraße 34, 81371 Munich, Germany
e-mail: martin.hodulak@mnet-online.de

W.F.E. Preiser et al. (eds.), *Building Performance Evaluation*,
DOI 10.1007/978-3-319-56862-1_9

9.2 Tradition and Innovation

"Beer is the world's most widely consumed and likely the oldest alcoholic beverage, it is the third most popular drink overall, after water and tea" (Wikimedia Foundation, Inc. 2016). The long history and overwhelming popularity of beer also applies to Germany, and particularly here to the regions of south-eastern Bavaria. These have the highest density of breweries as well as the allegedly oldest breweries worldwide, dating back as far as 1040 AD (Wikimedia Foundation, Inc. 2016). Most beers are still produced according to the German Reinheitsgebot, i.e., purity law, which was adopted in 1516, and may be regarded as probably the world's oldest existing food quality regulation.

German regional breweries are well aware of their product's heritage and regard it as their unique selling point, especially in global markets. The Bischofshof brewery's worldwide export of beer increased fourfold between 2004 and 2014. According to its managing director, this success is in large part due to their marketing of the Bavarian lifestyle, culture, and traditions, rather than just marketing the beer. Shipments to China frequently contain Bavarian dirndls and lederhosen alongside the brewery's beer bottles and crates. They help to promote Bavarian lifestyle on occasions such as the Chinese Oktoberfest in Quingdao. In addition to a well-targeted export strategy, the brewery's success can be attributed to its history, reaching back one thousand years, numerous quality labels, several world championship titles, and probably its most prominent customer, Pope Benedikt (Bayerische Staatszeitung 2015).

The production of beer itself relies more on innovative rather than traditional virtues even though, the brewery stresses, it needs both. Less than half a dozen global corporations dominate today's world beer market. In order to stay competitive on national and international markets, local and smaller breweries frequently merge to maintain a critical size, they license their successful brands and are increasingly engaged in developing new products. At the same time, they constantly optimize their processes and modernize their production facilities. In recent years, the Bischofshof brewery established marketing and production services for a number of regional microbreweries to help increase their efficiency and make better use of their own administration and production capacities. The brewery erected a new brewing house in 2006 at their Regensburg site, which substantially reduced their energy and water consumption (Bierwelten 2016).

9.3 Situation and Project Approach

The Bischofshof brewery site in Regensburg has been in use since the 17th century. After the renewal of the bottling plant with logistic facilities in 1992, and with the brewing house as the heart of the brewery in 2006, the administration buildings were next in line to be renovated or replaced. Triggering factors for refurbishment were the poor pre-war building quality and the outdated cellular offices, along with changes in organizational culture and work processes. The following were among the goals defined by management:

Company goals with relevance to the project:

- Provide optimal support to increase customer satisfaction and enthusiasm
- Pursue further process optimization, digitization, and reduced archiving
- Foster informal communication and knowledge flow

Project goals:

- Reduce the number of scheduled meetings through ad hoc meetings and cross-departmental links
- Increase visibility of management workplaces to improve communication
- Provide a new work environment that makes staff feel more at ease

The management decided against a seemingly obvious and common approach. This would have involved visiting best practice examples of contemporary office solutions and then choosing the one that seemed to be most appropriate. Instead, they chose an approach for their future workplaces, which is rather common in the planning of production facilities. This particular kind of planning is based on use of state-of-the-art technologies, scientifically-based and carefully engineered processes, and on detailed requirements as specified by the operating specialists. The engineering approach seems appropriate for the design of facilities where goods are produced. As these goods are tangible, their production processes and logistics are visible, and they can easily be tracked, analyzed, and optimized. As for offices, the engineering approach has other challenges. Office workplace design is, in large part, about communication, collaboration, and knowledge exchange among people. The issues at stake are intangible and the processes within the workflow are not visible. Moreover, since office workplaces are about people, the soft and qualitative topics prevail. Thomas Allen and Gunter Henn (Allen and Henn 2007) describe this challenge in "The Organization and Architecture of Innovation". Both authors developed workplaces optimized for information and knowledge flow. Basically, their concepts derive from two sources of knowledge. The first source is scientific evidence based on surveys and research. The second source is the everyday experience of the future users and their specific needs and requirements concerning their work environment and workstations.

A similar approach was used for developing the Bischofshof workplace concept. The brewery commissioned the Stuttgart-based Fraunhofer IAO (Institut für Arbeitswirtschaft und Organisation) to provide scientific knowledge. The IAO frequently initiates research projects about workplace design and is constantly conducting respective surveys. In 2002, the Institute published a survey on the self-perceived productivity and motivation of workers in different office types. Surprisingly, one of Germany's most popular office types, the double occupancy cellular office, was ranked as being among the least motivating and least productive (Spath and Kern 2003). This underscored the management's decision to develop the future offices to their own needs and specifications, rather than to adapt best practice solutions.

In order to define the specific user needs and requirements, the author's practice chose an approach initially developed over 60 years ago by the architectural firm of Caudill Rowlett Scott (CRS) in Houston, Texas (see also Chapter 18 by Parshall and Fonseca, Hellmuth, Obata + Kassabaum (HOK). In 1994, HOK acquired CRSS Architects, formerly CRS. CRS recognized the need for asking questions and translating the findings into specifications prior working on solutions (Pena and Parshall 2012). To them, it was clear that they could only design spaces that would provide optimum support to the building's users and their work if they understood exactly what the future users needed. The users they talked to, however, were seldom in a position to provide that information. When asked about their needs for future workplaces, they would either describe their current work environment, or repeat what they heard from their colleagues. This was clearly not a way to establish the basis for future workplace design. So, the required analysis should not consist of asking future users what they want, but rather what do they want to do within their future facility.

9.4 Requirements

From the outset, it was mutually understood that the proposed offices would not be an improvement or further development of the existing situation, but a rather new concept reflecting a new company culture under the new management. The staff's involvement and participation were regarded as the project's pivotal element and essential for its success (see Fig. 9.1). The involvement should result in a maximum input of the staff's operative knowledge for the project's benefit. Furthermore, it should help to strike the right balance between tradition and innovation. And, finally, the staff's active participation - and thus its influence - should support the outcome's acceptance.

9.4.1 Process and Methodology

The employees' participation consisted of three sets of interviews and workshops. In a strategic workshop at the project start, the management defined goals, stated restrictions, and discussed the agenda for the following interviews. The next day, most employees attended one of the four scheduled focus interviews. After the gathered data was structured, analyzed, and first concepts were developed, the

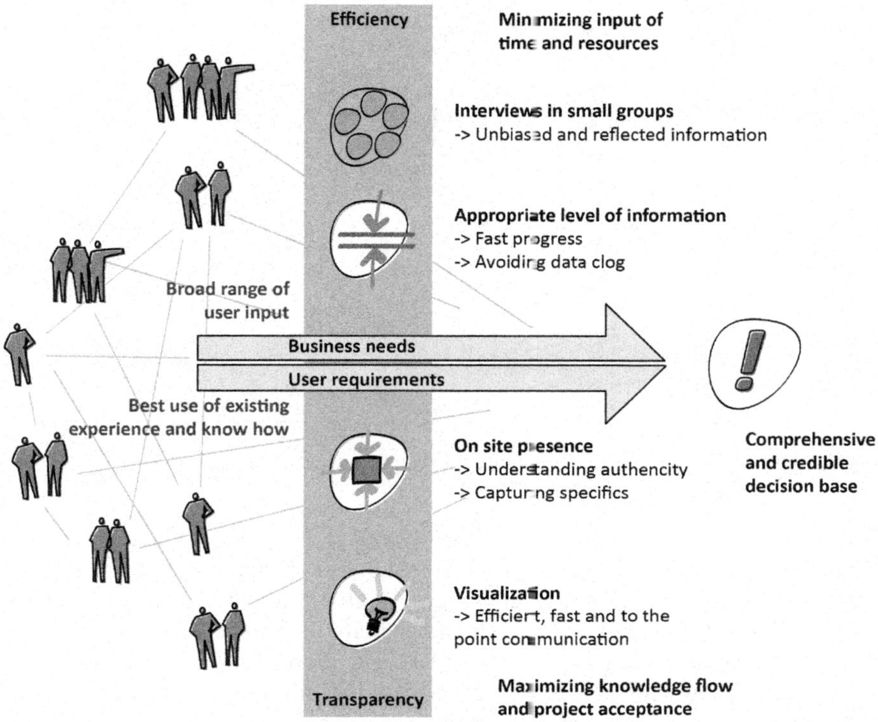

Fig. 9.1 Methodology chart. *Source* Author based on Hodulak and Schramm (2011)

management and representatives of each department met in a half-day consensus workshop. All the major decisions were made in this workshop. From this point on, the workplace requirements and concept were not altered, even though the office layout was frequently adapted and modified. All workshops and interviews were conducted in small groups and focused on qualitative aspects. Workshop information was simultaneously recorded visually on index cards and charts to keep the information flow transparent for all involved.

9.4.2 Engineering Innovative Workplace Concepts

"Most designers love to draw, to make 'thumbnail sketches', as they used to call these drawings. Today, the jargon is 'conceptual sketches' and 'schematics'. Call them what you will, they can be serious deterrents in the planning of a successful building, if done at the wrong time – before programming or during the

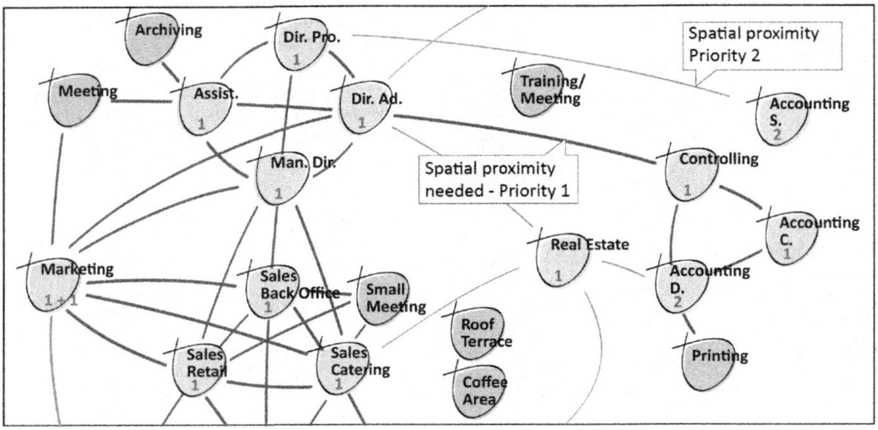

Fig. 9.2 Requirements for spatial proximities, excerpt, annotated. *Source* Author

programming process" (Pena and Parshall 2012, p. 20). The need for a distinct separation of programming and design was a constant topic throughout all interviews and workshops, as it was tempting to discuss the transformation of the old malt silo rather than analyze work styles and requirements. The interviews followed a standardized agenda. Tasks, work styles, and processes were discussed first, followed by information on functional proximities and concluding with requirements on workspace and work environment.

Based on their tasks, distinctive work profiles, e.g., management, administration and sales force, were identified. Corresponding properties of workplaces were defined as those that would best support their respective activities, such as individual and undisturbed work, collaboration or information exchange. Communication relationships, requirements for spatial proximities, and the resulting clustering and zoning of functions and workplaces were mapped on charts (see Fig. 9.2).

In the final stage and based on all previous considerations, workspace typologies were developed. Empirical studies on the effectiveness of open and closed workplace concepts were kept in mind, but all programmatic concepts were discussed and tested against the stated goals, needs, and requirements.

9.4.3 Implementing Traditional Elements

Increasing customer satisfaction and making employees feel more at ease were two of the project's main goals. Apart from functioning well, the new offices also had to respond to the staff's expectations as well as the customers' anticipated look and feel. Both were discussed in the course of the workshops. Visibility, the need for privacy, the openness and personalization of workspaces, and the centralization of printers were the most controversial of the discussed topics. It became apparent that the new offices could not be purely functional, but that some elements of the old office would have to be implemented as part of the concept.

9.5 Workplace Concept

The new offices maximize openness and lines of sight among all employees. Located on two levels, connected by an open staircase, and only separated by glass partitioning, most workplaces are visually connected (see Fig. 9.3). Even though noise can be blocked out, by closing the glass door, all members of the staff remain visible at all times. Most paperwork is digitized, archiving is reduced to one sideboard per employee, and individual printers are replaced by two shared printers. Individual offices of the field sales force were abandoned. Since the sales representatives only come in a few days each month they work in the cafeteria, in lounges, or meeting rooms.

The timberwork structure of the malt silo became part of the office design. The warm color and the texture of the aged wood create a positive contrast to the functional white and greys of the office furnishings. A number of historic artifacts were included: old models, products, and furniture such as the baroque wardrobe in

Fig. 9.3 Open space work area. *Source* Brauerei Bischofshof

Fig. 9.4 Bräustüberl. *Source* Brauerei Bischofshof

the reception area. The largest historic artifact is the Bräustüberl, a wood-clad pub interior, which was dismantled in the old building and set up in the new modern interior (see Fig. 9.4).

9.6 Post-occupancy Feedback

One year after the staff moved in, the Bischofshof project leader and the author met to review the office concept's performance. Among other things, the following topics were discussed.

- *Where do you see the major differences between the old and new situation?* We have moved closer together. The previous office building had a hotel-like feel, with long corridors and closed doors. Breaking down walls resulted in more spatial openness. But even more importantly, it resulted in more openness among the people. We see each other in the spaces around the coffee area, and the pigeonholes where we meet are in constant use. This might be partly due to the field sales force, who abandoned their office desks and now use the meeting spaces, bistro tables, and bar stools instead.

- *Were you able to increase work efficiency as one of the project goals?* We have no empirical evidence about this. It is my impression that there are more short ad hoc meetings. Our work styles also seem less formal. And we have definitely reduced barriers. We stop by, meet, and talk.
- *In the course of the programming we discussed the principle of visual connections among workplaces.* That works really well. Now and then, I realize that I don't need to make a phone call. I communicate via the line of sight.
- *How do you span the gap between tradition and innovation?* We wanted to keep the building and its construction in the tradition of a former production building, and we used modern tools and equipment wherever possible. Each employee has his own laptop, enabling him to work wherever he happens to be. We provide Wi-Fi on site and employees use mobile phones. Most employees - but not everyone - are now more mobile in their work.
- *What were the major challenges?* Prior to the move, each employee had their own printer. When we suggested shared printers, most employees were worried that these would result in longer distances, waiting times, and piles of documents being left behind. This issue was highly emotional; however, the concerns did not come true. A further concern was the use of glass partitioning and the resulting transparency of the workplace. We had to communicate early on in the project that glass partitions were only planned to separate work from circulation areas. Each employee would still have a solid wall at the back of his/her workspace.
- *Was the initial concept changed in the course of the project?* We changed the workplace seating from back-to-back to face-to-face. Employees were very concerned regarding wall-oriented seating. They wanted to sit with a wall to their back and facing their colleagues.
- *How have the employees felt and reacted since they moved in?* Their reactions have been very positive. During the brewery's open day, far more visitors than we had planned wanted to visit the new offices. Most of the employees wanted to show their new workplaces. We have also seen that employees are more conscious about the way they dress. Our receptionist is a nice example, demonstrating far more pride in the way she dresses and welcomes visitors. This applies to most employees who pay more attention to their appearance within the open space.
- *What are the elements epitomizing tradition within the new work environment?* It is the material and some elements. The original timber frame construction was kept as far as possible. The Bräustüberl was reconstructed in its original form and is another traditional element. It is our most favorite meeting room.
- *Was the investment of time and fees in programming well spent?* The interviews and workshops were a way of taking the employees' concerns and worries seriously and into consideration, well ahead of the start of planning. But they also contributed their experiences and expressed their interests. As a result, we had made almost no changes to the planning and realization.

9.7 Conclusion

In the first client meeting, the general manager mentioned a feature of the old office, prior to his start at the Bischofshof brewery. Three lights were mounted just above each of the office doors. The green light signaled, "come in", the red light stood for "do not disturb" and the white light was the signal for "I am out of the office". For the general manager these three lights epitomized the old corporate culture characterized by rigid hierarchical and departmental structures, regulations, and control.

Under the new management organization and processes, but furthermost, the corporate culture changed dramatically. The old offices were prestigious, spacious, individual, and very popular with the staff who got used to them over the decades. Given the choice, many employees would have seen the preferred solutions in the established concept. However, in the course of defining the staff's requirements and needs, the mismatch of office concept and company culture became evident.

In the project review, as a part of the BPE approach and its holistic view of the building life cycle (see also Chap. 1), the project lead on client side and the author regard the successful merging of tradition and innovation as a combination of three principles. Firstly, the concept is based on credible, scientific evidence derived from studies and surveys. Secondly, the staff and their everyday experience are fully reflected in the workplace program and conceptual design. And thirdly, the origins, the heritage, the regional context, and local involvement were not only all taken into account, but are a substantial part of the concept. The Beer Queen and the Bishop of Regensburg, who blessed the new Bischofshof workplaces, were also among the local officials and celebrities attending the opening ceremony for the new offices (see Fig. 9.5).

Fig. 9.5 Opening ceremony. *Source* Brauerei Bischofshof

References

Allen, T. T., & Henn, G. (2007). *The organization and architecture of innovation*. Burlington: Elsevier.

Bayerische Staatszeitung. (10.03.2015). Retrieved from http://www.bayerische-staatszeitung.de/staatszeitung/leben/detailansicht-leben-in-bayern/artikel/wie-bayerisches-bier-die-welt-erobert.html. Accessed 02 Feb 2016.

Bierwelten. (02.02.2016). Retrieved from https://www.bierwelten.de/brauereiverzeichnis/details/brauerei-bischofshof-ek.html. Accessed 02 Feb 2016.

Hodulak, M., & Schramm, U. (2011). *Nutzerorientierte Bedarfsplanung*. Berlin: Springer.

Pena, W. M., & Parshall, S. A. (2012). *Problem seeking: An architectural programming primer*. New York: Wiley.

Spath, D., & Kern, P. (Eds.). (2003). *Office 21—Push for the future*. Köln: Egmonts vgs.

Wikimedia Foundation, Inc. (02.02.2016). Retrieved from https://en.wikipedia.org/wiki/Beer. Accessed 02 Feb 2016.

Author Biography

Martin Hodulak is a facility programming and workplace strategy consultant, providing services to government and commercial clients. Over the past 18 years, he has been leading and conducting more than 70 projects on facility programming, workplace strategies and workplace design. Among them are commissions for automotive, pharmaceutical and IT industries, as well as for governmental institutions and higher education facilities His projects range from small-scale workplace designs to large programs on the urban design scale, such as the conversion of the Berlin Tegel Airport.

Apart from project work, he has been refining and teaching methodologies for programming and workplace design. He co-authored the first guidebook on programming for the German market (Hodulak, M. and Schramm, U. *Nutzerorientierte Bedarfsplanung* Springer, 2011) and published a comparative survey on corporate global workplace standards. Martin has lectured on programming and workplace design, and is published in national and international conference proceedings. He was trained as an architect at the University of Karlsruhe and the University of Bath and received his Ph.D. from the University of Stuttgart. In recent years, he has worked as a project leader, senior consultant and managing director for various architectural and consulting practices.

Chapter 10
Building-In-Use Assessment: Foundation of Workspace Psychology

Jacqueline C. Vischer

10.1 Introduction

User feedback studies, in which occupants provide information on performance of the building in use, are an important source of knowledge. User feedback studies aim to assess how buildings and building systems affect the comfort, effectiveness, and well-being of building users. As studies become more numerous and sophisticated, and information is more readily available on the effects of the built environment on users, feedback mechanisms have evolved to inform all stages of building programming, design, construction, and occupancy, known as building performance evaluation (Preiser and Vischer 2015). The building performance evaluation framework incorporates early notions of post-occupancy evaluation and links the information gleaned from users to decision-making at each stage of the building delivery cycle. The post-occupancy research focus on building occupants' feedback yields findings that shed light on the operation and management requirements of existing buildings, generate new knowledge about the human use of space, and inform key decisions during the design and construction process.

10.2 Assessment Tools

One active area of user feedback research focuses on environments for work, specifically office buildings. Since 1989, a range of measurement tools have been developed in the form of user surveys, instrument measurements of ambient conditions, and qualitative techniques—many of which are designed with a view to practical diagnosis of building performance leading to problem correction and

J.C. Vischer (✉)
University of Montreal, 2453 rue Ryde, Montreal, QC H3K 1R5, Canada
e-mail: vischer.biu@gmail.com

W.F.E. Preiser et al. (eds.), *Building Performance Evaluation*,
DOI 10.1007/978-3-319-56862-1_10

user-related improvements. This chapter provides an overview of ideas, concepts, and observations that have evolved from amassing, analyzing, and applying large amounts of feedback gathered from office building occupants over a period of almost three decades in countries all over the world. Together these ideas form a coherent theory of user-building interaction which can be applied to improving worker comfort and productivity, providing cost-effective accommodation, supporting technology-supported mobile work, and mediating rapid organizational change.

Building-In-Use (BIU) Assessment, devised in the 1990's, is one of the first tools for collecting reliable user feedback in environments for work. It aimed to standardize data gathered from building user surveys to ensure that user feedback could usefully be applied to diagnosing building performance (Vischer 1989). A short standardized questionnaire collects feedback in the form of users' ratings of building conditions and features. Collecting standardized data enables the construction of a database from which typical patterns of user response to office environments are calculated. Individual building scores are compared to database norms to provide a context for assessing the meaning of users' ratings of their workspace, and to indicate whether it is superior or inferior to a "typical" office building workspace. The evolution of this tool, its use in a wide range of work environments, and the rich variety of study outcomes led to many of the concepts and constructs discussed in this chapter.

Space for work, or workspace, is increasingly diverse. Whereas office planning was once based on simple division of workspace into large rooms containing rows of desks and a few private offices for managers, contemporary work environments include a range of individual and shared spaces, communal areas and amenities, and access to sophisticated electronic tools (Gillen 2006). Consequently, companies increasingly apply quality as well as cost criteria to workspace design to invest in environments that actively support workers' tasks (Vischer and Malkoski 2015). Research indicates that workspace design and management affect not only how people feel about their job, but also work performance, loyalty, engagement, and ultimately the value of human capital to the organization. The premise of BIU Assessment is the dynamic and interactive relation between users and space: i.e. the user's environmental experience includes the consequences of her behavior in that environment, and her experience of the environment is itself transformed by the activities she is performing (Vischer 2008a).

10.3 Satisfaction and Productivity

Early post-occupancy studies attempted to assess the success and failures of a building through users' satisfaction ratings, and they often still do. This approach asks building users to identify what they 'like' and 'dislike' about their work environment on the implicit assumption that self-reported satisfaction with individual features is a de facto measure of building quality. The logic implies that if users like their workspace and are satisfied the workspace is successful, whereas if they are dissatisfied, the building is not performing or has somehow failed.

Substantial knowledge of users' preferences has emerged from workspace satisfaction research. For example, findings show that office workers are typically dissatisfied with 'open plan' offices, whether this is due to noise levels, distractions, lack of privacy, or the sameness of 'cubicles' (Davis et al. 2011). However, the prevalence of this finding has not prevented employers from favoring open plan layouts—in part because workstations are cheaper to construct and more flexible to reconfigure than a traditional cellular office layout, and in part because more desks and equipment can be fitted into open plan layouts. As has been argued elsewhere, whether workers like or dislike workspace features pertains more closely to happiness research than to understanding how effectively the physical environment supports work (Vischer 2008a). More complex models of user-environment interaction, e.g., how well people can perform tasks, access needed tools, engage in appropriate communication, and identify territory, are needed to guide inquiry into workplace performance, that is, the effectiveness of workspace whose explicit objective is to support the performance of work. BIU Assessment connects workspace features with worker effectiveness: a performing workplace is designed to optimize worker productivity, so users' judgment of whether their space does or does not support their work is a better diagnostic measure than whether or not they like it.

10.4 Levels of Productivity

Workspace design influences productivity at three identifiable levels: these are individual, group, and organizational productivity Each category denotes a variation in scale of environmental influence (Vischer 2006).

- Individual productivity is typically assessed at the scale of the individual workpoint, through data on how the micro-environment—specifically environmental conditions such as lighting and visual conditions, variations in temperature and humidity, furniture ergonomics, and noise privacy—influences individual task performance. Data analysis measures effects on speed and error rates as well as on incidence of illness and absenteeism.
- The productivity of groups sharing workspace, such as a teamwork environment, is typically evaluated in terms of the quality and quantity of group processes, such as rate of innovation, number of creative ideas, and speed of decision-making. Teamwork is affected by the design and layout of the team workspace, including access, circulation, and ambient conditions such as noise. Group processes are affected by workgroup size and the relative accessibility of team members. Other environmental determinants of workgroup effectiveness include the positioning of work areas and shared space, and access to tools and equipment.

- A third level of productivity is a function of an organization's accommodation, that is, its overall work environment, including appearance and location, workspace and amenities, communications tools and technology, and the ways these are used. Accommodation choices support the organization's business objectives and affect competitive advantage to varying degrees. As Fig. 10.1 shows, the quality of support that the organization obtains from its accommodation can range from highly positive—actively supporting work—through neutral and poor, to highly negative (Vischer 1996).

In a positive organization-accommodation (O-A) relationship workers' tasks are facilitated, and in a negative O-A relationship workers' time and attention are lost dealing with adverse environmental conditions. The O-A relationship is dynamic and evolving as firms become attuned to the benefits of adjusting and updating workspace in response to changing technology and business processes. Locational advantages and access, as well as amenities such as fast elevators, convenient bathrooms, adequate parking, and attractive eating areas all affect organizational effectiveness.

While evidence accumulates that workspace design influences workers' effectiveness, accounts of workspace change suggest that employees resist 'social engineering' solutions where employers envision a work environment aimed at eliciting maximum productivity (Vischer 2009). Such an approach violates the socio-spatial contract, the implicit social contract between worker and employer that promises to provide a certain level of workspace quality in return for the worker's energy, effort, and knowledge (Vischer 2005). Contract violations, often unpremeditated, cause workers to feel devalued and increase their resistance to workspace change. In many organizations, moving workers out of private offices and into open workstations is a socio-spatial contract violation: the status, confidence, and responsibility that the employer communicates through allocating private enclosed workspace is undermined by allocating the same open workstation to everyone.

Managers who consider supportive workspace to be an investment in their workforce require evidence to guide their environmental design decisions.

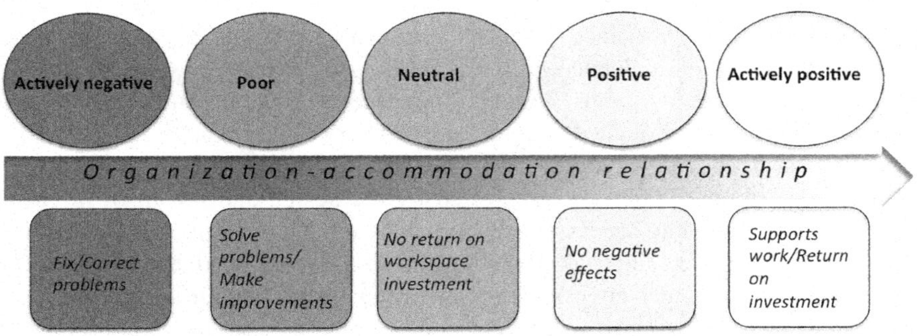

Fig. 10.1 Stages in the Organization-accommodation relationship. *Source* Author

Like information technology, workspace can and should be a tool for performing work. To design workspace as a tool for work, information is needed on what workers do, how they perform tasks, and ways in which they are—and are not—helped by their workspace. Ways of measuring users' feedback such as BIU Assessment are diagnostic tools whose findings apply to all three categories of productivity: individual task performance, teamwork effectiveness, and organizational accommodation.

10.5 Workspace Comfort

Collecting, interpreting, and applying complex feedback from users has generated a tri-partite model of workspace comfort that goes beyond simple user preferences and satisfaction ratings. Basic to this model is the concept of functional comfort, which connects user satisfaction with worker productivity by defining effective or successful workspace in terms of degree of environmental support for occupants' tasks and activities. BIU Assessment measures levels of functional comfort for a given workspace, and provides a diagnosis of workplace effectiveness that captures the impact of workspace features on work performance at the different scales of productivity. As shown in Fig. 10.2, functional comfort is one of three constituents of workspace comfort.

Physical comfort, defined in terms of meeting building codes and published comfort standards, ensures that people feel healthy and safe in the buildings they occupy. Without physical comfort there can be neither functional nor psychological comfort. Both physical and functional comfort are affected by psychological comfort: people's sense of belonging, territory, and environmental control, often expressed as the need for privacy.

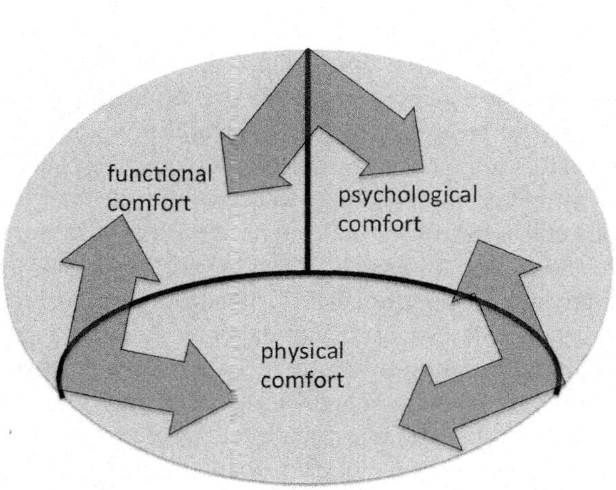

Fig. 10.2 Tri-partite model of workspace comfort. *Source* Author

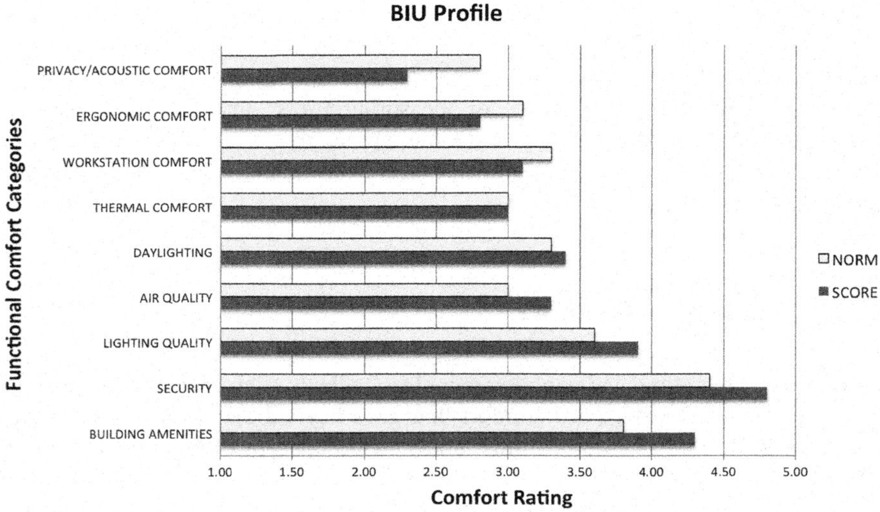

Fig. 10.3 Building-In-Use profile indicating intervention priorities based on differences between building score and database norm. *Source* Author

The functional comfort approach has been applied to diagnosing workplace performance in numerous office buildings. Figure 10.3 displays how users' functional comfort ratings of workspace features are analyzed with reference to database norms, providing a profile of workspace strengths and weaknesses that delivers diagnostic information to designers, planners, and managers. In this example, acoustic comfort is most in need of correction, while security and building amenities are supportive of workers' tasks. A functionally comfortable workspace is a performing workplace: employers receive a return on their workspace investment in terms of increased employee effectiveness (Vischer 2008b).

Workspace diagnosed as functionally uncomfortable slows down work, increases worker fatigue, and leads to workspace stress. In unsupportive environmental conditions—a negative O-A relationship—workers use their energy and attention to solve environmental problems. Consequently, task performance is compromised, energy for creative thinking and innovation is reduced, and the value of its human capital is not realized by the organization (Vischer 2007). Workspace stress occurs when elements of the physical environment interfere with the attainment of work objectives. Stressors that interfere with task performance, motivation, and social relationships "influence physiological processes, produce negative affect, limit motivation and performance, and impede social interaction" (Evans and Cohen 1987, p. 107). Today's workspace cannot be designed as a one-time, final, and permanent ergonomic support for all office tasks, but rather needs to be adaptable and negotiable to provide ongoing support to users. Workers need the skills and opportunities to engage with and adjust their environment over time and with changing task requirements in order to optimize functional comfort and cope with workspace stress.

Measuring levels of functional comfort in a building provides a diagnosis both of more stressful/least comfortable and of less stressful/more supportive workspace conditions. Reliable occupant feedback identifies all workspace environments somewhere along the continuum ranging from functionally comfortable and supportive of work to dysfunctional and stressful.

10.6 Psychological Comfort

Using BIU Assessment to measure functional comfort in buildings all over the world has shed light on the complex psychological layers that affect workers' relationship to their physical environment. Psychological comfort is a function of a sense of belonging, i.e., territorial appropriation, along with loyalty and commitment to the organization, and sense of privacy and environmental control, all of which are mediated by the socio-spatial contract and the behavioral expectations it implies (Augustin 2009; Vischer 2005).

Territory, whether of the individual or the group has psychological value both as space for one's work and as symbolic of one's place in the organization. Underlying these is a human behavioral schema expressed in terms of the personalization and appropriation of space: marking territory and constructing boundaries of social and environmental control. The introduction and use of new technology and sophisticated communications tools also affect workers' notions of territory. Territorial boundaries are not simply physical elements that enclose space: territoriality signifies sense of privacy, social status, and control. When people move out of private enclosed offices into open plan workstations, studies show they judge their environment more negatively, citing lack of privacy, acoustic conditions, and confidentiality problems (Brennan et al. 2002; McElroy and Morrow 2010). These reasons are given irrespective of whether or not their work is confidential, or whether or not they need to be alone to perform tasks effectively. Complaints about lack of privacy abound in studies of workspace change, independent of physical characteristics such as furniture configuration and partition height.

Workers' sense of privacy is connected to environmental control on at least two levels: mechanical or instrumental control, and control over process or empowerment (Vischer 2012a). Instrumental control refers to mechanical actions, such as chairs and worktables that are raised and lowered, cabinets and tables on wheels, operable windows, switchable lights, and a door. Evidence indicates a positive psychological impact from instrumental control in situations where employees are informed and even trained to make use of the controls available (Vischer 2012b). An important form of environmental control is the opportunity for personalization. Behaviors such as placing symbolic objects, family photographs, plants, and posters in individual and team workspace increase sense of belonging, loyalty, and morale (Elsbach 2004).

Opportunities for employees to participate in workspace decision-making increase control over process and environmental empowerment, which both affect psychological comfort. Studies have shown that worker participation in the workspace design process has a positive effect on people's response to and feelings about their workspace. People who are informed about workspace-related decisions, and who participate in decisions about their own space, are more likely to have feelings of belonging and territorial ownership. This enables workspace stress reduction through positive coping with environmental demands and encourages workers to find ways of solving their environmental problems.

10.7 Future Research

Considerable knowledge has accrued from using BIU Assessment both to assess building performance and to study the complexity of workspace psychology. The future of user feedback and its role in building performance evaluation requires strong theoretical frameworks that will lend greater coherence to existing knowledge, generate fruitful research, and create supportive work environments in office buildings.

While occupants' satisfaction ratings provide data on their likes and dislikes, satisfaction studies generate little information about environmental support for task performance, adding value to business processes, or why owners and managers should invest in workspace improvement. Generating diagnostic data on building performance through measuring how well the environment supports work generates findings that can be applied to decisions about how and when to intervene to solve environmental problems and effect improvement through removing, replacing, or transforming workspace features. Consequently, building interventions can be prioritized and appropriately scaled, workers' tasks are performed better, team communication and decision-making is more effective, and the organization is more productive. In addition, improved psychological comfort through empowerment helps enhance the creation and dissemination of organizational knowledge.

Companies that value human capital want to understand how new knowledge accrues in their organizations and how to distribute and share knowledge. Workspace plays an important role in these processes (Vischer 2010). Worker productivity in the knowledge economy is less a matter of improving speed and accuracy of routine tasks and increasingly a function of generating new ideas, being creative, working effectively in teams, and producing knowledge that adds value to the organization. While measures of functional comfort provide indicators of effects on productivity, other productivity indicators such as reduced illness rates, increased speed and accuracy of task completion, and even rates of generating new ideas, also measure workspace effectiveness.

Psychological comfort, the feeling of belonging, is an important predictor of employee retention and reducing costly staff turnover. More extensive measurement of territorial behavior and appropriation at work will yield improved knowledge of how and why environmental features affect employees' sense of privacy and how to meet privacy needs without compromising information exchange and team collaboration. Better understanding of territoriality, privacy and environmental control mechanisms through feedback from occupants will help organizations determine their returns on investment in workspace environmental quality in terms of recruitment and retention of high-quality employees.

Finally, users' workspace comfort interacts with sustainability and the 'green' qualities of commercial buildings (see Chaps. 16, 19, and 20). Sustainable building features, such as natural ventilation, water recycling, and passive cooling technology, affect physical, functional, and psychological comfort of users, and research provides evidence of behavioral changes as a result of sustainable design features. Some studies indicate a positive effect on users' psychological comfort as people are proud of working in sustainable buildings and feel empowered to make behavioral decisions, while others show little evidence of sustainable buildings providing more supportive workspace. There is also some evidence that giving occupants a more active role and responsibility—environmental empowerment—for changing their behavior in environmentally sustainable buildings is a necessary condition for success.

10.8 Conclusions

The environmental psychology of workspace is a rich and diverse field of study that is still growing. As human beings in all parts of the world spend increasing amounts of time in environments for work, the effects of the physical environment on occupants' performance, health, and morale needs to be better understood. The knowledge yielded by feedback from occupants through measurement tools such as BIU Assessment informs employers' decisions as well as corporate investments in the work settings they create, and assists and improves the building industry as designers, facilities managers, leasing agents, and construction professionals draw on it. Business managers also seek evidence of how workspace decisions affect their personnel as companies become more agile by implementing ongoing workspace change, often dispersing teams to more than one geographic locale. Using BIU Assessment to systematically collect reliable feedback from building users has yielded a rich mine of knowledge about building performance and a significant contribution to understanding the user-environment relationship.

References

Augustin, S. (2009). *Place advantage: Applied psychology for interior architecture*. New York: Wiley.

Brennan, A., Chugh, J. S., & Kline, T. (2002). Traditional versus open office design: A longitudinal study. *Environment and Behavior, 34*(3), 279–299.

Davis, M., Leach, D., & Clegg, C. (2011). The physical environment of the office: Contemporary and emerging issues. In G. P. Hodgkinson & J. K. Ford (Eds.), *International review of industrial and organizational psychology* (Vol. 24, pp. 193–238). New York: Wiley.

Elsbach, K. D. (2004). Interpreting workplace identities: The role of office decor. *Journal of Organizational Behavior, 25*(1), 99–128.
Evans, G. W., & Cohen, S. (1987). Environmental stress. In D. Stokols & I. Altman (Eds.), *Handbook of environmental psychology* (Vol. 1, pp. 571–610). New York: Wiley.
Gillen, N. M. (2006). The future workplace, opportunities, realities and myths: A practical approach to creating meaningful environments. In J. Worthington (Ed.), *Reinventing the workplace* (2nd ed., pp. 61–78). Oxford, UK: Architectural Press.
McElroy, J. C., & Morrow, P. C. (2010). Employee reactions to office redesign: A naturally occurring quasi-field experiment in multi-generational setting. *Human Relations, 63*(5), 169–191.
Preiser, W. F. E., & Vischer, J. C. (Eds.). (2015). *Assessing building performance* (2nd ed.). New York: Routledge.
Vischer, J. C. (1989). *Environmental quality in offices*. New York: Van Nostrand Reinhold.
Vischer, J. C. (1996). *Workspace strategies: Environment as a tool for work*. New York: Chapman and Hall.
Vischer, J. C. (2005). *Space meets status: Designing performing workspace*. Oxford, UK: Taylor & Francis/Routledge.
Vischer, J. C. (2006). The concept of workplace performance and its value to managers. *California Management Review, 49*(2), 62–79.
Vischer, J. C. (2007). The effects of the physical environmental on work performance: Towards a model of workspace stress. *Stress and Health, 23*(3), 175–184.
Vischer, J. C. (2008a). Towards a user-centred theory of the built environment. *Building Research & Information, 36*(3), 231–240.
Vischer, J. C. (2008b). Towards an environmental psychology of workspace: How people are affected by environments for work. *Architectural Science Review, 51*(2), 97–108.
Vischer, J. C. (2009). Applying knowledge on building performance: From evidence to intelligence. *Intelligent Buildings International, 1*, 239–248.
Vischer, J. C. (2010). Human capital and the organizational environment. In A. Burton-Jones & J. C. Spender (Eds.), *Oxford handbook of human capital* (pp. 477–498). Oxford, UK: Oxford University Press.
Vischer, J. C. (2012a). Managing facilities for human capital value. In K. Alexander & I. Price (Eds.), *Managing organizational ecologies* (pp. 34–47). New York: Routledge.
Vischer, J. C. (2012b). The changing meaning of workspace: Planning space and technology in the work environment. In S. Mallory-Hill, W. Preiser, & C. Watson (Eds.), *Enhancing building performance* (pp. 89–98). Oxford, UK: Wiley-Blackwell.
Vischer, J. C., & Malkoski, K. (2015). *The power of workspace for people and business*. Melbourne, Australia: Schiavello.

Author Biography

Dr. Jacqueline C. Vischer is an Environmental Psychologist specializing in environments for work and Professor Emeritus at the University of Montreal. She is a founder of the field known as workspace psychology. She has published numerous articles and several books, including *Environmental Quality In Offices* (1989), *Workspace Strategies: Environment As A Tool For Work* (1996), *L'Évaluation des environnements de travail: la méthode diagnostique* (co-author Gustave-Nicolas Fischer, 1998), and *Space Meets Status: Designing Workplace Performance* (2005). In addition, she has co-edited two books with Wolfgang Preiser. In 2004 Vischer received the Lifetime Career Award from the Environmental Design Association and in 2011 she was honored with the Nancy Vincent McClelland Merit Award from the American Society of Interior Designers.

Dr. Vischer is Professor Emeritus at the University of Montreal, where she directed the Interior Design Programme and founded the New Work Environments Research Group (Groupe de recherche sur les environnements de travail). As principal in her consulting firm, Buildings-In-Use, Vischer has advised a wide range of organizations internationally on managing workspace comfort, designing innovative workspace, and planning workspace change. Many of her writings are available at—and can be downloaded from—www.jacquelinevischerbiu.com

Chapter 11
Evaluating Workplace Constructs Using Computerized Techniques of Space Syntax

Linda N. Nubani

11.1 Introduction

Previous literature on space syntax showed how workspace layout generated boundaries that created relationships of accessibility and visibility (Rashid et al. 2005). These measures in turn regulated occupants' behavior and activities. Previous research also measured the impact of spatial layouts and various organizational constructs such as employees' performance, satisfaction, and face-to-face communication levels. However, there is little research that documents whether these techniques are implemented in professional practice. Within this chapter, the author discusses and compares the visual properties of six different semi-government and private offices in Dubai regarding their terms of intelligibility and their expected level of face-to-face communication among employees. The author provides a comparison between these offices and mainstream workplace concepts using space syntax techniques. The goal of using these techniques is to establish a systematic and an objective way in describing the relationship between organizational constructs and office layouts.

11.2 Current Workplace Standards in Dubai

Since cost and time are the two top concerns among all clients, the implementation of the proposed techniques will expedite the design process. Not only will it be able to meet clients' timeframes but it will also help them understand ways to improve both efficiency and performance among employees.

L.N. Nubani (✉)
Michigan State University, East Lansing, MI, USA
e-mail: NubaniLi@msu.edu

W.F.E. Preiser et al. (eds.), *Building Performance Evaluation*,
DOI 10.1007/978-3-319-56862-1_11

Dubai has become visible on the World's map with the vision of its leader Sheikh Mohammad Bin Rashid Al Maktoum. In 2014, he expressed his objective for Dubai Plan 2021 to turn it into "the city of happiness" and to make it the preferred place to live, work, and visit (Bashir 2014). In Dubai, foreign nationalities exceed 75% of the population. Expatriates come from over 110 countries, which makes the city more challenging specifically for the architecture and design practice. Duffy expressed Europe's challenges in the 1970s with designing IBM's open offices across the continent when following American office design practice might stimulate negative social tension among employees (Duffy 2015).

11.3 Economic Boom and Diversity in Dubai

Because of its cultural diversity, Dubai has been an oasis for foreign investments since 1999. In that year, Dubai Internet City, an Information Technology Park, was inaugurated to host top IT organizations such as Microsoft and Google. A year later, a Media Park was established to house leading organizations such as CNN and Reuters and 84 contemporary towers. Trade licenses within these two office parks include community fees that are used for upkeep. The fees also allow access to networking events, lease of conference spaces at reduce rates, and free use of break areas that support employees' productivity at work (see Figs. 11.1, 11.2, and 11.3).

Due to business parks such as these, both economy and population have been boosted tremendously. The population in 1999 was approximately 859,000 compared to approximately 2.5 million in 2016. The number of trade licenses issued in Dubai reached 4343 in 2012 compared to 19,000 trade licenses in 2015 (Staff 2012). Commercial licenses alone accounted for 74% of the total licenses. Since local laws require an office space to be attached to each license issued, a rough math indicates that there is room for 5000 office projects to be designed per year.

11.4 Workplace Design Standards

Generally speaking, offices constructed after 1999 in Dubai are shell and core. Although shell and core offices give clients the freedom to prepare the spatial layouts that support their objectives, they must absorb all the expenses that come with constructing their new office space. For example, the MEP (mechanical, electrical, and plumbing) work including the fire alarm and sprinkler systems, add 60% more to the budget. The absence of these services increase the amount of time required to prepare architectural drawings and to obtain permits within sixty days. The typical grace period given to clients by landlords is also 60 days.

Fig. 11.1 Prayer rooms are provided throughout the commercial facilities. *Source* Author

With both budget and time pressures, clients are urged to find designers who have the know-how and the capacity to cut down the design process to one week! This creates two problems: first, clients utilize online images to setup a theme for their office in order to expedite the design process. However, finding an existing layout that supports the required circulation can't simply be 'Googled!'.

Fig. 11.2 Photo showing employees taking their break within their office vicinity. *Source* Author

Second, changes at the site occur more than 70% of the time since clients did not allocate adequate time to the design process. This is good news to contractors as variations in relocating walls means additional income.

Fig. 11.3 Photo showing different break areas for employees in new office developments. *Source* Author

11.5 Architects Integrating Space Syntax in Their Practice

Space syntax is a group of theories founded by Hillier and Hanson that examine the effect of the environment on human behavior (Hillier and Hanson 1984). A set of techniques and programs have been developed since 1960s that can assist architects, planners, and designers to evaluate the implications certain properties of the environment have on psychosocial constructs.

In an opening address for the International Symposium on Space Syntax in 1997, Lord Norman Foster acknowledged the importance of using space syntax in the design process in his statement: "Although I myself am far removed from the academic world, it excites me to know, from the perspective of my own very demanding environment in architectural practice, that the techniques they have pioneered actually work." (Foster 1997, pp. xvii–xxii).

11.6 Importance of Interactions at the Workplace

There are two means to arrive at innovations: analytical power to create ideas (Heerwagen et al. 2004) and interacting with other employees to maximize the quality of the work produced (Schon 1991). Previous research indicated that face-to-face communications were unplanned and relied heavily on the location of the employee within the workplace (Tooren 2011). During an unplanned encounter, employees got instant feedback along with visible expressed emotions that served as further assurance. Employees were also provided with emotional support to assist them with their difficulties (Gutwin and Greenberg 2001). Further research indicated that face-to-face interactions resulted in the cognitive load on employees being much lower. Moreover, increased amount of shared knowledge among employees lead to higher productivity and performance (Reagans and Zuckerman 2001).

Employees naturally have the habit to scan other employees while walking within the office and grab the opportunity to converse with those who appear to be available (Backhouse and Drew 1992). Clearly, face-to-face interactions are primarily influenced by the spatial configuration of the workplace. Tanaka (2002) measured and recorded gaze and eye movement of employees by using a wearable apparatus. Specifically, Tanaka compared the amount of movement employees made in an effort to converse with colleagues while looking at their PCs and the location of their desks within the office layout. The results showed links between the layout of the office and concentration level.

Much of the earlier research centered around the notion of open plan office versus cellular private office in their discussion of the effect of office layout on communication and face-to-face interaction (Haynes 2008). Boutellier et al. (2008), for example, found that employees sitting along highly integrated corridors reported higher levels of communication or face-to face interaction. Stryker et al. (2012) revealed that both visibility and availability of informal and formal meeting spaces played a significant role in accomplishing complex tasks. They defined high visibility as those employees who sit closer to shared spaces such as corridors or break areas whether the employee is in a closed office or in an open cubicle. In other words, proximity among employees was necessary in establishing a rapid contact among them, however, too much proximity resulted in distractions.

In the recent past, few methodologies emerged from the space syntax literature that could objectively examine the impact of spatial configuration on face-to-face interactions using three measures: distance, visibility, and integration. Rashid et al. (2005) used axial line analysis techniques in examining visibility co-presence, i.e., the number of people located within the visual field of a route, and face-to-face interaction in analyzing four US federal buildings. In all four case studies, the result was that people preferred interactions that happened around individual workspaces.

11.7 Intelligibility in the Space Syntax Literature

A group of architects and cognitive psychologists examined various building configurations and circulation systems using axial line analysis and visibility graph analysis (VGA) (Natapov et al. 2015). Specifically, they looked at intelligibility where resulting values enable them to compare different buildings objectively. Simply put, intelligibility results from the correlation of the layout's local property, i.e., connectivity, to the layout's global property, i.e., integration. Environments with high levels of intelligibility were easier to navigate and easier to understand (Haq and Zimring 2003; Peponis et al. 1990).

The idea of VGA was first presented by Turner and Penn (1999) as an extension to space syntax literature specifically pertaining to the work of Benedikt in 1979. Benedikt fully provided measures that mathematically described properties of an isovist such as area, perimeter, and occlusivity (Benedikt 1979). An 'isovist' is the amount of area visible around oneself. For example, compare the amount of visible area around employee 1 and employee 2 in Fig. 11.4a, b. Using VGA, Turner and Penn (1999) were able to explore the relationship among all the isovists existing in an environment at an urban scale and at a building scale. They discussed the relationship between global isovist properties of the Tate Gallery in Millbank. There they traced people movement by taking into consideration obstructions on the floor such as statues and other objects.

The visibility graph is created by setting a grid of points by identifying visual properties as seen from each grid point within the space, and subsequently, the indivisibility connections each grid point has is calculated. Connectivity represents the number of direct visual connections a grid point has, and integration shows the number of steps required to access every grid point in space. To elaborate, if employees from their seats in the marketing department are able to see their colleagues by standing at the copier's location next to their seats, i.e., defined as intermediate location, then one can say that the employee's seat is located on a highly integrated node. Given such a location, the employees can communicate with their colleagues with ease. However, the number of visual connections employees have directly from their seats indicates a grid point that enjoys a much higher degree connectivity. To elaborate, Fig. 11.4c. shows a color coded grid generated in the entire office layout to indicate how connected each employee is to the rest of the layout. Colors range from light grey as extremely connected to dark grey as the least connected. In this particular plan, employee 2 is more connected than employee 1. Note also how the six managers sit in the least visually connected spaces in the plan, and they are a few steps away from being visually connected with the staff. The level of privacy, distraction and face-to-face encounter can now be simulated using these techniques.

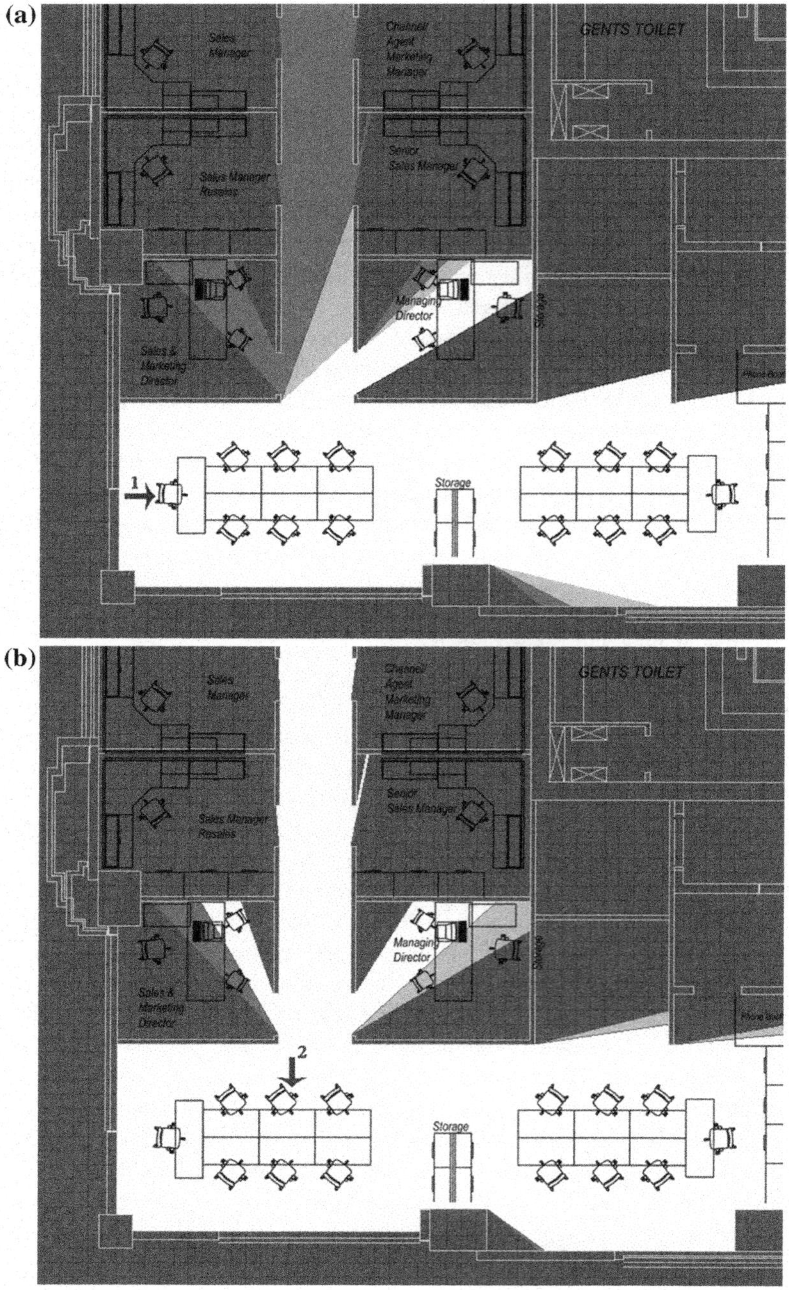

Fig. 11.4 **a** An isovist (polygon) drawn 360° around employee 1. **b** An isovist (polygon) 360° around employee 2. **c** Color coded grids in the office layout indicating connectivity levels of all employees. *Source* Author

(c)

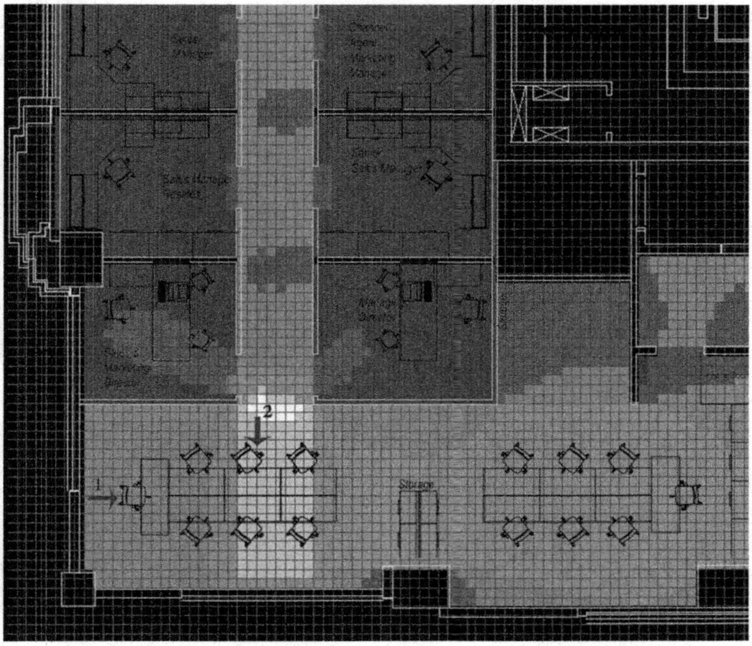

Fig. 11.4 (continued)

11.8 Space Syntax in Design Practice in Dubai

The idea of bridging research and practice in the design sector is rare in Dubai. In her practice, and to meet her client's needs in commercial interiors, the author began to offer free of charge simulations using space syntax techniques. The absence of knowledge and use of the widely used assessment methods such as Building Performance Evaluation or Post-Occupancy Evaluation (Preiser et al. 2015) in the city gave the author a competitive edge, especially since space syntax can help clients understand the impact of early design decisions on several building performance aspects such as safety, privacy, productivity, distractions, and overall satisfaction. Space syntax as a technique promises to use less time in achieving the goals of the client. This will eventually reduce changes at the site as discussed earlier in this chapter.

The six case studies selected for this chapter represent semi-government and private medium to large scale offices. The six floor plans displayed in Tables 11.1 and 11.2 are labeled Plan 1 to Plan 6. From the space syntax literature, the author focused on the measure of intelligibility due to its objectivity and its link with levels of communication. Intelligibility values shown in Tables 11.1 and 11.2 were prepared using the Depthmap program developed by Turner (2001). Plans 1 to 3

Table 11.1 Summary of office layouts with high intelligibility values. *Source* Author

<table>
<tr><td>FLOOR PLAN</td><td>Graph showing connectivity values where lighter shades of grey are highly connected spaces and dark shades are the least connected spaces</td></tr>
<tr><td>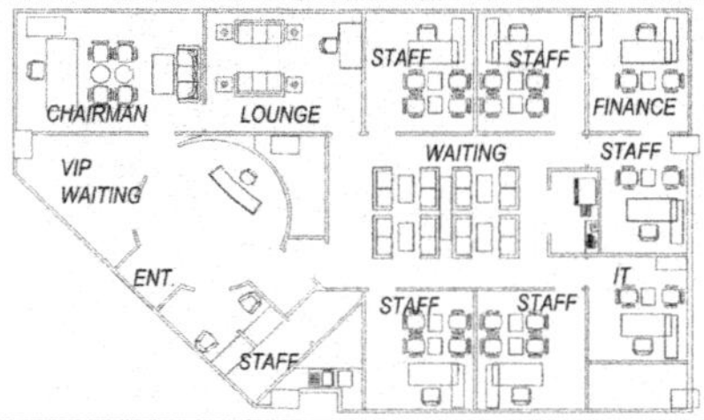
</td><td>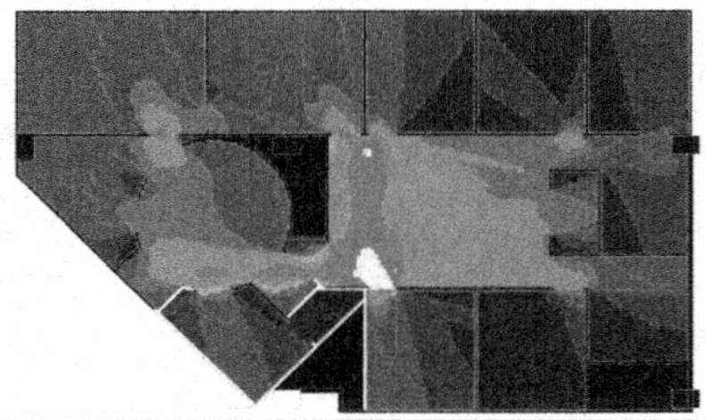</td></tr>
<tr><td>PLAN 1</td><td>INTELLIGIBILITY VALUE 0.88</td></tr>
<tr><td>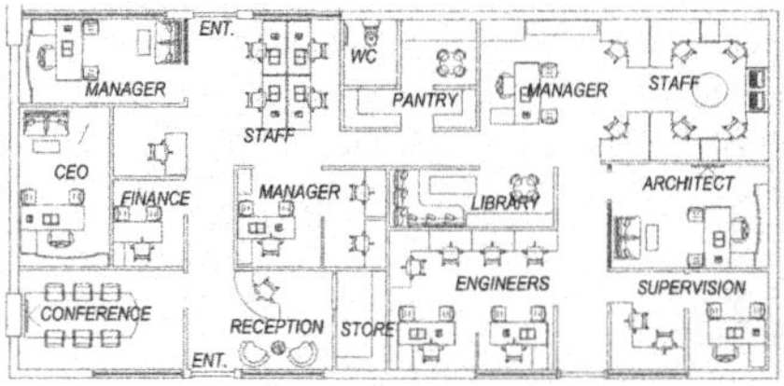
</td><td>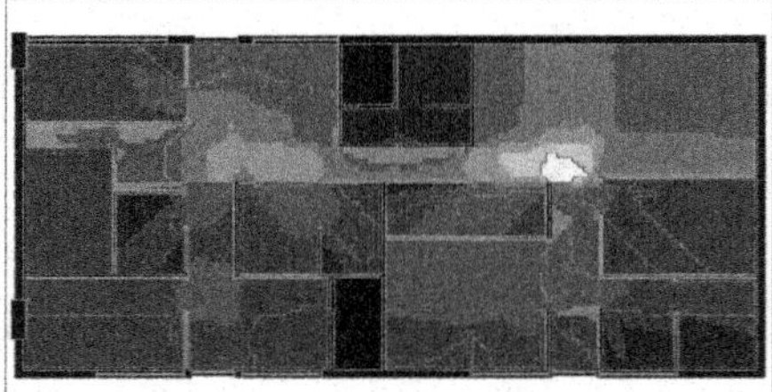</td></tr>
<tr><td>PLAN 2</td><td>INTELLIGIBILITY VALUE 0.67</td></tr>
<tr><td>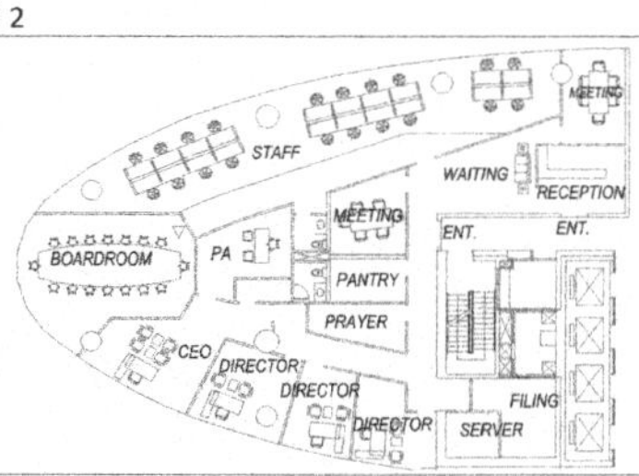
</td><td>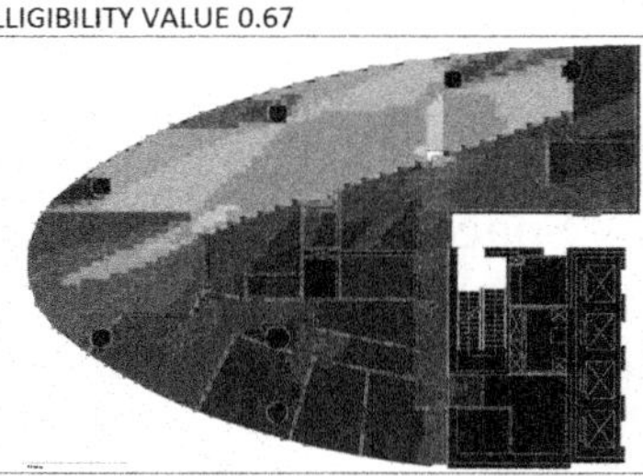</td></tr>
<tr><td>PLAN 3</td><td>INTELLIGIBILITY VALUE 0.62</td></tr>
</table>

represent plans characterized with high intelligibility. Plans 4 to 6 are office layouts with low intelligibility. The results from the analysis are as follows:

- Plan 1 is a semi-private office entity that offers express services to VIP clients. The client wanted a very high level of communication and constant eye contact with all visitors without any exception. One can see from the graph shown in Table 11.1 that most parts of the office including the chairman's office enjoy a large amount of visibility, with the exception of the finance officer. To increase intelligibility values within the plan, two entries were introduced from the Chairman's office. The purpose of this configuration was to give VIP Customers immediate attention.

Table 11.2 Summary of office layouts with low intelligibility values. *Source* Author

FLOOR PLAN	Graph showing connectivity values where lighter shades of grey are highly connected spaces and dark shades are the least connected spaces
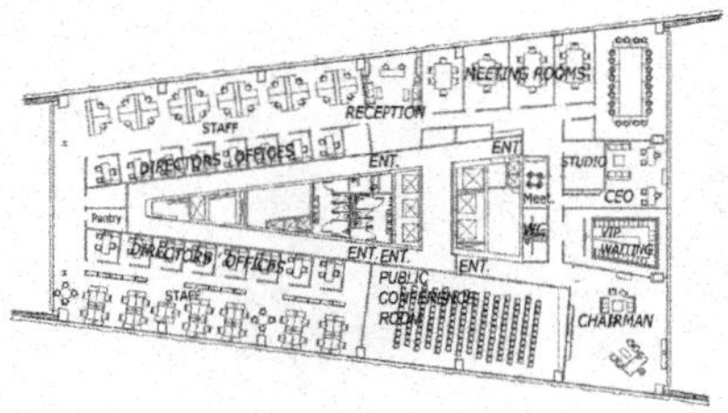	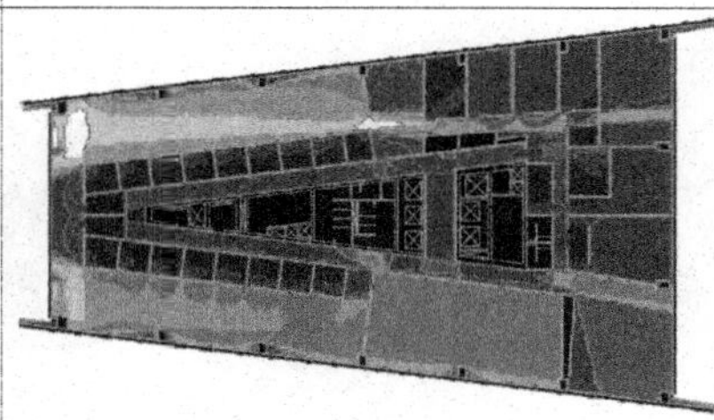
PLAN 4	INTELLIGIBILITY VALUE 0.34
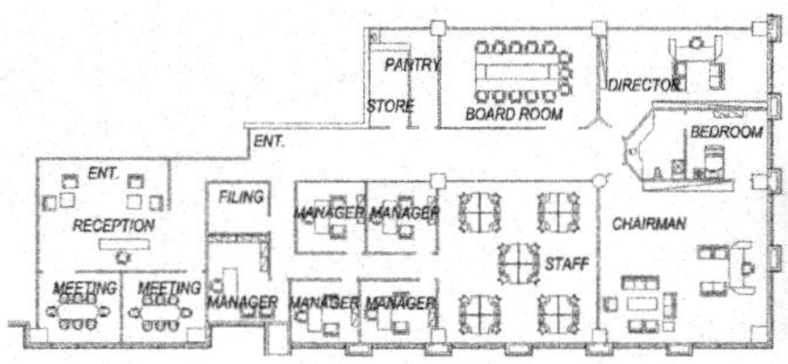	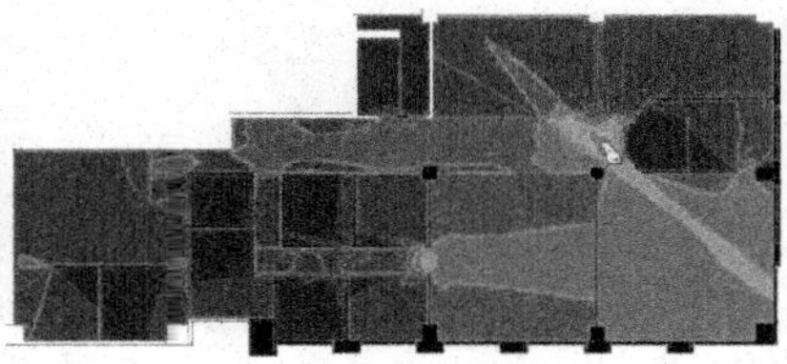
PLAN 5	INTELLIGIBILITY VALUE 0.28
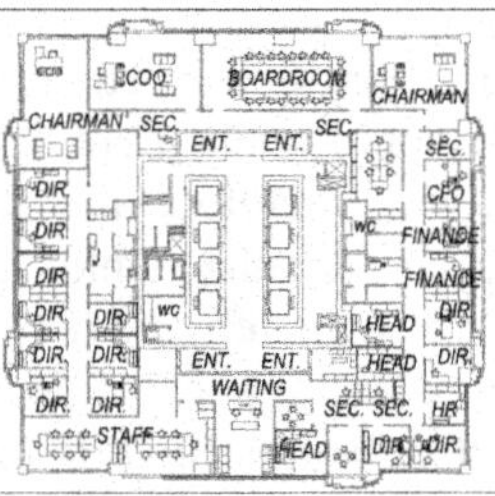	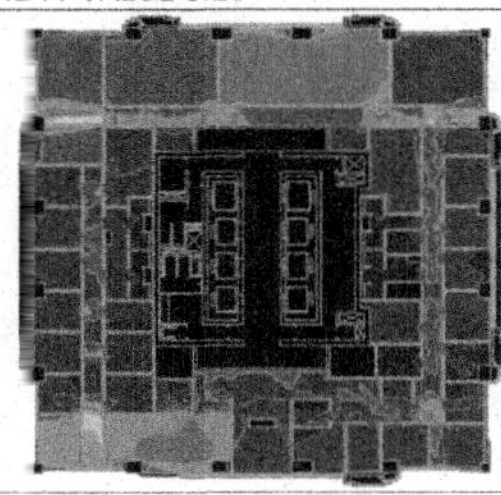
PLAN 6	INTELLIGIBILITY VALUE 0.17

- Plan 2 is a mid-scale contracting company where employees and senior managers are in constant interaction with each other in order to stay updated with project details. Note how the break area and the library are scattered to increase intelligibility within the office. While the CEO's office is close to the main entrance, he has the least visually connected office, however, he is close to the most integrated corridor in the layout. In other words, this layout supported the client's need to lower the level of distractions within the CEO's office, and give everyone else high level of encounter opportunities.
- Plan 3 is an office that finances semi-government developments. Note how symmetrical this office is in terms of visual connections. Directors enjoy minimal distractions, while the rest of the staff enjoy high levels of communication.

Fig. 11.5 Photo of the Office plan 5 showing a long hallway connecting the reception at one end with the chairman's office at the end of the hall. *Source* Author

- Plan 4 is a semi government entity that is somewhat segregated. Senior management is not easily accessible. Senior management spends more time outside the office as opposed to regular employees, and therefore the number of encounters must be limited. Moreover, the entrance of the public conference venue is directly accessible from the elevator lobby, thereby adding more privacy to employees.
- Plan 5 represents a real estate developer where the client needed privacy for his employees. Figure 11.5 shows the main hallway of this office. Note how the hallway is defined by one way tinted glass partitions. The staff sit behind one side of the partition and the boardroom is located behind the other side of the partition. This office was designed with the intent to keep employees focused on their work, and to minimize interactions in the office.
- Plan 6 is a real estate developer where segregation is encouraged and therefore, the plan is intentionally designed with low intelligibility values in mind. Here, the CEO and Chairman's offices are completely isolated from the employees' areas by a series of doors. If these doors were relocated, the value of intelligibility in this case will increase.

11.9 Conclusion

The objective of the author was to encourage the implementation of space syntax techniques in professional practice. The literature review indicated two things: First, face-to-face interactions and casual encounter among employees in the workplace increase communication levels, and they reduce cognitive load among employees. Second, intelligible office layouts as measured by using space syntax techniques increases interactions. These results were documented widely within the space syntax literature, however, they were rarely applied in practice, and especially in workplace design. The author demonstrates to the reader that implementing these techniques in the author's practice in Dubai expedited the design process in the first place, and supported the client's needs. The author further suggests surveying users six months after the occupancy in order to test whether the predictions made earlier during the design process were met. If these predictions were met, then users' ratings as related to their spatial location might be objectively measured using these techniques. This in turn will be a powerful method to be implemented within the building performance evaluation (BPE) process.

References

Backhouse, A., & Drew, P. (1992). The design implications of social interaction in a workplace setting. *Environment and Planning B: Planning and Design, 19*(5), 573–584.

Bashir. (2014, December 17). *Mohammed bin Rashid launches Dubai Strategy 2021*. Retrieved from http://www.wam.ae/en/news/emirates-international/1395273975190.html. Accessed 1 July 2016.

Benedikt, M. L. (1979). To take hold of space: isovists and isovist fields. *Environment and Planning B: Planning and Design, 6*(1), 47–65.

Boutellier, R., Ullman, F., Schreiber, J., & Naef, R. (2008). Impact of office layout on communication in a science-driven business. *R&D Magazine, 38*(4), 372–391.

Duffy, F. (2015). Buildings and their use: The dog that didn't bark. In W. F. E. Preiser, A. T. Davis, A. M. Salama, & A. Hardy (Eds.), *Architecture beyond criticism: Expert judgment and performance evaluation* (pp. 128–132). New York: Routledge.

Foster, N. (1997).Opening address. In M. D. Major, L. Amorim, & F. Dufaux (Eds.), *Proceedings of the first international space syntax symposium* (Vol. 3, pp. xvii–xxii). London: Space Syntax Laboratory, The Bartlett School of Architecture, University College London.

Gutwin, C., & Greenberg, S. (2001). A descriptive framework of workspace awareness for real time groupware. *Computer Supported Cooperative Work (CSCW), 11*(3), 411–446.

Haq, S., & Zimring, C. (2003). Just down the road a piece: The development of topological knowledge of building layouts. *Environment and Behavior, 35*(1), 132–160.

Haynes, B. (2008). The impact of office layout on productivity. *Journal of Facilities Management, 6*(3), 189–201.

Heerwagen, J. H., Kampschroer, K., Powell, K. M., & Loftness, V. (2004). Collaborative knowledge work environments. *Building Research & Information, 32*(6), 510–528.

Hillier, B., & Hanson, J. (1984). *The social logic of space*. Cambridge: Cambridge University Press.

Natapov, A., Kuliga, S. Dalton, R., & Hölscher, C. (2015). Building circulation typology and space syntax predictive measures. In K. Karimi, L. Vaughan, K. Sailer, G. Palaiologou, & T. Bolton (Eds.), *Space syntax: Proceedings of the tenth international space syntax symposium* (Vol. 30, pp. 1–16). London: Space Syntax Laboratory, The Bartlett School of Architecture, University College London.

Peponis, J., Zimring, C., & Choi, Y. K. (1990). Finding the building in wayfinding. *Environment and Behavior, 22,* 555–590.

Preiser, W. F. E., Davis, A. T., Salama, A. M., & Hardy, A. E. (Eds.). (2015). *Architecture beyond criticism: Expert judgment and performance evaluation*. New York: Routledge.

Rashid, M., Zimring, C., Wineman, J., Flaningam, T., Nubani, L., & Hammash, R. (2005). The effects of spatial behaviors and layout attributes on individuals' perception of psychosocial constructs in offices. In A. Van Nes (Ed.), *Proceedings of the fifth international space syntax symposium* (pp. 71–88). Amsterdam: Delft University of Technology.

Reagans, R., & Zuckerman, E. W. (2001). Networks, diversity, and productivity: The social capital of corporate R&D teams. *Organization Science, 12*(4), 502–517.

Schon, D. A. (1991). *The reflective practitioner*. Alderschot: Ashgate Publishing.

Staff. (2012, May 7). *New Dubai trade licenses up 27%*. Retrieved from http://www.emirates247.com/business/economy-finance/new-dubai-trade-licences-up-27-2012-05-07-1.457555. Accessed 1 July 2016.

Stryker, J., Santoro, M., & Farris, G. (2012). Creating collaboration opportunity: Designing the physical workplace to promote high-tech team communication. *IEEE Transactions on Engineering Management, 59*(4), 609–620.

Tanaka, R. (2002). Future workplace design. *Displays, 23,* 41–48.

Tooren, W. (2011). How to create office layouts that stimulate collaboration? *United Academics Journal of Social Sciences,* 2011(7), 7–23.

Turner, A. (2001). Depthmap: A program to perform visibility graph analysis. In J. Peponis, J. Wineman, & S. Bafna (Eds.), *Proceedings of the third international space syntax symposium* (Vol. 31; pp. 1–9). Atlanta: A. Alfred Taubman College of Architecture and Urban Planning, University of Michigan.

Turner, A., & Penn, A. (1999). Making isovists syntactic: Isovist integration analysis. In M. D. Major & L. Amorim (Eds.), *Proceedings of the second international space syntax symposium* (Vol. 1, pp. 1–9). Brasilia: Universidad de Brasilia.

Author Biography

Linda N. Nubani holds a Ph.D. in Architecture (Environment and Behavior/Criminology) from the University of Michigan at Ann Arbor (2006), Masters degrees in architecture are from the University of Michigan (2003) and the University of Cincinnati (2001). Dr. Nubani moved to Michigan in 2016 to accept a teaching and research position with Michigan State University. Prior to that, she lived in Dubai, UAE from 2004 to 2016 where she taught at the American University in Dubai. In 2004, she co-founded APID, the Association of Professional Interior Designers in Dubai. APID is the first professional association of interior design in the Middle East. In 2006, she co-founded the Green Buildings Conference that has become very active in many countries around the Gulf region. In 2008, she co-founded the Italian Architecture Magazine Compasses in Dubai. Also in 2008, Dr. Nubani founded an interior design firm that she called "Mazarii Interiors". The word Mazarii implies "Green Practice" in the native language of UAE. In the same year, she became a certified professional interior designer, and in 2014, she became a certified Civil Defense and Life Safety consultant. Dr. Nubani has designed more than 70 projects in the Middle East, France, and Spain. The highlights of her work include: designing the largest smoking lounge in the World at Dubai Intl. Airports; completing more than 40 office projects for the private and the government sector; renovating homes for celebrities and royal families; and, winning the annual light competition for the National Day of the UAE. Her research interests include environment and behavior, space syntax, post occupancy evaluation, and crime prevention through environmental design.

Chapter 12
Social Interactions in Work Environments: Expanding Building Evaluation

Mille Sylvest

12.1 Introduction

This chapter aims to provide a viable way of understanding the social aspects of life in the physical environment, as well as to relate this understanding directly to the evaluation of the environment.

This is done through distinguishing between aspects related to social space—such as privacy, individual space requirements in relation to surrounding others, distances to social areas, and territoriality generally used within environmental psychology and building evaluations on one hand—and the ecological and situated aspects of formal and informal interactions, collaboration, and communication on the other. The focus of the method presented at the end of the chapter is to evaluate in situ social activity patterns in a way that is useful for architectural practice.

Humans are an integral part of the socio-physical environment, engaging simultaneously with different physical settings and other humans that co-inhabit these settings. This co-inhabitation means that the natural human situation is inherently a social one. This is especially true for today's employees of various companies and organizations, who are surrounded by a number of co-workers and managers in their everyday work life, all of whom they have to cooperate, collaborate, and interact with.

M. Sylvest (✉)
3XN Architects, Kanonbådsvej 8, 1437 Copenhagen, Denmark
e-mail: mil@3xn.dk

W.F.E. Preiser et al. (eds.), *Building Performance Evaluation*,
DOI 10.1007/978-3-319-56862-1_12

12.2 From Social Space to Social Interactions

Research into social space forms a large part of the early research undertakings within environmental psychology. Then and now, social aspects are often mentioned within environmental psychology and its related fields. These social aspects, however, are mostly space and distance related, and thus, from within a framework of situated social activity, environmental psychology often appears to be lacking a viable understanding of the situated social aspect of life in the physical environment.

The different social aspects used in environmental psychology and the building evaluation literature can be divided into three categories:

1. Social Space Aspects, referring to the individual centered nature of the experience of being more than one person in an environment. They include topics such as privacy, space requirements etc.
2. Social Interaction Aspects, referring to interactional activity between two or more individuals in the environment.
3. Social Aspects of Design and Evaluation, referring to evaluation methods and considerations, and the societal aspects that affect design and evaluation thinking (see Fig. 12.1).

Social Space Aspects traditionally receive the most focus within environmental psychology and the evaluation literature. These aspects include issues relating to a person's distance to others: personal space, crowding, territoriality, privacy, and disturbances among others (see Bechtel and Churchman 2002; Gifford 2014; Bechtel 1997), rather than social interaction and activity pattern opportunities provided by the environment. Although important aspects of a humane environment the Social Space Aspects often constitute the entire social aspect of environmental psychology publications. As a result, research within the field generally has an underlying basis in the physical aspects of the socio-physical environment and often deploys an individual perspective, rather than a social one. This becomes a one-dimensional perspective that often disregards the social interactions that are at the foundation of human co-habitation within the surrounding physical environment.

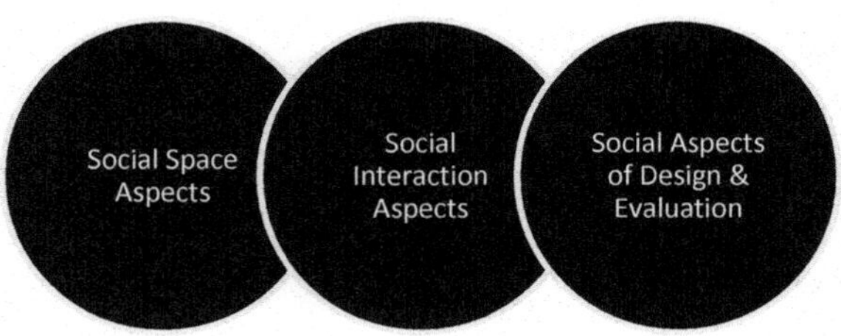

Fig. 12.1 The three social aspect categories used in environmental psychology and building evaluation literature. *Source* Author

12.3 Building Evaluation

Excellent work has been done in recent decades in order to develop viable evaluation methods and ensure there is an opportunity for relevant knowledge to be fed back into future design projects. The development of first Post-Occupancy Evaluation (POE) (Preiser et al. 1988) and later Building Performance Evaluation (BPE) (Preiser and Schramm 1997; Preiser and Vischer 2005; Mallory-Hill et al. 2012) historically has been undertaken in order to aid the creation of more humane and well-functioning buildings for the end-user, more sustainable user-building interactions, and buildings that fit the organization in question and thus help create optimal work-flows, productivity, and employee retention. However, with a primary focus on technical and maintenance issues, as well as space requirements and general user satisfaction, there are important areas that still need to be evaluated in order to gain a truly holistic picture of human life in the built environment. These areas include social activity patterns among the users and the opportunities the built environment offers in terms of social interactions, collaborations, informal meetings, and a positive social climate.

With the rise of knowledge and information based work settings, it is imperative to ensure well-functioning buildings that support social interactions and collaborative activities in order for knowledge workers to share this information. Often completely absent from building evaluations, a social interaction perspective has yet to be incorporated into the BPE process.

With a focus on social interactions in the built environment, this chapter takes a non-traditional approach to building evaluation by expanding the areas considered necessary and important to evaluate. Following this line of thought, the argument is to enable the design of increasingly more humane buildings, by emphasizing the need to include within a building's evaluation how it performs in relation to social climate, employee interactions, and collaboration.

12.3.1 Social Perspectives in Building Evaluation

In the first edition of *Assessing Building Performance* (2005), Preiser and Vischer state that BPE "links diverse phenomena that influence relationships between people, processes and their surroundings, including the physical, social and cultural environments" (p. 4). However, evaluations of building performance—in relation to social activities, such as collaboration, informal interactions, and communication—are generally only included in evaluations of innovative office spaces, as exemplified by Hodulak (2012) and Kato et al. (2005). The focus on social interaction appears rather limited within other types of building evaluations, where little attention is given to the social interaction aspects of the everyday life of the users within the evaluated settings.

However, based on the human condition as social co-habitants, the social interaction aspect is an important evaluation feature in any building or designed environment. Social interactions are not simply higher order aspects in a 'Maslowian' sense. Rather, social interactions, collaborations, and a positive social climate are crucial aspects of productivity and efficiency on one hand, and of employee satisfaction, retention, stress related issues, and social inclusiveness on the other. As such, social interaction aspects are related to both organizations and employees, and should be included in any evaluation of building performance. This is also a point partly made by Windsor (2005), who argues for the necessity of including a social-organizational perspective into all stages of BPE.

12.3.2 An Increasing Social Focus

In a chronological literature review conducted of the six major contributions to POE and BPE published during the last three decades (see Preiser et al. 1988; Baird et al. 1996; Federal Facilities Council 2001; Preiser and Vischer 2005; Mallory-Hill et al. 2012; Preiser et al. 2015) the author found a total of 411 word entries mentioning any kinds of social aspects. Out of these, 121 were categorized as social space aspects, 214 as social interaction aspects, and 76 as social aspects of design and evaluation (see Fig. 12.2). Where newer editions of the publications were available, first editions were intentionally chosen for this literature review in order to be able to track the development related to the focus on social interactions over time.

What is interesting to note is that despite a limited focus on social activity within building evaluations, recent publications exhibit at least some indications that there is a heightened interest in the social interaction perspective. When categorizing various contributions to these major publications within the evaluation field, it becomes evident that however limited the focus on social interaction aspects might be, it is increasing. Taken from the early POE publications to the more recent BPE

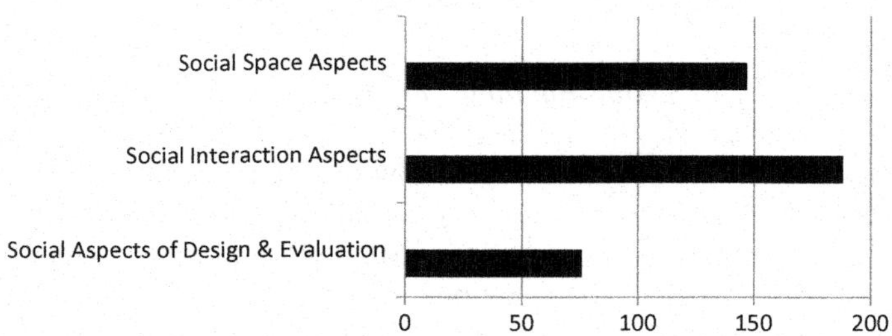

Fig. 12.2 The 411 word entries of six major POE/BPE publications categorized under the three social aspect categories. *Source* Author

publications, there is a significant rise in authors mentioning aspects relating to social activity, as well as how to design for these aspects of life in the built environment (see Fig. 12.3).

One exception, as Fig. 12.3 shows, is Baird et al. (1996), who provide an early focus on social interactional aspects within building evaluation. However, this focus seems to disappear again, only to reappear in an increasing fashion in the latest publications within the field. These publications include several chapters that are at least partially focused on social interactions. Furthermore, it becomes evident that interest in interactional aspects is increasing when looking at examples, such as Mallory-Hill and Westlund (2012) who refer to investigations by Heerwagen and Wise (1998), Heerwagen and Zagreus (2005), and Pyke et al. (2010). They note that: "Other variables that are suspected to play a role in perceived satisfaction include aspects of design that enhance social experience, aesthetics and beauty" (Mallory-Hill and Westlund 2012, p. 175). This verifies the observed development towards a more socially oriented focus within the field of building evaluation.

It is sometimes unclear what is meant by social experience or social activity in evaluation-method descriptions. In many contributions to BPE, the social interaction aspect is limited to the mere existence of collaborative space, or distances to formal or informal meeting areas, rather than to interactive activity patterns and observations. Furthermore, with the exception of the previously mentioned investigations by Hodulak (2012) and Kato et al. (2005), the existing social aspects in building evaluations are rarely evaluated in situ, but instead through a very limited part of otherwise interesting and relevant POE surveys.

What is needed is not only a broad focus on social aspects and social interactional activity in building evaluations, but also that this focus is placed on observed and situated, context specific social activity. To that end, the following sections present an evaluation method rooted in focused ethnography. Through this method,

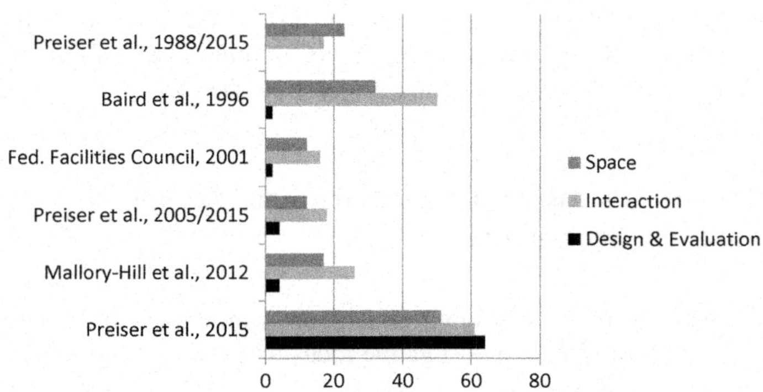

Fig. 12.3 The different social aspects and their frequency within the evaluation literature. *Source* Author

the aim is to expand the areas traditionally evaluated within POE and BPE, and thus, make possible a more holistic picture of how well buildings live up to modern-day demands regarding collaboration and team-based work.

12.4 Evaluating Social Activity: A Multi-method Approach

When evaluating social activity in the built environment, it is advantageous to apply a multi-method approach that combines existing evaluation methods with an ecological perspective on in situ social interactions within the setting, and thus creates knowledge regarding actual, everyday social interactions among users. In addition, a rooting within focused ethnography ensures that the overall evaluation method is commensurable with the time constraints of architectural practice, and thus viable and useful in practice.

The approach to evaluating social activity patterns and interactions within the situated, ecological framework presented here does not include carrying out a traditional POE or BPE, but rather a qualitative exploration and investigation of how social activities can be evaluated and included into practice-oriented building evaluation methods.

12.5 *Focused Ethnography*

Focused ethnography is a feasible method for commercial architectural practice, where project time and resources are insufficient for prolonged in-depth field work. The focused method narrows the scope of the field study by reducing observation time as well as by entering the field with a specific focus informed by background knowledge about the setting in question (Wall 2015). The method is therefore an attempt to comply with both the epistemological requirements of ethnography on one hand, and the constraints of applied projects on the other.

12.6 Method Testing in Three Northern European Work Environments

The high standard of most built environments in Scandinavia and a large part of Northern Europe allows for an investigative focus on the social aspects of everyday life in work settings. Even though it would still be both relevant and useful, an in-depth focus on social interactions is not necessarily possible in other parts of the world, where users and evaluators are still forced to focus on basic design related aspects, such as lack of daylight for office workers, standards related to air quality, etc.

Three buildings were tested with the presented method: Tangen Polytechnic School (Norway), Stadshuis Nieuwegein (Netherlands), and Ørestad College (Denmark), all designed by the Danish architecture firm, 3XN architects (see Figs. 12.4, 12.5 and 12.6). Although varying in use and user groups, the three

Fig. 12.4 Tangen Polytechnic, Kristiansand, Norway. *Source* © 3XN/Adam Mørk

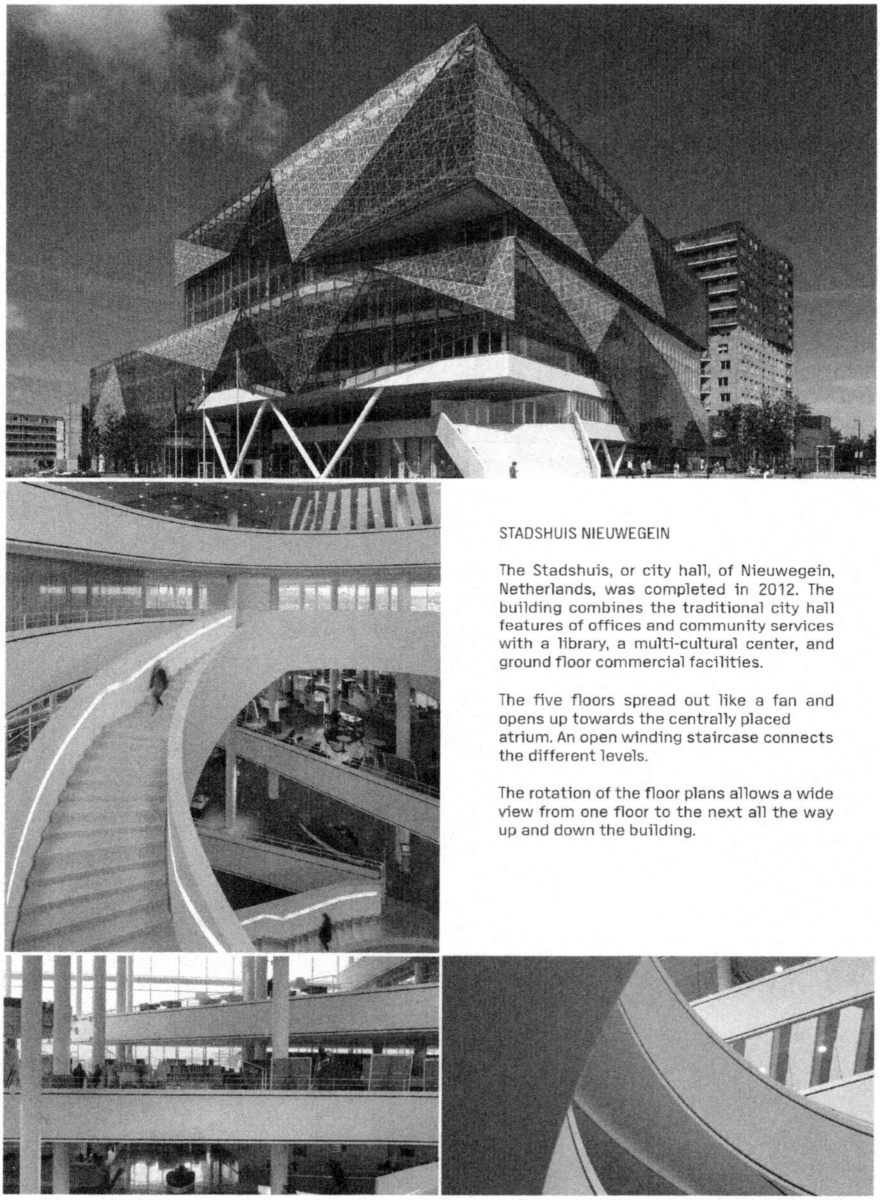

Fig. 12.5 Nieuwegein City Hall, Nieuwegein, Netherlands. *Source* © 3XN/Adam Mørk

Fig. 12.6 Ørestad College, Copenhagen, Denmark. *Source* © 3XN/Adam Mørk

buildings exhibit fundamental similarities related to the firm's design traditions: open, transparent spaces; communication opportunities through visual contact between the floors; and chance meetings in the open atrium, and on the central staircases.

Table 12.1 The different methods included in the evaluation of social interactions in the built environment. *Source* Author

Building observation	*Informant observation*
Observations of building and general user activity patterns	Observations of the activity patterns of each informant during a workday
• To provide information before informant observations	• To determine where informants engage in social interactions
• To inform interview guide	• To understand reasons for choosing certain places for interaction
Semi-structured interview	*Activity mapping*
Interviews with each informant	Individual and place centered Activity Mapping of informants.
• To gain in-depth understanding of reasons behind observed social activity patterns and use of building	• To determine the different types of social interactions in the spaces
• To gain information about informants experiences in the building	• To understand relationship between material and social contexts of the space
Photographic recording	*Video recording*
Photographic recording of spaces and interactions	Video recording of specific and complicated settings
• To visualize the interaction types in different areas	• To determine the social use of crowded environments
• To visualize the design of different areas	• To analyze types of interactions in the setting

The multi-method approach tested consists of building and informant observations, activity mapping, semi-structured interviews, as well as photographic and video-based recordings of specific areas (see Table 12.1). The different methods are included on the basis of how well they serve evaluations practices on one hand, and architectural practice on the other.

12.7 Investigation Methods

As a focused ethnographic method, only two consecutive weeks including a single evaluator is spent on site in each building. A total of 21 informants, distributed evenly between the three buildings are included in the informant observations, activity mappings, and interviews described below.

12.7.1 Building Observations

The first part of this overall method consists of building observations of both the physical aspects of the building and its layout and functions, as well as of general activity patterns among the users. These observations serve as a familiarization with the building in question and the different movement patterns within it.

12.7.2 Informant Observations

Informant observations are conducted with one informant per day, by continuously following the informant around throughout his or her entire workday. The observational role deployed here is one of participant observer (Ackroyd and Hughes 1992).

The informant observations allow the gathering of knowledge about the actual everyday social activity patterns of the informants during their workday. This knowledge on activity patterns, types of interactions, and places these occur, is supplemented with the conversations that arise throughout the day with each informant. This produces valuable insights into the everyday life in the building in question.

This role as participant observer enables contact with users who are not informants, as the visible nature of the role combined with days spent in different departments and areas of the building allows for users to be approached regarding their perceptions, stories or anecdotes about the building. In addition, this allows for relatively easy recruitment of any additional informants.

During the days of building and informant observations, field notes are produced in the form of comprehensive descriptions on site throughout each day. These field notes take the form of descriptions of the informant's activity patterns and movements within the building, summaries of conversations with the informant, his or her thoughts about or experiences in the building, as well as any stories he or she would tell about the building or the everyday life within it.

The activity of writing down field notes, while the observed informants are occupied doing individual work or engaged in meetings, is useful in several ways. Firstly, it allows for the registration of very recent or simultaneous events related to both informant observations and the activities happening in the surrounding environment. Secondly, the activity of recording field notes means that the informant feels less observed during his or her work, as the researcher appears occupied with the notes.

12.7.3 Activity Mapping

The method of activity mapping, or what is generally termed 'Behavioral Mapping' within environmental psychology, applied here, includes a form of individual-centered mapping, focusing specifically on social interactions. Printed floor plans allow the recording of where and with whom informants engage in social interactions during the observation day, as well as the social or work-related nature of these interactions. Since individual-centered activity mapping necessitates following particular informants over time and throughout locations, the method proved well suited as an integral part of the informant observations.

12.7.4 Semi-structured Interviews

The insights produced during observations and activity mapping of social interactions is used to inform an interview guide for interviews conducted with each informant during the day of individual observation. These interviews serve as a method to describe reasons for and thoughts about the observed activities within the building, and take the form of semi-structured interviews. After the preliminary building observations, the overall structure of the interview guide is adjusted in order to fit the building in question. In addition, the guide is adjusted each day, in order to allow the individual informants to elaborate on observations or topics covered throughout the day of informant observation.

Interviews are recorded using a small audio-recording device, and transcribed in order to become available for coding and analysis.

12.7.5 Photographic Recordings

Photographic recordings of areas, specific rooms, and observed activities and situations are conducted during observations. These photographs are used in two ways. On the one hand, they serve as visualizations in order to be able to return to the settings when in doubt about the layout or design of a certain setting or area. On the other, they serve as visualizations of situations, activities and use of specific places or areas, and as such, they can be included in the analysis in combination with field notes, activity maps, and interview transcriptions.

12.7.6 Video Recordings

Areas within the building that are busy or crowded, and thus complicated to observe in real time, are video-recorded in order to become available for analysis. The video recordings are carried out as in situ observations of everyday social practices in the setting (Pedersen et al. 2012). A partial transcription of demarcated social interactions is produced and coded in combination with field notes and interview transcriptions.

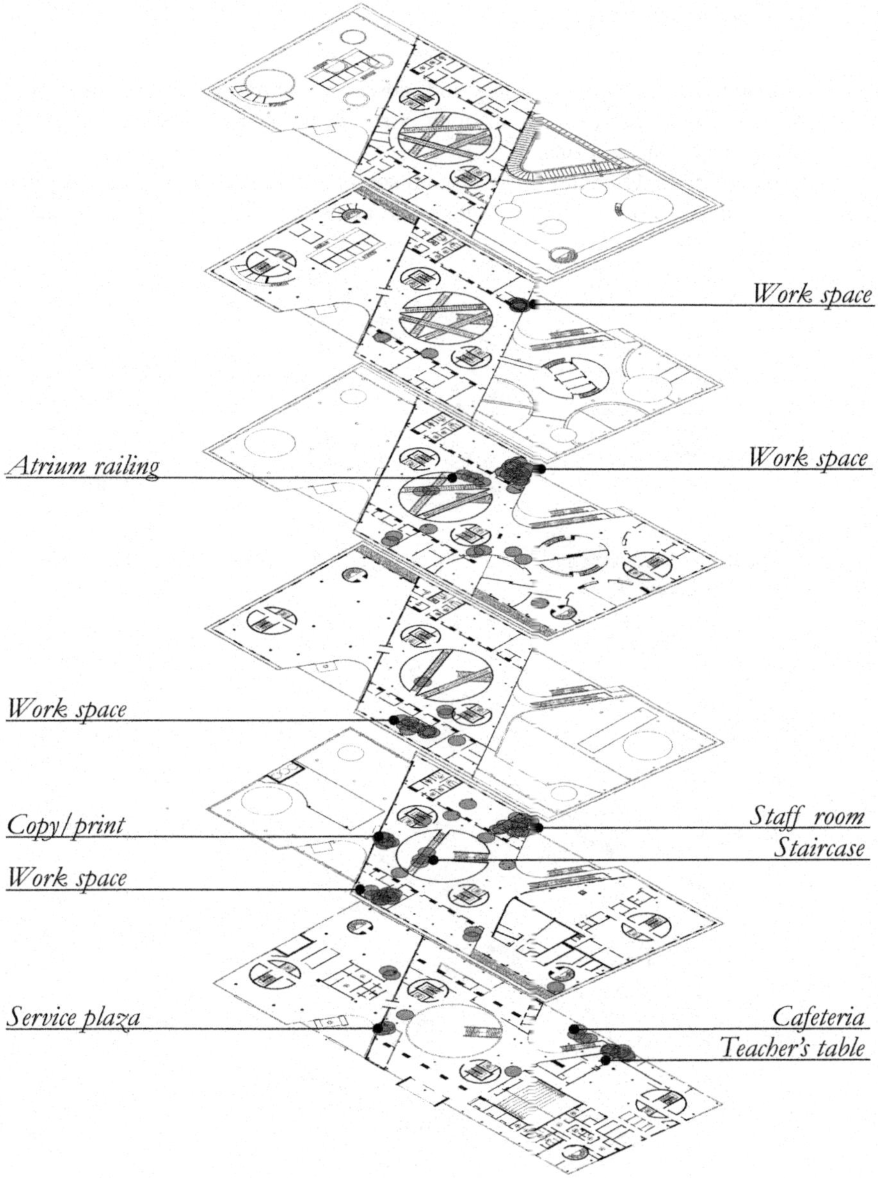

Fig. 12.7 Observed social activities and their placement within Tangen Polytechnic. *Source* 3XN/Author

12.8 Case-Study Findings

By testing the social evaluation method described above on three case-study buildings, it becomes clear that both open, visually transparent spaces and closed rooms afford social interactions.

Interactions generally center on or around social junctions that provide adequate interactional space within the setting; work areas and rooms, office kitchens,

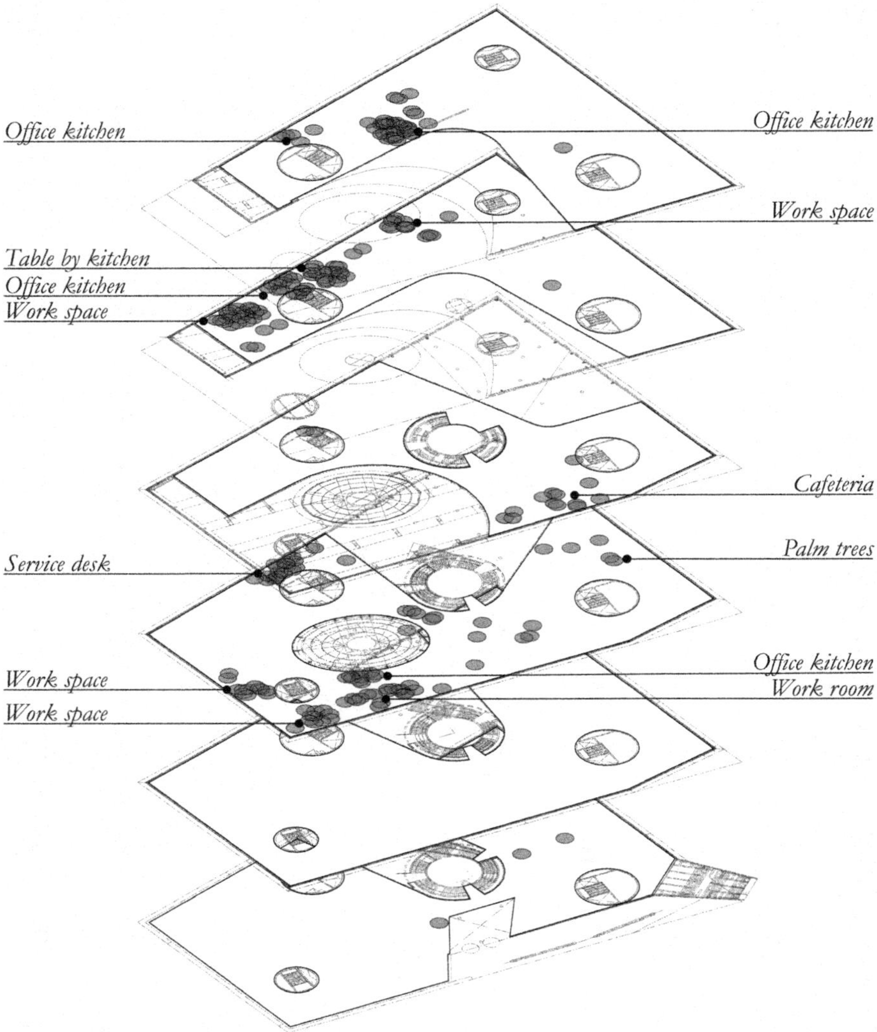

Fig. 12.8 Observed social activities and their placement within Nieuwegein City Hall. *Source* 3XN/Author

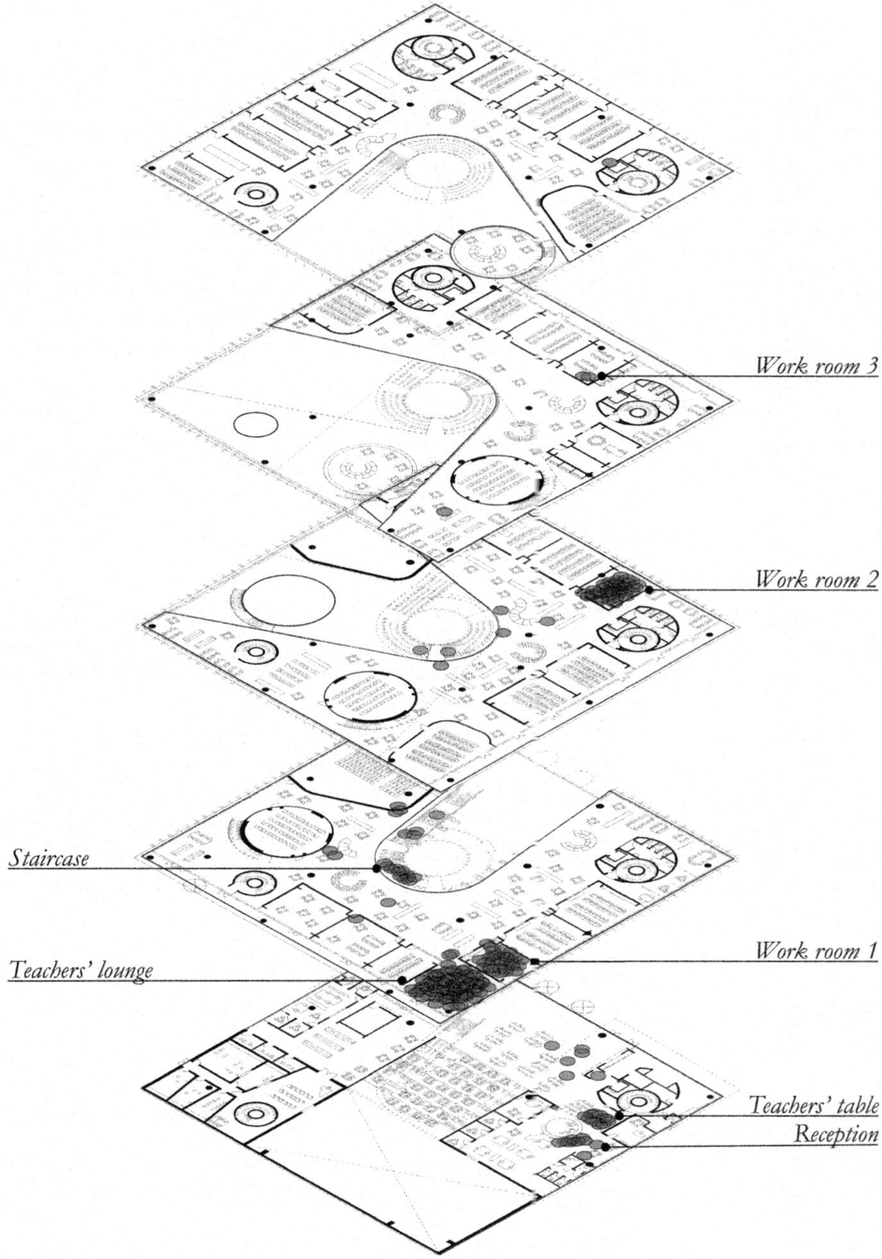

Fig. 12.9 Observed social activities and their placement within Ørestad High school. *Source* 3XN/Author

communal atrium spaces, and stair landings (see examples of observed social activities in Figs. 12.7, 12.8, 12.9). These interactions are encouraged by the spatial diversity of visually open areas with opportunities for more privacy.

The open design and central placement of the atria enhance the visual contact within all three investigated buildings. Hence, the atria increase opportunities for random meetings and social interactions among users by providing opportunity for cross-spatial interactions and social views. However, the visual contact to and from the staircases and all floors surrounding the investigated atria also occasionally lead to feelings of insecurity or unpleasantness, due to the level of personal visibility within the space.

Furthermore, the findings show that social interactions are dependent on both overall organizational culture and more locally developed cultures within specific areas or departments of the organization. These cultures restrict the possible actions within any given space while also providing the users with a choice of suitable setting to work or interact in, depending on their mood or given task.

12.9 Conclusions

Social interactional aspects need to be evaluated in order to gain a truly holistic picture of human life in the built environment. To this end, a method is presented that has proven viable and useful in evaluating in situ social activity patterns within three work environments in Northern Europe.

The combination of building and informant observations, activity mapping, and visual recordings on one hand, and the in depth interviews regarding reasons and thoughts related to the observed actual activity patterns on the other, results in a reliable and useful collection of data on social interaction patterns in the built environment. Furthermore, while the methods used prove useful related to evaluations practices, the relatively short time span of focused ethnography proves to be efficient and commensurable with the time constraints of architectural practice.

References

Ackroyd, S., & Hughes, J. A. (1992). *Data collection in context*. London: Longman Group.

Baird, G., Gray, J., Isaacs, N., Kernohan, D., & McIndoe, G. (1996). *Building evaluation techniques*. Wellington: McGraw-Hill.

Bechtel, R. B. (1997). *Environment & behavior: an introduction*. Thousand Oaks: Sage.

Bechtel, R. B., & Churchman, A. (Eds.). (2002). *Handbook of environmental psychology*. New York: Wiley.

Federal Facilities Council. (2001). *Learning from our buildings: A state-of-the-practice summary of post-occupancy evaluation*. Washington, DC: National Academy Press.

Gifford, R. (2014). Environmental psychology matters. *Annual Review of Psychology, 65*, 541–579.

Heerwagen, J. H., & Wise, J. A. (1998). Green building benefits: Differences in perceptions and experiences across manufacturing shifts. *Heating, Piping, and Air Conditioning, 70*(2), 57–63.

Heerwagen, J. H., & Zagreus, L. (2005). *The human factors of sustainable building design: Post occupancy evaluation of the Philip Merrill Environmental Center*. Berkeley: University of California Press.

Hodulak, M. (2012). Programming spaces for innovation. In S. Mallory-Hill, W. F. E. Preiser, & C. Watson (Eds.), *Enhancing building performance* (pp. 98–102). Oxford: Wiley-Blackwell.

Kato, A., Le Roux, P. C., & Tsunekawa, K. (2005). Building performance evaluation in Japan. In W. F. E. Preiser & J. C. Vischer (Eds.), *Assessing building performance*. Oxford: Elsevier (republished 2015, London: Routledge).

Mallory-Hill, S., Preiser, W. F. E., & Watson, C. (Eds.). (2012). *Enhancing building performance*. Oxford: Wiley-Blackwell.

Mallory-Hill, S., & Westlund, A. (2012). Evaluating the impact of green building on worker productivity and health: A literature review. In S. Mallory-Hill, W. F. E. Preiser, & C. Watson (Eds.), *Enhancing building performance* (pp. 167–178). Oxford: Wiley-Blackwell.

Pedersen, M., Klitmøller, J., & Nielsen, K. (Eds.). (2012). *Deltagerobservation – En metode til undersøgelse af psykologiske fænomener*. Copenhagen: Hans Reizels Forlag.

Preiser, W. F. E., Davis, A. T., Salama, A. M., & Hardy, A. (Eds.). (2015). *Architecture beyond criticism—Expert judgement and performance evaluation*. New York: Routledge.

Preiser, W. F. E., Rabinowitz, H. Z., & White, E. T. (1988). *Post-occupancy evaluation*. New York: Von Nostrand Reinhold (republished 2015, London: Routledge).

Preiser, W. F. E., & Schramm, U. (1997). Building performance evaluation. In D. Watson et al. (Eds.), *Architectural design data* (7th ed., pp. 233–238). New York: McGraw-Hill.

Preiser, W. F. E., & Vischer, J. C. (Eds.). (2005). *Assessing building performance*. Oxford: Elsevier. (republished 2015, London: Routledge).

Pyke, C., McMahorn, S., & Dietsche, T. (2010). *Green building & human experience: Testing green building strategies with volunteered geographic information*. Washington, DC: U. S. Green Building Council.

Wall, S. (2015). Focused ethnography: A methodological adaptation for social research in emerging contexts. *Forum: Qualitative Social Research, 16*(1). http://www.qualitative-research.net/index.php/fqs/issue/view/51.

Windsor, A. (2005). User satisfaction surveys in Israel. In W. F. E. Preiser & J. C. Vischer (Eds.), *Assessing building performance* (pp. 140–148). Oxford: Elsevier (republished 2015, London: Routledge).

Author Biography

Mille Sylvest, PhD is a Behavior Researcher with Copenhagen based firm, 3XN Architects.

Her research focuses on social activity and wellbeing in the built environment and whether the intentions of the architect can be observed in the everyday life and social activity patterns among building users. As a part of 3XN Architects, she works closely with the architects in order to determine how best to inform the design process regarding social activity patterns and wellbeing among building users through evaluations, design tools, and research dissemination. In addition, she is responsible for the academic facilitation of the Science Forum, a BLOXHUB initiative aimed at bridging academia and practice through collaborative research projects.

Mille's research is funded by Innovation Fund Denmark and the Realdania Foundation.

She frequently lectures and presents her research at universities, architecture festivals and design weeks, as well as to companies around the world.

Chapter 13
Making POE Work in an Organization

Duncan Joiner

13.1 Introduction

A principal conclusion drawn by the author, Joiner, and Ellis from their experience of post-occupancy evaluation (POE) practised across a variety of New Zealand government agencies, was that POE would only prove its long term value when it had developed an operational data base, which was used successfully to influence facilities management, design and design policy (Joiner and Ellis 1989). The New Zealand Ministry of Works and Development (MWD) had begun developing techniques for POE in 1979 with an expectation that it would inform a data base for government facilities in particular, and a more general national data base on building performance.

Joiner and Ellis also concluded that the practice and methods of POE must be sympathetic to a host organisation's culture if it is to be sustained and usefully inform the design and management of its buildings. The idea was that the full value of POE would be realised if it became an integral part of organisations' facilities management and procurement processes. However, after 1988 New Zealand government services were decentralised, and most occupy premises leased from commercial owners. MWD was abolished, and its services were privatised, so the context for POE for government facilities changed dramatically.

Participatory walk through, and focus study POE methods developed and practised by MWD for government facilities have survived, and with government decentralisation they have developed as consulting services. Consultants have managed POEs for a range of private, public, and commercial facilities, and POE is listed as a service on the website of the Building Research Association of New

D. Joiner (✉)
67 Nicholson Road, Wellington 6035, New Zealand
e-mail: d.joiner@clear.net.nz

W.F.E. Preiser et al. (eds.), *Building Performance Evaluation*,
DOI 10.1007/978-3-319-56862-1_13

Zealand (BRANZ) and is also included in the New Zealand Institute of Architects list of services. Where POE continues to succeed, it is clear that consultants have been aware of the principles identified in the early reviews of the MWD POE programme.

13.2 Background

MWD developed POE in response to pressures from the New Zealand Government for improved accountability of the Ministry's architectural services. POE was developed with the help of researchers from the School of Architecture at Victoria University of Wellington. Advice was taken from other experts, notably Professor Robert Shibley, who was at that time directing a POE program on behalf of the US Army Corps of Engineers, one of the few government agencies in the world then successfully operating a POE program (Shibley 1985).

In 1988 MWD was converted into a state-owned enterprise, and by the mid-1990s its design branch had become a private sector corporation. Commercialization meant that it could provide design and property management services for private sector clients. Equally, most government agencies were free to commission a private firm as an alternative to the former Ministry. There was no longer a central government agency with responsibility for the design, construction and management of buildings for the government. The context in which the MWD POE program developed had disappeared.

The research input to the MWD POE program is well documented (Daish et al. 1983; Kernohan et al. 1992). Apart from developing the techniques for POE, the MWD also learned a good deal about the organizational aspects of operating a POE program. From that initial experience, it could be seen that in applying POE effectively to the improvement of building design quality that these organizational factors require as much attention as the POE techniques themselves. The second part of this chapter describes how with understanding of the essential organizational factors, the POE model has developed and spread after the MWD context was removed.

13.3 Establishing and Maintaining POE

MWD was, until April 1, 1988, New Zealand's government department responsible for undertaking major public works projects. The operating units of MWD consisted of seven district offices, with a Head Office in Wellington (see Figs. 13.1 and 13.2). Architects in the MWD offices worked with government departments on the development of their building programs, design and construction contract management. They also worked directly with property management and maintenance

Fig. 13.1 MWD Head Office Building in Wellington. *Source* Author

Fig. 13.2 MWD District Office in Thorndon, Wellington—one of seven MWD District Offices. *Source* Author

staff. They held, therefore, continuous knowledge and experience about the performance of government buildings.

The POE program was initiated with the twin aims of improving accountability, and generating information that would contribute to the improvement of existing buildings and to create policy for future design. An innovative feature of the POE process was its participatory nature. The research team from Victoria University showed in a comprehensive literature review (Daish et al. 1983) that most building evaluation had been done against physical rather than use criteria. Although they found an extensive body of theory on behavioral methods of evaluation, most of these were untested and at any rate did not meet their need for methods which were quick and pragmatic, but also scientifically rigorous. Since user participation had been established as a key principle in MWD's POE strategy, the researchers had to develop their own process and techniques for user involvement.

Two techniques in particular are characteristic of the POE developed in New Zealand. One is the "Walkthrough" or "Touring Interview" method, whereby small groups of users, designers, builders, managers, and other concerned parties walk through the building observing, commenting, and evaluating. Walkthroughs were planned, managed, and facilitated by a Task Group, which is responsible for putting together a report on the Walkthrough POE. The second technique is the Focus Study, a more concentrated evaluation which examines in detail, again with user participation, a particular problem or success identified in a building.

Three other principles guided the development of MWD's POE techniques. Firstly, techniques were intended to be robust, that is, capable of being used by a variety of people, generalists as well as specialists. Secondly, POE was to be a "bottom-up" rather than a "top-down" process, which could be initiated locally according to need. Thirdly, POE needed to be able to clearly demonstrate its own cost benefits.

Early reviews of the POE program showed that it had been extremely successful. The POEs had yielded good information for upgrading or fine-tuning the buildings under evaluation, and they had generated positive attitudes and increased awareness among participants with respect to environmental issues and aspects of building use. Where POEs had been less successful was in generating information of long-term value in the design of future buildings, and in stimulating a demand for increasing the number of POEs to be done. There was a problem of creating and sustaining demand for POE in an organization where its utilization was a matter of choice.

The reasons given for not initiating POEs were that it was too expensive and time-consuming. The existing POE product was thought to be well-founded and essential to the longer-term objectives of building a data base, and making an input to policy. At the same time, there was an argument for adding to the range of POE techniques some which were even quicker and cheaper than the participative walkthrough, and which would have greater appeal, particularly to the busy architect. For, it was the architects in the district offices who were showing the most reluctance to use POE. In implementing these strategies, four types of organizational change measure were devised: structural, motivational, informational, and technical.

- A structural measure involves a permanent or semi-permanent change. Two structural measures were decided on by MWD. The first was to establish a special POE unit at Head Office. The second was to re-define in a formal way the services offered by MWD to its clients, to make POE a structural part of those services. A structural measure provides a continuing stimulus to the result it is designed to bring about.
- Motivational measures attempt to change people's attitudes through persuasion, reward, incentive, or disincentive. MWD had been doing this in relation to POE since the program was introduced, in the form of "brushfire" tours of the district offices. But the problem with motivational measures is that the results do not necessarily last. Attitudes may revert if the desired change is not compatible with individual's work aims as they perceive these, or if individuals are replaced by others.
- Informational measures present rational arguments that are expected to make sense to those receiving them. Their effectiveness is limited if the logic of these arguments does not appeal to or fit with current objectives of the audience. Nevertheless, full information is essential, and MWD decided to spell out more clearly the aims and benefits of POE by improving the literature they produce for potential users.
- Technical measures involve developing new or improved techniques which will make it easier for those concerned to achieve the desired aim. The research team at Victoria University was commissioned to develop new techniques better suited to the needs of working architects.

In relation to these four types of measure, it was clear that a supply marketing strategy for POE, which aimed to sell an existing product more effectively, would gain more from the first two, while a demand strategy would benefit most from new techniques and from information about those techniques. However, MWD's experience suggests that all four types of measure are most effective when acting in concert.

13.4 POE as an Organizational Process

Ellis and Joiner (1985) developed the idea of design quality being negotiable. They argued that design quality depends on the effective management of resources: not just physical and technical resources, but human resources too. Furthermore, the whole notion of design quality lacks relevance if not applied to a building in use; in such cases it depends on the definitions of quality held by all those involved, users as well as planners and designers. Establishing design quality through the processes of designing, occupying, and using a building is in this sense "political" (Ellis 1984), and is characterized in varying degrees by social negotiation about standards and about perceptions and definitions of quality. "Good physical design is a necessary but not a sufficient condition for design quality" (Ellis and Joiner 1985).

Participative POE brings users and other parties together in negotiation with each other in direct relation to the physical environment, and MWD harnessed POE to serve the organizational process of negotiating design quality.

13.5 POE for Design and Facilities Management

MWD's experience was that POE was useful for facilities management as well as for design, and there was plenty of evidence that this was being realized in other places as well, including North America and Europe. While the POE movement was led by large public sector organizations like MWD (Shibley 1985) the private sector was gradually following. In the office field, several furniture companies began offering some version of POE as a service to their customers, as were a few interior design and space planning practices. Although as a whole, the architectural profession was slow to follow this trend. As MWD found in its own organization, architects appeared to be fixated by the notion that their business was the production of new buildings, and were either unable or unwilling to shift to the idea of providing a continuing service to their clients.

The MWD response to this perceived gulf between architects' normal activities and POE was to re-design POE to bring it closer to the everyday needs of architects. As a result, POE was turned into a more flexible set of tools for use in a variety of situations. In this form, it was more closely integrated with the design process, and its appeal to architects and to their clients increased.

13.6 Building a Data Base

A large public sector design practice provides for its clients the advantages of scale. One of those advantages is the scope a large organization has for maintaining and deploying specialist resources, such as a unit for accumulating POE data and translating it for future use by the organization.

Reviews of some of the POE reports produced by MWD showed that the way information was recorded and presented was crucial to its usefulness in future situations. Ensuring that the quality and depth of information is collected consistently during the POE process itself required, in the MWD's experience, a specialist overview, preferably from the same specialist responsible for accumulating and translating that information into a form for future use.

The final form or product that such design information should take was another issue under debate. One possibility was to follow the example of the US Army Corps of Engineers, with their very thorough Design Guide Publication Series

(Shibley 1985). Another was to integrate POE data into a computerized information system which was being developed. From the MWD experience, there were different requirements for information from POEs—for immediate use in facilities management or design, and for providing longer-term guidance across a range of building projects. Appropriate data storage for these requirements was not resolved by MWD.

13.7 POE After MWD

When MWD was restructured and privatized, the most logical organization to develop and maintain a central POE data base for government facilities disappeared. Furthermore, in the same deregulation of government services, individual government agencies were required to take over and manage (often through out-sourcing) the design and management of their buildings. They all started to work and plan independently with no common thread of building quality for government facilities mediated through one agency.

Therefore it is not surprising, that a national building database of information from POEs has not been developed. However, BRANZ has recently commissioned a collation of 5500 best and worst aspects of buildings from the records of some 170 POEs. Approximately 5500 sets of keywords have been interpreted from these POEs and are used in the collation to identify stakeholder groups' statements about their experiences of buildings (BRANZ 2016).

Although the idea of a central database for New Zealand's government did not survive, the recent development by BRANZ shows promise of carrying the idea forward. In the meantime, however the POE process itself has survived and expanded. It has been successfully used in on-going programs by some of the central government agencies, and is being used by a variety of commercial organizations as well including retail, offices, medical facilities, and workshops, and there has been interest from maritime industries in POEs of passenger and crew accommodation in ships (Joiner 2006, 2007).

It is clear that the walkthrough POE process is adaptable to the requirements of a wide variety of organizations. This is indicated by the success of its continuation beyond the central government agency for which it was developed. But case studies of the use of POE for courthouses demonstrate some characteristics of making POE work in an organization which perhaps account for its continuing use and success (Joiner 1996). After two courthouse evaluations, the Ministry for Justice recognized that what they had initially regarded as a research-based activity was actually a useful service that consultancies could deliver to them (see Fig. 13.3). Contracts to undertake subsequent courthouse POEs were then won on a competitive basis from other tenderers. What had happened was that the Ministry for Justice had taken ownership of the POE process. Tender documents for the later POEs set out a clear structure for each study and a list of headings for the reports. The organization had become clear about why it wanted POE. This is just one example of making POE

Fig. 13.3 Courts buildings in Wellington. *Source* Author

work in an organization, but it is experiences like these with tendering processes for POE which also converted architects to the idea that POE is something that their clients will want and that their clients think that they can do for them in association with design activity.

13.8 Conclusion

From MWD's early experience it was clear that there can be few universal rules for the successful organization of POE. But those aspects of MWD's experience which were singled out by Joiner and Ellis (1989) are still relevant to its ongoing value. POE, like other institutional implants, must be sympathetic to the host culture. Architects and consultants who are successfully assisting organizations with POE understand supply and demand strategies, and are tailoring POE to meet their clients' needs. They have successfully recognized the structural, motivational, informational, and technical aspects of making POE relevant to organizations.

It is still worthwhile to reiterate the principles laid down by Shibley (1985). First, start simply. In retrospect, MWD might have started even simpler than it did, by building on architects' day-to-day evaluative activity, rather than introducing a concept perceived as something distinct from this. Second, know your organization. Here it is clear that consultants who are working successfully with multiple evaluations are taking the trouble to do this. Shibley's third principle states the need to develop multiple levels of application. Multiple levels of application have been the key to the adaptability of POE and its ongoing value to a wide range of organizations and their long and short-term demands for building information.

References

Building Research Association of New Zealand (BRANZ). (2016). Best and worst aspects of buildings. Retrieved from: http://www.branz.co.nz/cms_display,php?st=18sn=287. Accessed 21 Nov 2016.

Daish, J., Gray, J., & Kernohan, D. (1983). *Post-occupancy evaluation of government buildings: Final report*. Wellington: School of Architecture, Victoria University.

Ellis, P. (1984). Making buildings work: Should designers get involved in politics? In J. A. Powell, I. Cooper, & S. Lera (Eds.), *Designing for building utilisation* (pp. 217–229). London: Spon.

Ellis, P., & Joiner, D. (1985). Design quality is negotiable. In K. Dovey, P. Downton,0. & G. Missingham (Eds.), *Place and placemaking: Proceedings of the PAPER 1985 Conference*, Melbourne. Melbourne: PAPER Association.

Joiner, D. (1996). POE practice. In G. Baird, J. Gray, N Isaacs, D. Kernohan, & N. McIndoe (Eds.), *Building evaluation techniques* (pp. 85–89). New York: McGraw-Hill.

Joiner, D. (2006). Evaluating the design of passenger and crew accommodation on ships. In *Proceedings of the international symposium on marine design* (pp. 77–84). London: Royal Institution of Naval Architects.

Joiner, D. (2007). User feedback in ship design. In *Proceedings of the international conference on human factors in ship design, Safety and Operation* (pp. 87–95). London: The Royal Institution of Naval Architects.

Joiner, D., & Ellis, P. (1989). Making POE work in an organisation. In W. F. E. Preiser (Ed.), *Building evaluation* (pp. 299–306). New York: Plenum Press.

Kernohan, D., Gray, J., Daish, J., & Joiner, D. (1992). *User participation in building design and management*. Oxford: Butterworth-Heinemann.

Shibley, R. G. (1985). Building evaluation in the mainstream. *Environment & Behavior, 17*, 7–24.

Author Biography

Duncan Joiner PhD(London) BArch(Hons)(Auckland) FNZIA RIBA LDINZ, is a registered and chartered architect. He was educated at the School of Architecture at the University of Auckland, and the Bartlett School of Architecture, University College London. He has a PhD from the University of London.

For over 40 years Duncan has combined creative design and research in government and commercial design practices, in the design and research consultancy practice he and his partner established in Wellington, and in his academic appointments. He was New Zealand Assistant Government Architect (Design) for ten years from 1978, and Chief Architect in Works Consultancy Services Ltd from 1988 to 1992. Duncan has held the positions of Head of the Wellington Polytechnic School of Design, inaugural Pro Vice-Chancellor of the Massey University College of Design, Fine Arts and Music, and Professor of Design.

Chapter 14
POE for Organizations with a Repetitive Building Type

Thierry Rosenheck

14.1 Introduction

This chapter is about post-occupancy evaluation (POE) methods developed while working for an agency in the U.S. Government. The views expressed herein are the author's alone and are not necessarily those of the U.S. Government.

The identity of the agency is kept anonymous intentionally to focus on POE methods and facilitate imagining how they would fit any other organizations. Still, a brief generic description of the agency is useful: it manages an international portfolio; its facilities fit a repetitive building type, which are typically located on campuses.

The POE Program at this organization started in 1992, and was initially modeled after POE methods for the US Postal Service Facility Department (1990) developed by Jay Farbstein. At different stages, the program involved others: Craig Zimring and his team from Georgia Tech (1997, 2001) helped develop the first interactive POE database; and Gerald Davis, Francoise Szigeti, and the team of the International Centre for Facilities (1993, 2001) helped develop an organizational profile using American Society for Testing Materials (ASTM) Serviceability Scales.

Other indirect influences came from Zeisel (1984) whose 'design development spiral' diagram lays out a vision for a continuously changing built environment transformed by constant testing from new knowledge about itself (see Fig. 14.1).

Similarly, Preiser (2001) diagrammed the Building Performance Evaluation (BPE) process model showing a continuous loop relationship between 'evolving

T. Rosenheck (✉)
US Government, Washington, DC, USA
e-mail: rosenheckte@yahoo.com

W.F.E. Preiser et al. (eds.), *Building Performance Evaluation*,
DOI 10.1007/978-3-319-56862-1_14

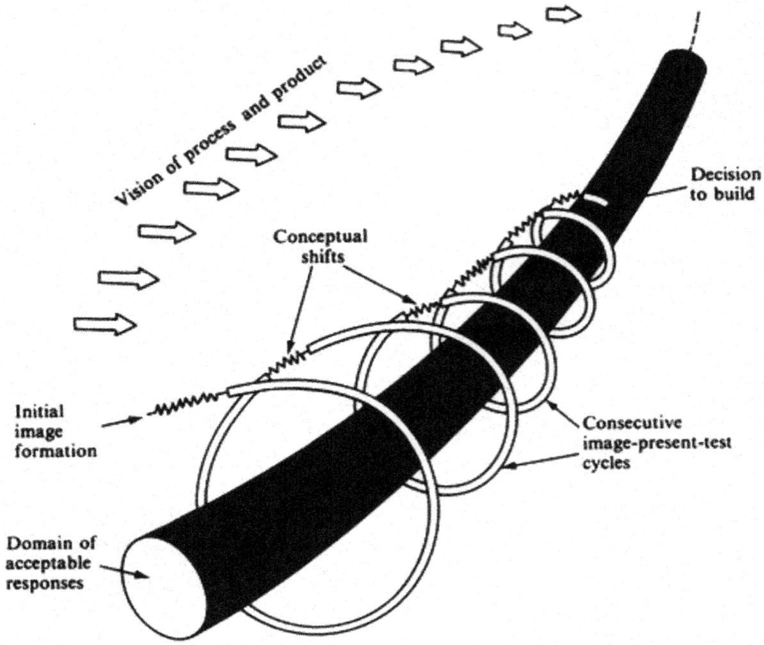

Fig. 14.1 Design development. *Source* Zeisel (1984)

universal design performance criteria' and 'feedforward' responding to six project delivery phases and actual performance (Fig. 14.2).

The methods practiced by the POE Program are described in three distinct sections:

- Pre-visit preparations
- On-site visit field survey
- Post-visit data recording and analysis

Following these is a section on feedforward to describe the implementation of POEs—potentially the most challenging aspect of the Program. Finally, the concluding section discusses the future challenges of the POE Program.

14.2 Pre-visit Preparation

A POE is initiated one to two years after occupancy of a recently completed campus project. Six to ten POEs are usually conducted yearly, two on the same trip. Once the campuses are identified, the administration is contacted and four online surveys are sent to capture a preliminary assessment of building performance. One survey

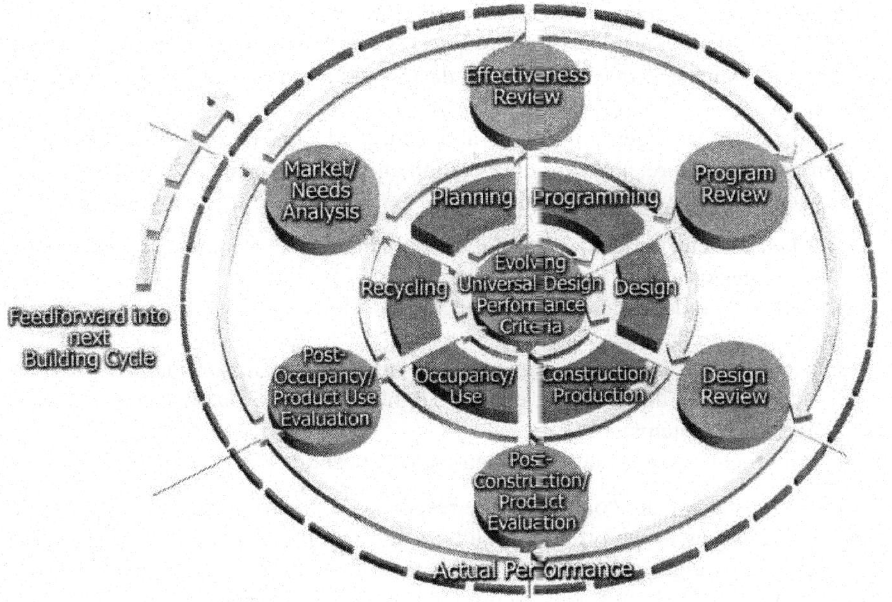

Universal design evaluation: process model with evolving performance criteria.

Fig. 14.2 Building performance evaluation process model. *Source* Preiser (2001)

goes to all staff, and three go to subject matter experts (SMEs): Facility Manager (FM), Administrative Manager, and Security Manager.

Online surveys prepare the POE team by providing a snapshot of occupant satisfaction with the facilities at the campus. The general staff survey is the primary survey, and is divided in five sections:

1. User Profile—demographics
2. Site Characteristics—location, quality of the exterior
3. Workstation Characteristics—ability to support the work of the respondents
4. Shared Spaces and Amenities—i.e. conference rooms, cafeteria, parking
5. General Descriptions—the only open ended questions: what occupants like most, least, and what features they would like repeated in a similar, future facility

The number of respondents per campus varies from 25 to 80 or more depending on campus size. The other three surveys are essentially one-on-one to management SMEs. SMEs are asked how well, from their expert perspective, facilities and systems meet their intended purposes. The SMEs also have open-ended questions about what they like most, least and what features they'd like to see repeated in future facilities. A brief description of each survey focus is provided below:

- Administrative Manager—questions about the adequacy of parking for visitors and staff, effectiveness of meeting spaces for different events, etc.
- Security Manager—questions on the design and the performance of the security systems.
- The Facility Manager—questions about the performance of the facilities, systems, infrastructure, and ease of operation and maintenance (O&M). This survey is wide-ranging with four sections:

  1. Management—adequacy and ease of maintenance, training, commissioning of systems;
  2. Systems—adequacy of performance for various systems such as mechanical and electrical;
  3. Facilities—ease of maintenance and operation at major facilities such as adequacy of the heating, cooling, lighting, power, data/telecom, finishes, furniture systems;
  4. Comments—three standard open-ended questions.

When the surveys are completed, the responses are distributed at headquarters to various disciplines to analyze the results of the facilities' performance. A typical team has six to eight participants. The core team includes a POE manager, and mechanical, electrical, and construction engineers. The in-house team is not dedicated to the POE Program. Instead, team members rotate with each POE visit. The disciplines will vary and include interior designers, programmers, and other specialists. Typically, a senior architect-consultant joins the team to provide a practitioner's perspective, and aid in analyzing data and recording findings.

A standard itinerary is reviewed with all stakeholders. An itinerary that indicates the roles and contributions of team members is developed. The campus management reviews the itinerary, adjusts it to fit their calendar and assigns a point of contact (POC) on site to facilitate the POE's activities. Prior to departure, the POE team meets with the professionals who were involved in the initial design and construction of the campus to be surveyed. The POE team assembles and reviews project documents, highlights of the projects, and project delivery issues that might have occurred. In addition, other preparations may include asking campus staff to gather information for the team, such as energy consumption and performance data on specific systems.

14.3 Site Visit

All campuses in this agency follow a similar model that supports both public and back-of-house functions. A fence or wall surrounds the campus, which have entry access-pavilions that screen visitors, staff, and vehicular traffic at the perimeter. Inside the grounds, the campus has an open space, one Primary Office Building (POB), and ancillary support facilities such as maintenance shops, storage, utility building.

Most of the POB's focus is on offices; however, there are also public areas and several levels of back-of-house areas. Public spaces might include lobbies, multi-purpose rooms, and reference centers that greet visitors. Areas for conducting public business have a waiting room similar to what is found in a bank. Most of the offices utilize an open-plan layout with systems furniture. Interspersed in the workspace are shared spaces, such as conference rooms of various sizes, training rooms, workrooms, storage, and kitchenettes. The POB has amenities, with a multi-functional atrium itself surrounded by other amenities, such as a cafeteria, fitness center, health unit, and bank teller.

Site visits typically take 3–4 days and start with a 2-day multidisciplinary walkthrough and end with team members conducting their surveys individually with one-on-one interviews focused on their team members' respective disciplines. Most of the data generated is gathered during the walkthrough interview. The data falls in two categories: soft and hard metrics. Hard metrics are used to measure the systems and infrastructure performances; soft metrics are used to gauge how well the workplace or amenities meet their intended purpose. Being multidisciplinary, the team is adept in managing these differences.

Metrics to measure workplace functionality are not readily available. In most cases, the 'soft metric' feedback data captured in walkthrough interviews is sufficiently reliable at a 'common sense' level. For instance, a heavy door is difficult for an elderly to open—observations can corroborate this feedback, etc. Positive feedback is also gathered, such as having an outdoor terrace adjacent to the cafeteria. As this feedback is obtained, the team takes notes, photos, and indoor environmental-quality measurements, such as light, sound, and air-quality readings for temperature, relative humidity, and CO2 levels. A new protocol for the technical team includes spot measurements and capturing data from the Building Automation Systems (BAS) for additional information on energy consumption to analyze for efficiency.

The standard walkthrough begins outside the campus in public spaces, and then proceeds through the access-pavilions, the grounds, ancillary facilities, then the POB. Once in the POB, the team visits the shared spaces, then proceeds from workspace to workspace and interviews available section-heads, as well as observing and recording the activities of visitors and staff.

Types of interviews vary: most happen during the multidisciplinary walkthrough and are informal, with team members asking 'what works well and what does not work well in supporting the work you do.' Other interviews include one-on-ones with SMEs, and group interviews with staff that are selected from different divisions of the organization. These interviews generate insightful discussions on a wide-range of topics, such as work environments, organizational culture, and general responses about the campus' architecture. Other topics can also provide clarification to responses from the online surveys. At the end of each day, the team debriefs by reviewing major findings of that day. These sessions are critical; they focus the team on common issues, and prioritize significant items to be reported to headquarters. Prior to departure, the team out-briefs the campus management with a summary of POE findings. Before heading to the next POE, the team captures all findings in a quick-list. The number of findings per POE ranges from 100 to 180 or more.

14.4 Post-visit Data Recording and Analysis

14.4.1 Recording the POE Database

A distinguishing feature of the POE Program is its database, a repository of findings, and a powerful tool for analysis. The building block of the repository is made up of single POE findings, themselves made up of the elements described below:

1. Headline—a description of the finding
2. Description—a description of the physical aspects that can be measured or located
3. Observation—actions that describe system performance or human activities in that setting
4. Requirement Analysis—reference to contractual requirements, guidelines or standards related to the particular finding.
5. Evaluation Analysis—the evaluation in which 2 and 3 above are compared to 4
6. Recommendation—description of action to take in response to the evaluation in 5 above
7. Photos or drawings—illustrations of what is discussed in 2 and 3 above

Tags and Indices—Beyond capturing the fundamental signature of a project, such as location, the year built, etc., the POE team tags finding for easy retrieval and analysis. That way, findings deemed to be "one-offs" can be differentiated from patterns and trends. The database contains many tags. The ones below are more relevant:

1. Facility—i.e. maintenance-shops, POB, parking
2. Space Type—i.e. lobby, utility closet, cafeteria
3. Discipline—up to three disciplines the finding is related to, i.e. architecture, programming and/or construction
4. Keyword—up to three can be selected; i.e. layout, acoustics and/or privacy
5. Recommendation category—further described in the Analysis section below

14.4.2 Analysis

The POE Program's process of analysis has evolved, and while some methods remain constant, the Program pursues improvements. Not all analyses used are discussed in this chapter, and only the more relevant highlights are described.

The potential for higher-level statistics is present but not yet fully exploited. The online survey responses from the general staff provide a large data set, particularly when aggregated from several POEs with data from over 2500 respondents. Its analysis uses descriptive statistics. In addition, the three open-ended questions from each survey provide considerable qualitative data that require content analysis.

This type of analysis is labor intensive and has been rarely attempted. Clearly, the online surveys contain data that can be further analyzed. The POE Program is searching for software to realize this. By far, the most valuable analysis is to examine several POEs at a time. At convenient intervals, the POE Program conducts "POE LookBack" analysis of several recently completed POEs, with the objective of identifying patterns.

14.4.3 Recommendation Categories for Implementation

Once the basic elements of a finding are identified, additional evaluation is conducted to establish whether the finding observed is related to project delivery issues at programming, design, construction, commissioning, or could be related to training/O&M. This evaluation categorizes the recommendation into five categories for the implementation, or feedforward phase:

1. Criteria—repetitive findings are forwarded to a working group charged to manage and update Standards and Guidelines.
2. Project Execution—repetitive findings are sent to related discipline offices.
3. Deficiency—findings are sent to the campus FM for corrective action.
4. Study—recommendations for studies are forwarded to the offices affected by the issues; or to be undertaken by the POE Program.
5. For Information—not for action, but for future monitoring.

14.5 Feedforward

The feedforward phase follows analysis and shifts the focus to implementing actions that inform the next generation of projects. Because different institutions organize their stakeholders according to their unique internal organizational cultures, the feedforward process must reflect each organization's make-up, and cannot be easily generalized. In this organization, the POE Program does not have authority to implement recommendations. Instead it must rely on communicating its findings to stakeholders for the implementation of recommendations. There are several avenues available:

1. Trip reports
2. Presentations
3. "LookBack" reports
4. Stand-alone studies
5. Participating on design reviews
6. Briefing senior management
7. Contributing to Guidelines 'Best Practice' reports.

14.6 Discussion and Observations

The POE methods as described in the previous sections have become reliable and constant. Below is a discussion of what has worked well, or can improve the POE Program. The type of POE conducted is an 'indicative POE' (Preiser 2001); and fits with this organization's repetitive building type. A 3–4 days POE gives the right cost/benefit balance and provides a wide-angle gaze on workplace performance, and the feedback can inform future projects more rapidly.

Field methods—observations need to be coupled with stronger metrics. Technical instruments can measure precisely, but measuring work productivity is still elusive. Use of sensor technology is promising, although in an international context, this may trigger unanticipated ethical questions.

Data Analysis—the level of analysis done is adequate to prepare and conduct the POEs, and to indicate patterns of findings. However, more powerful software to conduct inferential and content analyses will reveal more from the recorded data.

Database—the repository of findings is a powerful tool but has been lacking an intuitive interface that would enhance utilization of a database with over 6500 findings from over 47 POEs collected since 2007. The database is growing, and its interface is constantly modified to be more user friendly to the entire organization.

Team Composition—the POE Program does not have a dedicated team of multidisciplinary professionals, and must rotate its team members. There is a trade-off between rotating and using a dedicated team. In rotating teams, the POE knowledge gained is disseminated and integrated throughout the organization's staff. With a dedicated cadre, cumulative POE experience is increasing. It is the POE Program's desire to build a dedicated core to address re-occurring issues, and also include rotating specialists.

Feedforward—a POE is potentially complete without including feedforward; and there is no standard feedforward to fit all organizations. But ignoring the opportunity for feedforward makes for empty POEs, especially in the context a repetitive building type. The POE Program must continue to develop and adapt its process with subsequent changes in the organization.

In the context of the POE Program's organization, the POE Program created its own 3-D model (Rosenheck 2008) as a guide to show how its POE process, through the feedback loop and using knowledge management/database, can influence the six BPE phases of project delivery, and the organization's guidelines and procedures (Fig. 14.3).

POE practice and organizations could benefit from a better understanding of feedforward. Clearly, this deserves further study, especially through the use of POE reports and best practices findings. POE Reports are reporting methods that are frequently re-examined to make them more effective. One approach is to associate personas with the findings, as if re-told by the occupants of the campuses. Findings related to O&M would be told by the FM; those with privacy issues would be told the management. Reporting is a work in progress, with the intent to make findings come alive in a story-telling form. Best Practices are POE findings being used to

Fig. 14.3 Modeling Post-Occupancy Evaluation in context of organizational learning. *Source* Author (2008)

illustrate best practices for published guidelines. This pilot effort has the best prospects of linking lessons from past projects to future ones.

This chapter shows that POE methods have matured—from pre-visit preparation, site visit, and recording, to analysis. The methods have become accepted procedures in the context of BPE's six phases of project life cycles. Still, it is unclear how well feedforward is understood as an actionable and replicable procedure. This chapter describes the steps one organization has taken in this regard. However, more discussion and a better understanding of feedforward in the building evaluation practice could solidify POE's benefits to organizations, and enhance organizational learning.

References

Davis, G., Thatcher, C., & Blair, L. (1993). *Methods for setting occupant requirements and rating buildings: Serviceability tools manuals* (Vol. 1 & 2). Ottawa: International Centre for Facilities.

Preiser, W. (2001). The evolution of post-occupancy evaluation: Toward building performance and universal design evaluation. In Federal Facilities Council (Ed.), *Learning from our buildings: A state-of-the-practice summary of post-occupancy evaluation* (pp. 9–22). Washington, DC: National Academy Press.

Rosenheck, T. (2008). Modeling post-occupancy evaluation in context of organizational learning. In *Proceedings of the environmental design research association.*

Szigeti, F., & Davis, G. (2001). Appendix A—Functionality and serviceability standards: Tools for stating functional requirements and for evaluating facilities. In Federal Facilities Council (Ed.), *Learning from our buildings: A state-of-the-practice summary of post-occupancy evaluation* (pp. 63–78). Washington, DC: National Academy Press.

United States Postal Service Facilities Department (Ed.). (1990). *Post occupancy evaluation (POE) guidebook.* Facilities Department.

Zeisel, J. (1984). *Inquiry by design*. New York: Cambridge University Press.

Zimring, C. (1997, December). Database to aid embassy designer. *Urban Land* (p. 18).

Zimring, C., & Rosenheck, T. (2001). Post-occupancy evaluations and organizational learning. In Federal Facilities Council (Ed.), *Learning from our buildings: A state-of-the-practice summary of post-occupancy evaluation* (pp. 42–53). Washington, DC: National Academy Press.

Author Biography

Thierry Rosenheck is a registered architect and has been employed in the US Government since 1991. His work has included a mix of new projects, rehabilitations and initiating a POE Program in 1992. Thierry received his Bachelor's of Architecture from Howard University in 1972, and Masters of Architecture in Environment Behavior Studies from the University of Wisconsin-Milwaukee in 1987.

Chapter 15
US Army Command Headquarters: Evaluating Existing Buildings to Set Design Requirements

Greg Allen Barker

15.1 Introduction

Building evaluation can be a powerful tool when scrutinizing policies and standards an organization uses in its capital projects process. An appropriate and focused methodology can establish whether a criticism has merit and, if so, provide an empirical basis for changing said policies or standards. This chapter illustrates the use of building evaluation to evaluate and adapt a technical requirement used by the US Army in the design of its headquarter facilities. While the case study is focused on technical building requirements, the discussion will also provide the context of how such specific analysis might lead to further evaluations to connect the technical requirement to a successful user experience from the built environment.

15.2 Background

This chapter presents a case study in which design requirements were developed using the evaluation of an existing building in conjunction with computer modelling. The United States Army replaced three army command headquarters as part of the Base Realignment and Closure (BRAC) process. The scope of work included the development of design requirements used a number of methods, but this chapter will focus on the method for setting the gross building area based on net space requirements of each building's design program.

The identification of gross building area for this special evaluation was in response to Army Regulation 405-70 (AR 405-70), which provides a standard for the allowable gross building area of facilities based upon an allotted gross square

G.A. Barker (✉)
dba Greg Allen Barker, AIA, 120 Cerro Romauldo Ave, San Luis Obispo, CA 93405, USA
e-mail: gabarker@sbcglobal.net

W.F.E. Preiser et al. (eds.), *Building Performance Evaluation*,
DOI 10.1007/978-3-319-56862-1_15

footage per occupant plus a multiplier of 1.25 for the net requirements of select support spaces. Through the experience working with the regulation, project managers at the Army Corps of Engineers were convinced that AR405-70 did not result in sufficient space to meet the functional requirements of facility users. This scope of work was intended to provide an objective evaluation of AR 405-70 and make recommendations for design requirements that would result in adequate provisions of space.

These design requirements were to be applied to new headquarters for the following three Army commands:

- Training and Doctrine Command (TRADOC);
- US Army Forces Command (FORSCOM), responsible for delivering troop resources to the joint commands in different regions of the world, sharing a building with United States Army Reserve Command (USARC); and
- Army Materiel Command (AMC).

Their existing facilities were extremely different in terms of age and typology. TRADOC was spread among numerous buildings dating back to the mid-1800s at Fort Monroe, Virginia. AMC occupied a recent building constructed using pre-engineered components and assemblies at Ft. Belvoir, Virginia. FORSCOM was in a relatively recently completed, custom designed building at Ft. McPherson, Georgia.

15.3 Scope of Work

The development of Standard Elements for 4-Star Headquarters consisted of multiple objectives and methodologies. The Standard Elements to be used in the design requirements packages included the following:

- size, floorplate configuration, and column systems;
- building loss features and rules;
- standard space modules;
- features to support occupant functionality;
- cafeteria and fitness space requirements;
- LEED (Leadership in Energy and Environmental Design) Silver opportunities and guidance; and
- exceptions to criteria.

Multiple methods were employed in the development of the elements:

- structured observations of existing operations;
- group interviews using the Serviceability Tools and Methods;
- the development of conceptual floor plates to develop rules for efficiency;
- building loss analyses of the conceptual floor plates and a recently completed US Army administrative building.

The chapter focuses on these aspects of the scope and methods related to evaluating Army requirements for the allocation of administrative space, how the associated recommendations from the study were implemented, and their effectiveness in project delivery.

To follow up on the implementation and effectiveness of the recommendations, the design architects of the three completed facilities were contacted, asked about their experience working with the space requirements, and asked to provide relevant illustrations for their completed buildings. The architects of the headquarters for FORSCOM/USARC agreed to participate and were interviewed by telephone.

15.4 Issues Provoking the Evaluation

The allocation of administrative space controlled by the U.S. Army is governed by Chapter 5 and associated appendices to AR 405-70, with limited exceptions not relevant to this discussion. The regulation pertains to all administrative space, regardless of the echelon or mission of the facility, with some differentiation of private offices described below. Otherwise, a headquarters building is allocated administrative space the same as an office building that administers local capital projects, for example.

The basic space allocation for new construction is 162 gross square feet per occupant with the exception of commanders, deputy commanders, division heads, and branch heads who are authorized higher allocations based on rank and assignment. Additional allowances are provided for file areas, conference rooms, training rooms, reception areas, exhibit areas, copy rooms, and mail rooms. The 1.25 multiplier is then applied to the sum of these listed areas to calculate from the allowable net area to the allowable gross area of the facility. An anecdote to this information is that a federal administrator laid out the basics of these requirements in the early 1960s.

There were concerns among project managers going into the planning of the three army command headquarters, based on feedback on other recent Army administration buildings, that AR 405-70 would not provide sufficient space for the missions and staff complements of these commands. Specific concerns expressed by project managers included:

- the space allocations were more representative of norms for office space in the early 1960s than customary practice in the first decade of the 21st century;
- the army commands were top-heavy organizations, with typically command officers such as lieutenant colonels and colonels serving as analysts and relatively few enlisted and civilian personnel compared to most administrative facilities;
- the army commands had specialized facilities not envisioned when the regulation was developed, such as secure operations, information, and data centers.

The goal of these evaluations was to determine if AR 405-70 would result in adequate space allocations for the three Army command headquarters, and recommendations for addressing any shortcomings identified.

15.5 Methods

Some of the activities, used to develop the Standard Elements as a whole, contributed to the analysis of space allocation, while others were specific to this task. Structured observations of occupant operations and space utilization contributed to several aspects of the study's scope. These provided a comprehensive understanding of each command, the unique aspects of the work performed, and the features of their existing facilities that either contributed to or detracted from staff's ability to perform their work.

Two relatively parallel and complementary tasks were directed at evaluating AR 405-70. One task consisted of measuring the space of Sparkman Complex Building 9, located at the Redstone Arsenal in Huntsville, Alabama. Sparkman B9 was perceived to be the most recent example of an Army administration building closest in scale to the planned Army command headquarters. Even among themselves the operations of the three Army commands were not truly comparable except in the most general sense, so Sparkman B9 was recognized as the closest comparable building available with accepted limitations.

The parallel task was to develop hypothetical and high efficiency floorplates at the smallest and largest scales for anticipated and planned headquarter buildings. The small floor plate model was assumed to require 40,000 gross square feet (gsf) (3716 m^2) while the largest required 80,000 gsf (7432 m^2). To support other tasks in the scope, the model floorplates were developed with great attention to details minimizing the loss of any assignable area, e.g., coordinating the ceiling and structure grids. Figure 15.1 from the Standard Elements describes and illustrates the rules used to develop the model floorplates.

Models were developed and evaluated using the Vectorworks CAD application. Several scenarios were developed for the small and large floor plates to test the impacts of varying design decisions such as placement of vertical cores and coordination of grids at end bays. The proportions of enclosed offices, support space, and open office workstations were consistent with the design programs for the three facilities. Exterior walls were assumed to be eight inch (203 mm) thick. Figure 15.2 shows one such scenario for a small floor plate with interior improvements shown.

The model is composed of shaded zones, originally in color, related to different classifications of space and their order of measurement, as listed in Table 15.1.

The models did not include a number of the space measurement categories typically found in office buildings, as the goal was to demonstrate maximum net programmatic areas that could be accomplished using the basis for floor plate configuration. Features such as large atria, i.e., void areas; individual tenant

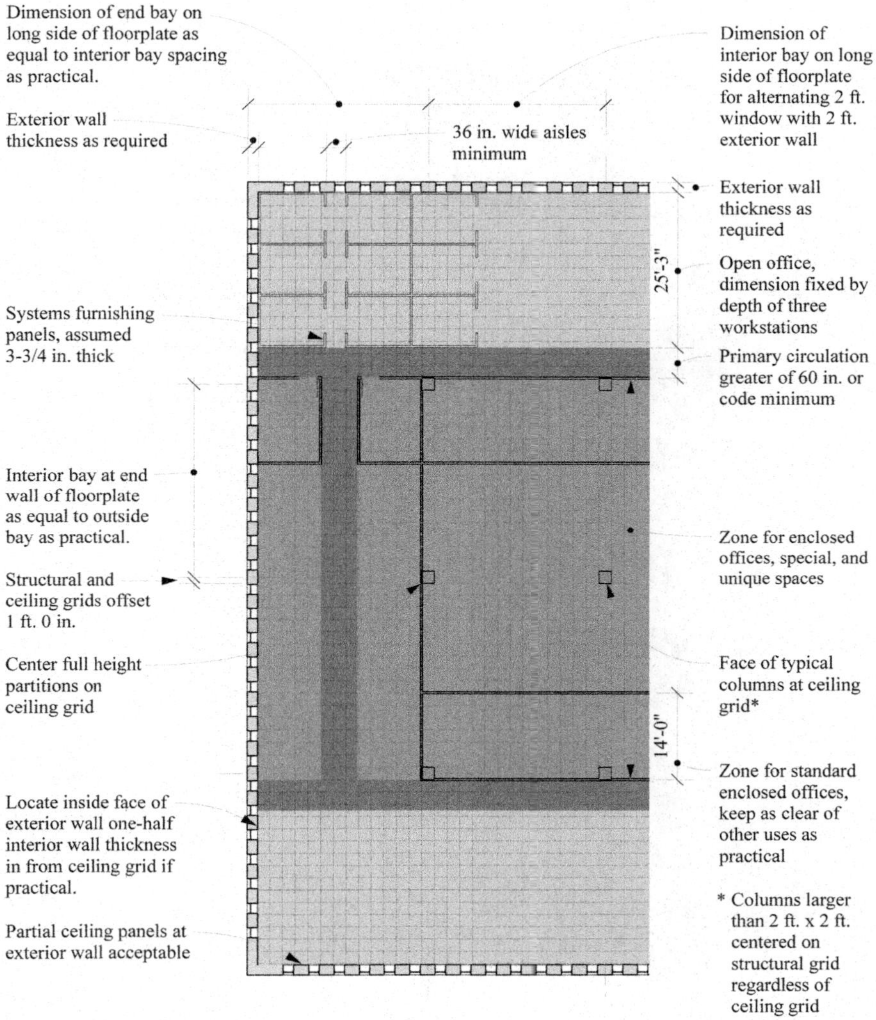

Fig. 15.1 Basis for floor plate configuration. *Source* Author

stairways between floors, i.e., occupant void area unassignable area; or perimeter encroachments were not included in the models, but included in subsequent measurements of existing buildings. In addition, exterior gross area was not listed as it is the resultant sum of the modelled spaces and exterior walls.

The Army Corps of Engineers provided DWG design files of Sparkman B9 that were imported into Vectorworks for measurement purposes. Polygons for each classification of space were drawn to scale and organized on design layers for each

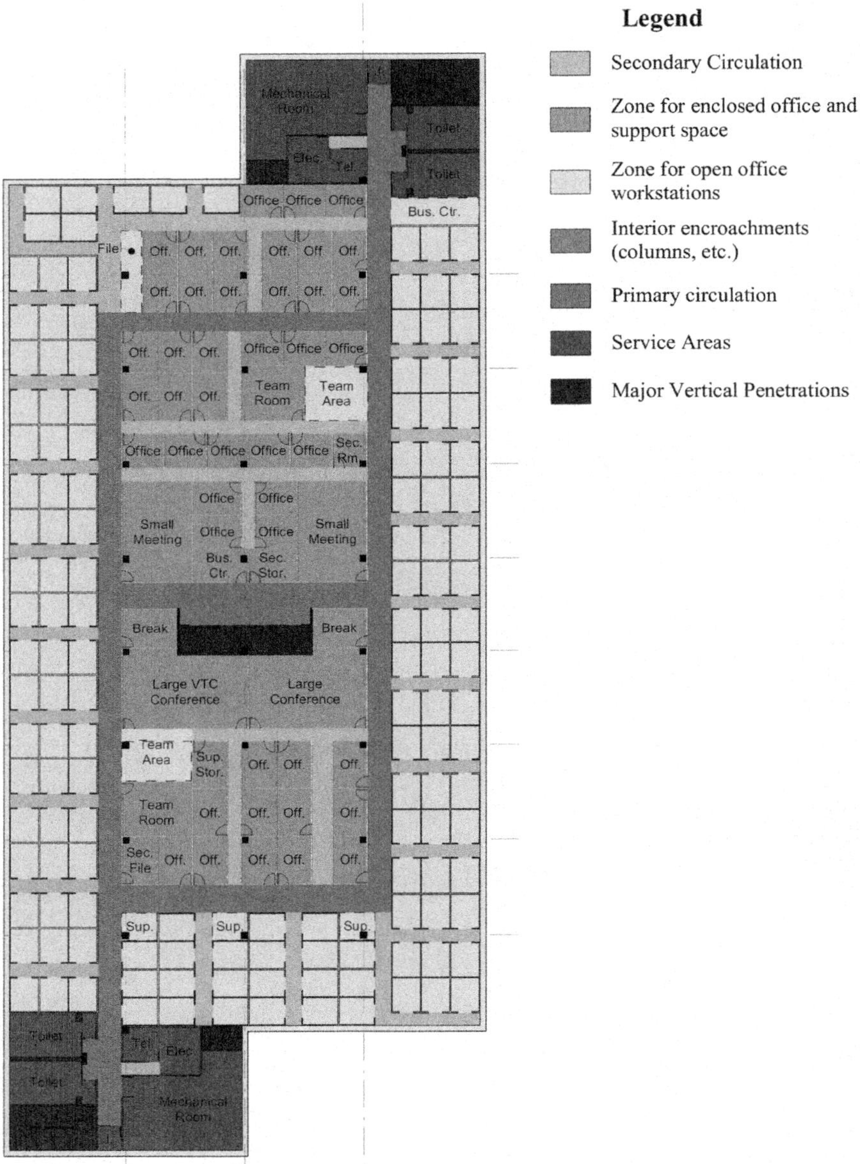

Fig. 15.2 Conceptual small floor plate with core elements at perimeter and interior improvements. *Source* Author

Table 15.1 Classifications of space modelled with measurement hierarchy. *Source* Author

Type of space	Rank in measurement sequence
Major vertical penetrations	4
Service areas	5
Primary circulation	6
Zones for enclosed offices, support space, and open office workstations (elements of plannable area)	8
Restricted area	9
Interior encroachments	10
Occupant void area	11
Secondary circulation	14

floorplate evaluated. Once a polygon was created the application was able to then provide the square footage as part of the object's data.

The polygons followed the taxonomy for space measurement provided by ANSI Z65.1-1980 Standard Method for Measuring Floor Area in Office Buildings. It should be noted that the American National Standards Institute subsequently collaborated with the Building Owners and Managers Association International (BOMA) to produce a substantially revised measurement standard: ANSI/BOMA Z65.1 (2010). The revised standard has shifted the emphasis in favor of the interests of real estate leasing agents, creating a new methodology with new categories for the classification of space supporting the method and its goals.

ANSI/BOMA Z65.1 now corresponds less with the standards and definitions for space measurement of other organizations. ASTM E2619M Standard Practice for Measuring and Calculating Building Loss Features That Take Up Floor Area in Buildings continues to evaluate *building loss factor* in a manner more relevant to building users seeking to maximize the functionality of their investment in facilities. American Institute of Architects' Document D101 The Architectural Area and Volume of Buildings emphasizes measurements to test scope and support preliminary construction cost evaluations.

The space measurements were tabulated in hierarchical order consistent with the standard. Table 15.2 tabulates the analysis for three scenarios: the large floorplate using 8 × 10 open workstations, the large floor plate using 8 × 8 open workstations, and Sparkman B9.

The table includes a column with each element's order in the measurement sequence, although it should be noted that there is discontinuity in the table because the 12th and 14th in sequence were broken out to emphasize comparisons between actual measurements and what would be allowed under AR 405-70. The column labelled "1" for each scenario represents the results of the measurements while the column labelled "2" illustrates what would be allowable under AR 405-70. The most significant findings are:

Table 15.2 Conceptual floor plate measurements. *Source* Author

Legend	Conceptual Floorplate:		COS (w/ 8X10 wkstations)							COS (w/ 8X8 wkstations)							Sparkman Bldg. B9						
	Note: When measuring or p-lining space the sequence of measurement and hierarchy is always from the outside in. Whether starting from outside gross or from programmable area, measure the elements in the order in the next column, "Measurement Sequence". Notice that 12th is listed above 13th, so p-line Assignable before Unassignable.	Measurement Sequence	Total Sq.Ft.		% of Outside (Exterior) Gross		Each Area at left is multiple of available area 1		Each Area at left is multiple of available area 2	Total Sq.Ft.		% of Outside (Exterior) Gross		Each Area at left is multiple of available area 1		Each Area at left is multiple of available area 2	Total Sq.Ft.		% of Outside (Exterior) Gross		Each Area at left is multiple of available area 1		Each Area at left is multiple of available area 2
	Area 1 available for NSF Program Requirement = Total of Assignable Area (Recommended). Measured to center line of enclosing walls and of enclosing furniture panels. Does NOT include Secondary Circulation.	12th	50,100		62.2%	A	1			49,780		61.8%	A	1			17,882		58.6%	A	1		
	Secondary Circulation 1. Includes all Secondary Circulation outside workstations and office modules. Measured to centerline of enclosing walls and to center line of enclosing furniture panels.	14th	9,257		11.5%	B	0.18			9,577		11.9%	B	0.19			6,469		21.2%	B	0.36		
	Area 2 available for NSF Program Requirement according to AR405-70 = Total of Assignable Area plus circulation adjacent to workstations. Measured to inside face of enclosing walls.	12th	55,094	15	68.4%	A		A	1.00	55,097	15	68.4%	A		A	1.00	23,385	15	76.6%	A		A	1.00
	Secondary Circulation 2. Includes only Secondary Circulation adjacent to full-height walls. Includes factor for area occupied by thickness of walls and thickness of furniture panels.	14th	4,263	14	5.3%	B		B	0.08	4,260	14	5.3%	B		B	0.08	966	14	3.2%	B		B	0.04
	Subtotal		59,357	13	73.7%	C	1.18	C	1.08	59,357	13	73.7%	C	1.19	C	1.08	24,351	13	79.8%	C	1.36	C	1.04
	Unassignable Area (currently not assigned, plus any remaining Building Loss Factor)	**13th**	0	12	0.0%	D	0.00	D	0.00	0	12	0.0%	D	0.00	D	0.00	0	12	0.0%	D	0.00	D	0.00
	Occupant Void Area, e.g. tenant stairwell	11th	0	11	0.0%	E	0.00	E	0.00	0	11	0.0%	E	0.00	E	0.00	0	11	0.0%	E	0.00	E	0.00
	Interior Encroachments (includes columns, earthquake bracing walls, etc. and enclosing walls)	10th	280	10	0.3%	F	0.01	F	0.01	280	10	0.3%	F	0.01	F	0.01	173	10	0.6%	F	0.01	F	0.01
	Restricted Area (restricted by regulation, code or lease, e.g.: access to door to roof, clearance at outside wall, etc. and enclosing walls)	9th	0	9	0.0%	G	0.00	G	0.00	0	9	0.0%	G	0.00	G	0.00	0	9	0.0%	G	0.00	G	0.00
	Plannable Area = total Inside Gross Area minus the five items below, which is a calculated number.	8th	59,637	8	74.1%	H	1.19	H	1.08	59,637	8	74.1%	H	1.20	H	1.08	24,524	8	80.3%	H	1.37	H	1.05
	(All of the items below are measured to dominant portion of the outside wall.)																						
	Perimeter Encroachments (e.g.: window sill, pilaster)	7th	0	7	0.0%	I	0.00	I	0.00	0	7	0.0%	I	0.00	I	0.00	70	7	0.2%	I	0.00	I	0.00
	Base Building Circulation on empty building, or **Primary Circulation** on an occupied or planned floor	6th	10,249	6	12.7%	J	0.20	J	0.19	10,249	6	12.7%	J	0.21	J	0.19	2,730	6	8.9%	J	0.15	J	0.12
	Service Areas	5th	5,616	5	7.0%	K	0.11	K	0.10	5,616	5	7.0%	K	0.11	K	0.10	1,375	5	4.5%	K	0.08	K	0.06
	Major Vertical Penetrations	4th	2,590	4	3.2%	L	0.05	L	0.05	2,590	4	3.2%	L	0.05	L	0.05	936	4	3.1%	L	0.05	L	0.04
	Void Areas (e.g.: atrium void)	3rd	0	3	0.0%	M	0.00	M	0.00	0	3	0.0%	M	0.00	M	0.00	0	3	0.0%	M	0.00	M	0.00
	Interior Gross Area (to dominant portion of outside wall). Design for blast resistance, so assume glass is less than half of height of wall, so measure to inside face of wall.	2nd	78,092	2	97.0%	N	1.56	N	1.42	78,092	2	97.0%	N	1.57	N	1.42	29,635	2	97.1%	N	1.66	N	1.27
	Exterior Gross Area	1st	80,510	1	100.0%	O	**1.61**	O	**1.46**	80,510	1	100.0%	O	**1.62**	O	**1.46**	30,533	1	100.0%	O	**1.71**	O	**1.31**

Typical Support Spaces are: conference room; copier/printer room; secure file storage; kitchenette/break area; supply/storage area.

TEAG, Inc. - Standard Elements for 4-Star Headquarters - February 23, 2007

- The hypothetical models used a strict discipline to minimize loss of assignable area require building gross area to be 1.61–1.62 times net area.
- Sparkman B9 had a building gross area 1.71 times net area.
- AR 405-70 would require net to gross ratios of 1.31–1.46. It would be likely that buildings designed to the regulation would not be capable of delivering the assignable area needed, thereby forcing compromises to adapt to the available space.
- The rules shown in the Basis for Configuration would result in an outcome about 6% more efficient than Sparkman B9 achieved.

The basis for floor plate configuration was in large part responsible for the efficiency differences between the models and Sparkman B9. The models strictly adhere to principles designed to achieve efficiency, including:

- coordinating office and workstation dimensions with structural and ceiling grids;
- avoiding perimeter encroachments;
- setting the depth of open office bays at a whole number of workstations plus code required width of secondary circulation; and
- maintaining a rectangular grid without curves or elements intersecting at acute angles.

Figure 15.3 illustrates how a lack of coordination between building and furnishing systems can result in unassignable space. For example, the left of middle column has unassignable space adjacent to a column with an awkwardly placed grouping of file cabinets. Additionally, Sparkman B9 was designed with two grids intersecting at a 45° angle. The inefficiency of this shows in the large triangle of

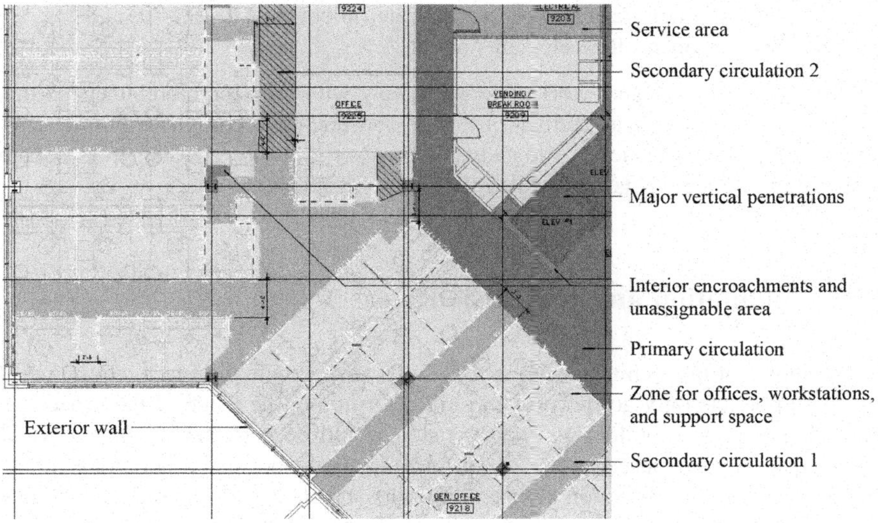

Fig. 15.3 Enlarged partial plan of sparkman B9 space analysis. *Source* Author

secondary circulation near the center of the figure. The measurement and comparison of difference typologies can reveal their costs in terms of additional space and project costs so that building owners can make informed decisions regarding the cost design options that may produce value in entirely difference domains such as aesthetics.

Based on the analysis, the Standard Elements recommended using a multiplier in the range of 1.61–1.70 for net-to-gross calculation of requirements for personnel and for special spaces.

15.6 Application

The project architect and director of interior design from Fentress Architects were interviewed regarding their experience working with the space requirements for the FORSCOM/USARC headquarters building.

The project was procured using a design-build process. Design-build teams responded to solicitations what provided an itemized space program with an allowable gross building area of 456,000 square feet (42,364 m^2). Neither of the interviewees recalled seeing the Standard Elements as part of the requirements package either during or after selection.

While most of the changes that were made after award of the project involved adjacencies and stacking, there were also significant additions and deletions to the scope, such as the addition of the large IT center. The basis for making these changes in scope had clearly been influenced by the Standard Elements. While AR 405-70 was referenced during the discussions of changes in scope, the bases used were:

- 1.25 times the net areas allowed for private offices in AR 405-70, and
- 1.6 times net open office elements added.

Army project managers worked cooperatively with the design architects so that each change in scope was realistic, flexible, and readily accomplished while continuing to stress efficiency. Ultimately, the project grew to 631,000 gross square feet, which the architects felt pushed the limits of the site.

15.7 Validation and Conclusion

The opinions of the architects who successfully worked with the space requirements for these projects only go part way toward determining the value of the process in the delivery of new buildings. The next step would be performance-based evaluations that include observations of space utilization and staff operational effectiveness. Such an evaluation program would have to be structured considering certain cautions.

The first would be the impact of churn over time. As an example, AMC is a relatively stable organization whose work processes change slowly in response to policy and stepwise with changes in information technology utilized in its work. TRADOC on the other hand is a highly dynamic organization with units meeting their mission, dissolving, and being replaced by new units in response to new issues on an ongoing basis. One might expect the operational goals of AMC to be highly similar several years after completion of its headquarters, while for TRADOC the primary operational requirement remaining stable over time would be the need for resiliency to change.

A second caution could be referred to as systemic demands placed on the buildings. While each was conceived as the headquarters for their respective commands, upon occupancy the three buildings became resources in one of the largest building inventories in the world. This is a macro-level equivalent to the prior caution: have the needs of the greater system resulted in increased demands upon the individual buildings that cannot be fairly compared with the original goals for each building?

A well-conceived performance evaluation program could probably account for these dynamics, and would be the next logical step to validate the value of setting space requirements based on the empirical evaluation of existing buildings and modeling as a comprehensive tool.

References

AIA Document D101-1995. (1995). *The architectural area and volume of buildings*. Washington, DC: American Institute of Architects Press.

ANSI Z65.1-1980. (1989). *Standard method for measuring floor area in office buildings*. New York, NY: American National Standards Institute.

ANSI/BOMA Z65.1. (2010). *Office buildings standard methods of measurement*. Washington, DC: BOMA International.

Army Regulation 405-70. (2006). *Utilization of real property*. Washington, DC: US Department of the Army.

E2619M. (2009). *Standard practice for measuring and calculating building loss features that take up floor area in buildings*. W. Conshohocken, PA: ASTM International.

The Environmental Analysis Group. (2007). *Standard elements for 4-star headquarters*. Mobile, AL: US Army Corps of Engineers.

Author Biography

Greg Allen Barker holds a Master of Architecture degree from the University of Illinois, U-C (1983). His undergraduate degree in Architecture is from California Polytechnic State University, SLO (1978). Mr. Barker has over 30 years of experience specializing in predesign consulting including needs assessment, system master planning, physical master planning, facility programming, post occupancy evaluation, and applied research for mostly governmental facilities including postal, correctional, court, public safety, border stations, vocational, and offices.

Mr. Barker is a registered architect in California, Nevada, and Oregon as well as an NCARB certificate holder. The American Collegiate Schools of Architecture, Amoco, and the University of Illinois have all presented Mr. Barker with awards for his research. He served as secretary and vice-chair of the Board of Directors of the Environmental Design Research Association. He is a member of the American Institute of Architects and serves on ASTM International subcommittee E06.25 on Whole Buildings and Facilities.

Part IV
Advances in Evaluation Methods

Preamble

Ihab M.K. Elzeyadi

Over the past 50 years building performance evaluation (BPE) has developed from a concept to a multi-method and multi-disciplinary process that impacts various disciplines of the building industry. With much advances in the way this profession designs, certifies, and appraises buildings; some lingering questions remain, such as: (1) what advances in evaluation procedures have been conceived, (2) how can these be applied to green-certified, sustainable, and buildings designed for well-being, and (3) how can these results be translated to advance the design process and better predict performance? The eight chapters in this section address these questions and provide new insights in building evaluation and performance methods. The goal of most building evaluations is to produce evidence-based of performance that would in-turn support decisions about planning, design, construction, management, and certification of future buildings. The following chapters tackle these problems by arguing that designers should refine evaluation processes and develop advanced comparative methods to be used on complex buildings that push the limits of performance. The chapter authors draw on examples from several successful BPE case studies affecting a variety of high performance building typologies from commercial offices to educational and healthcare settings.

In the first chapter of Part IV, this author discusses a multi-tiered process for the evaluation of high-performance buildings and LEED™ certified ones. This process follows an inductive approach of comparing design simulation and predictions to actual performance on the physical objective level as well as the subjective and symbolic levels. The chapter discusses results from a 36-months longitudinal multi-season POE assessment case study of a laboratory building that combined innovative methods and data collection protocols to measure energy use, indoor environmental quality, and occupant satisfaction providing continuous feedback loops that improve building design, delivery, and operation.

Becker argues for expanding the role of post-occupancy evaluation (POE) within the building performance evaluation cycle by accepting various forms of data collection techniques, such as qualitative, ethnographic, and anecdotal evidence and observations that can be continuously and quickly collected by practitioners in the field. He advocates POEs as a simple diagnostic tool carried out by facility managers and used within organizations for their continuous improvement. As in medicine, these can serve as clinical experiences that aid day-to-day decision-making and provide basis for more rigorous evaluation research that might be followed out by collaborative teams of academics and practitioners.

Parshall and Fonseca's chapter describes a comprehensive and feasible method that addresses scope, budget, and time commitments for an architectural firm's clients. They describe a five-step and four considerations process that suit many purposes. This approach goes beyond that of traditional architectural programming and evaluation to include risk management and environmental health concerns. An important consideration in the development of this method is economy of effort while achieving the greatest value for their clients.

Marans and Callewaert provide an interesting approach to evaluate a cultural change program aimed at promoting sustainability behavior and attitudes at a university campus. The evaluation is intended to inform and provide feedback to campus stakeholders to better the day-to-day operations and programs and to serve as a model for the use of behavioral research in addressing critical environmental issues within universities and other settings.

Mallory-Hill and Gorgolewski examine the gaps between predicted and achieved performance of nine green-certified buildings located across Canada. They assessed key performance indicators for green buildings under major categories of occupancy, energy use, water use, and indoor environmental quality. Data collection involved both qualitative and quantitative techniques.

Focusing on university campus architecture and planning assessments, Bain proposes a valuable use of POEs as a tool to help building owners make permanent decisions about the future of existing assets rather than looking back at how they performed after occupation. She proposes a Functional Assessment protocol with the use of a wide range of data gathering techniques to determine the future value of specific buildings for the long term. Best situated within the campus planning process, this type of assessment determines solutions to current issues and aid planning strategies for the future.

In their chapter Kato, Mori, and Kato, report on a comprehensive BPE of pediatric intensive care units in hospitals in Japan, along with comparing situations with those of the USA. Data collection was carried out in 13 hospitals using questionnaire survey, site visits, and content analysis of 30 blogs written by parents having children patients. They concluded their chapter by illustrating how BPE supports the planning, design, and management methodology in Japan's healthcare system.

Last but not least, Fay discusses the full cycle of a POE process by proposing methods of planning and conducting a POE, actively reporting the findings, and applying outcomes through the use of a collaborative design charrette. The design charrette, as a dissemination and application tool, presents an opportunity to engage with research that can inform future designs while also developing familiarity with the POE framework, methodologies, and building occupants perspectives.

Chapter 16
A Comparative Analysis of Predictive and Actual Performance of High Performance LEED™ Buildings

Ihab M.K. Elzeyadi

16.1 Sustainable Building Performance: A Systems Approach

There seems to be an implicit, and sometimes explicit, view that human comfort and building performance occurs in separate envelopes. Previous building performance studies (e.g. Elzeyadi 2012, 2015a, b) show that human satisfaction and comfort are multi-faceted and are affected by the fourfold components of the environment in its physical, physiological, psychological, and social attributes. While one can assume that individuals are impacted by the environment in different ways, their general achievement of overall satisfaction as well as how they operate and interact with their building are the result of both the process and product of their environment. The occupants' overall appraisal of its ambience is based on both how it is currently performing as well as their expectations and predictions of its performance (Preiser et al. 1988).

In general for sustainable buildings—and particularly LEED™ certified ones—the process of design, construction, and building performance need to be evaluated simultaneously. Many previous studies have focused on energy and resource performance of LEED™ buildings (Scofield 2009, 2013; Stoppel and Leite 2013). Other studies compared indoor environmental quality and occupant satisfaction of LEED™ buildings as compared to conventional ones (e.g. Altomonte and Schiavon 2013; Schiavon and Altomonte 2013, 2014; Newsham et al. 2009, 2013; Abbaszadeh et al. 2006). These studies have mostly focused on evaluating LEED™ buildings as products relying on either their actual consumption metrics or post-occupancy evaluation (POE) surveys of their occupants. Rarely have the process, goals, and predictive performance of these buildings been considered in the

I.M.K. Elzeyadi (✉)
Department of Architecture, School of Architecture & Allied Arts, 1206 University of Oregon, Eugene, OR 97403-1206, USA
e-mail: ihab@uoregon.edu

W.F.E. Preiser et al. (eds.), *Building Performance Evaluation*,
DOI 10.1007/978-3-319-56862-1_16

evaluation or compared to how they perform against their design intent. To that extent, both their physical performance, resource consumption, and occupant satisfaction of their indoor environmental quality need to be evaluated in a holistic approach. LEED™ certified buildings offer excellent cases to test the design process/product hypothesis due to the substantial documentation that they require to achieve their certification levels. LEED™ buildings differ from green buildings due to their substantial documentation and certification process that account for various metrics under different categories. While not all LEED™ certified buildings can be high performance and many non-LEED™ certified buildings can outperform certified ones, this chapter will discuss building performance evaluation of LEED™ buildings for available access to documentation and certification process (Turner and Frankel 2008). For example, LEED™ buildings require detailed reports on their credits achieved, LEED ™ scorecard, design charrettes, resource consumption simulation, and energy reduction prediction below local energy codes. In addition, the credits achieved or not achieved by a LEED™ certified building forces the design team to investigate sustainability on multiple scales from the site to the building and the indoor environment scales including materials used. It provides an ideal building evaluation paradigm to question how these certified buildings perform according to their hypothesized expectations, design process, as well as their final product.

16.2 Sustainable Building Performance as Place Experience

Building on early definitions of place experience and performance, sustainable building performance is defined as all the qualities of the place that are collectively perceived and evaluated by its occupants as affecting their needs, wants, and the tasks they perform, with minimal impacts on the resources of the global environment in both product and process. Following a systems perspective, qualities of a sustainable place experience can be grouped in intellectual taxonomies according to their levels of meaning to occupants (Rapoport 1988). This chapter adopts a previously developed building performance assessment framework developed by the author (Elzeyadi, in press) to conceptualize both building performance and occupants' experience in LEED™ buildings as a system. This conceptualization acknowledges the complex systems of interactions between people and their indoor/outdoor environment on three scales: (1) the micro-scale related to the individual and their indoor environmental quality (IEQ), (2) the mini-scale of groups and their building, and (3) the macro-scale, centered on the urban setting and site factors. The interaction of these various levels define the environment as experienced by the user following a nested model of place experiential qualities broken down to three qualities, instrumental, latent, and symbolic (see Fig. 16.1).

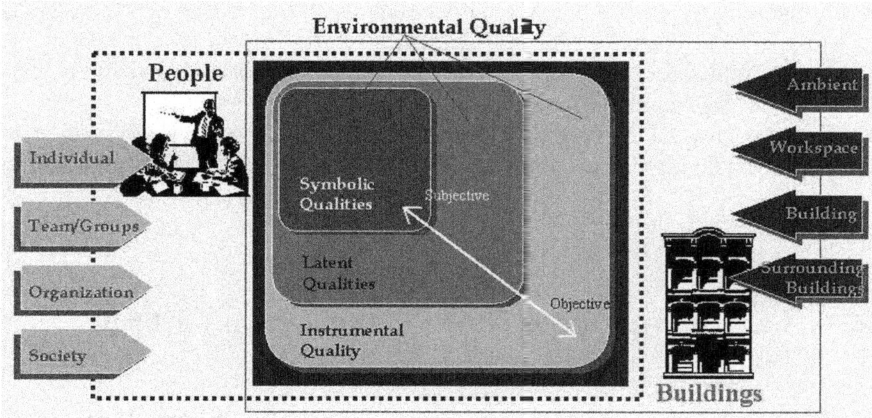

Fig. 16.1 Building performance and place experience. *Source* Author (2003)

1. Instrumental Level:
The instrumental level of a building performance represents everyday qualities that enable individuals and groups to perform their tasks, behave, and act appropriately and predictably in place (Preiser 2009; Preiser and Wang 2012). For example:

- On the micro-scale, it include qualities such as thermal comfort, indoor air quality, lighting, visual comfort, noise, and auditory comfort.
- On the mini-scale, it represents building resource performance, energy efficiency, accessibility, universal design, and operation practices.
- On the macro-scale it represents commuting behavior, walk score, street connectivity index, solar envelope, etc.

2. Latent Level:
The latent level of building performance represents subjective qualities of an environment that engages individuals and groups in place as well as provide subjective value to both occupants and their settings (Rapoport 1990). For example:

- On the micro-scale, it include qualities such as personal space, control, privacy, personalization, and access to views of nature.
- On the mini-scale, it represents building legibility, way finding, indoor décor, safety and security, energy, and recycling behavior.
- On the macro-scale it represents connections to the outdoors, local amenities within walking distance, community engagement, and access to the building.

3. Symbolic Level:
The symbolic level of building performance represents qualities of an environment that are related to higher-level meanings of symbols and artifacts that correspond to beliefs and combine both the objective and subjective dimensions of place (Zimring and Peatross 1997). For Example:

- On the micro-scale, it include qualities such as status, value, myths, aesthetics and poetics of place.
- On the mini-scale, it represents building image, transparency, visibility, performance dashboards of resource consumption, etc.
- On the macro-scale, it represents organizational value system, neighborhood fit, eco-district connectivity, livability, inclusive urbanism, and design for diversity factors.

16.3 Case Study: POE of a LEED™ Platinum Certified Building

The Robert and Beverly Lewis Integrative Science Building (LISB) brings world-class researchers together under one roof from a range of different disciplines. The \$65 million educational facility, which opened in October 2012 in Eugene, Oregon — USA, is home to strategic research clusters centered on interdisciplinary and integrative research missions. The 103,000 S.F. (9569 m^2) facility unites the sciences by stitching the adjacent science buildings into one complex (see Fig. 16.2). The ambitious goals of LISB are reflected in its forward-thinking design. The building employs a central daylit atrium design with extensive skylighting to connect an open layout of dry and wet labs, educational spaces, meeting rooms, and impromptu gathering spaces that builds on ideas of flexible and transparent inter-connected spaces.

Creating a science building with sustainable features was an imperative in designing the Lewis Building. This building is predicted to use 58 percent less energy than conventionally designed buildings of similar size and function. Energy savings features include natural ventilation in non-lab spaces, solar shading, daylight harvesting, night flush cooling, variable flow chemical fume hoods equipped with automatic sashes that close when not in use, and the extraction of heat from an adjacent utility tunnel. To reduce usage of potable water, the building reclaims

Fig. 16.2 The Lewis Integrative Sciences Building (LISB). *Source* Elzeyadi (2015a)

reverse osmosis treated water from a neighboring zebra fish research facility and uses this water to flush all urinals and toilets. All of the storm water on site is also collected and treated, and 28 solar hot water panels on the rooftop heat all domestic hot water. Operable windows with sensor-signals controls, daylighting, and access to views of nature are among the various sustainable design strategies used to earn the facility a LEED™ Platinum certification.

16.4 Comprehensive Building Performance Protocol for LEED™ Buildings

To validate the proposed performance evaluation framework (see Fig. 16.1), a comprehensive building performance evaluation protocol (BPEP) was carried out for 36 consecutive months for the LISB. The protocol assessed building energy and resource performance, indoor environmental quality (IEQ), and occupants' perception of comfort for a number of critical spaces in the case study building (see Fig. 16.3). The research team engaged with the design team, facility managers, and building users for 36 months from post-construction analysis to a detailed Post-Occupancy Performance Evaluation (POPE). This included:

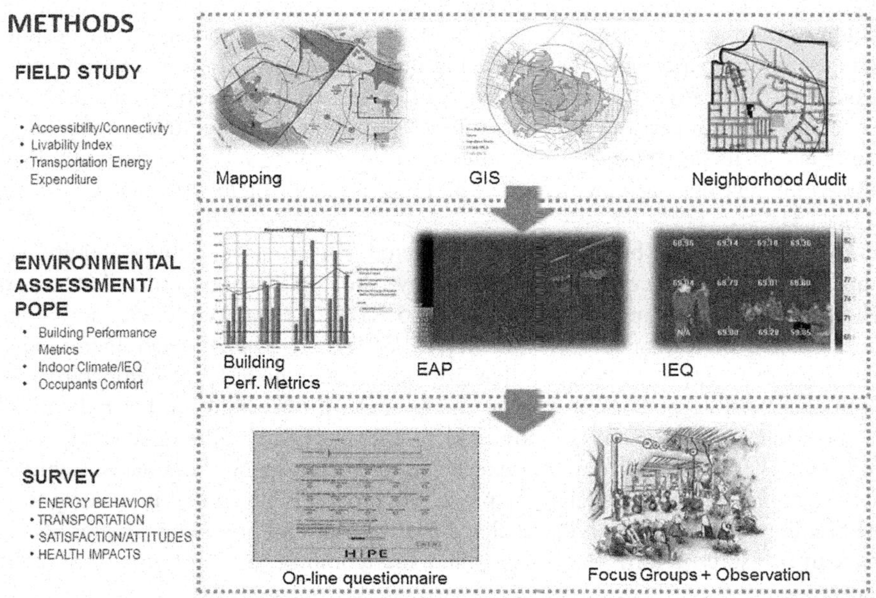

Fig. 16.3 A comprehensive POPE design and methods employed for the study. *Source* Author

1. Field study to assess the building and its context on the urban macro-scale for the instrumental, latent, and symbolic qualities.
2. Environmental assessment and POPE on the building scale, which included assessing both predicted and actual use of energy, water, gas, and steam, IEQ spatial analysis based on LEED™ credit achieved, and physical comfort metrics of the building thermal, visual, and acoustical qualities.
3. Occupants' surveys by employing questionnaires, interviews, and focus groups with the building users and traditional POE tools (see the Introduction by Preiser, Hardy, & Schramm). Spatial analysis and visualization of IEQ assessments relating the qualitative phenomenological and quantitative performance impacts of the studied spaces.

16.5 The Building Energy Performance

The building systems are extensively commissioned to ensure that they meet the design goals and predicted resource consumption. Following a three year post-occupancy monitoring procedure and additional commissioning, results show that the building was below its targeted predicted energy consumption for the first year by 2–3%, exceeding its predicted energy performance by 15% for year two and slightly above its predicted performance by 2–3% for year three. On average, the building energy performance is well within its predicted target levels despite changes in occupancy and malfunctioning of some systems controls over the past two years. This stresses the importance of continuous monitoring and analysis procedure for high performance and LEED™ buildings. The building is considered exemplary in its performance, exceeding most buildings in its categories with 36% energy savings over a typical code complying building of its size and type (see Fig. 16.4).

By analyzing the building energy metered performance without additional adjustments for any central plant efficiency, the total energy utilization index (EUI) of the building in year three was 169.5 KBTU/SF/Yr. While 2% more than the predicted performance of 165.68 KBTU/SF/Yr., it is 40% better than recent buildings of its size and type as well as 30% better than the baseline comparative of a building built to the current Oregon energy code. By applying further reductions in operational management and power plant efficiency, the building would further improve its energy efficiency and exceed its predicted energy efficiency savings. The 36 month performance analysis procedure proved to be crucial in making sure the building is performing to its predicted EUI metrics and providing a better bench mark to measure performance success and shortcomings both in terms of process and product.

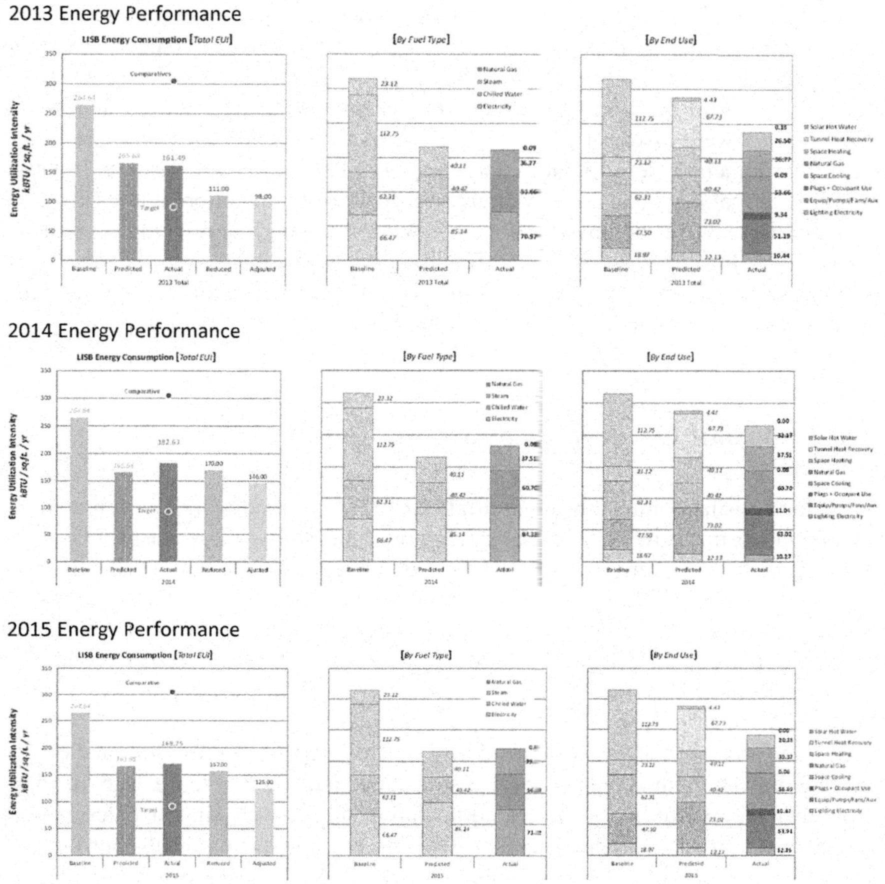

Fig. 16.4 Energy and resource consumption in EUI of LISB over 36 months period. *Source* Author

16.6 IEQ Assessments

Despite the importance of energy and resources savings of LEED™ buildings, proven impacts on occupants' health and performance from improved IEQ will be a game changer in adopting high performance buildings strategies and providing a big driver to design teams to select strategies based on proven evidence and human impacts (Elzeyadi 2015b). These impacts could vary from reduced Sick Building Syndrome (SBS) symptoms to increase indoor comfort of occupants due to better lighting, thermal conditions, and air quality. Evidence to support these claims has been mixed. Existing studies of occupant satisfaction and comfort in LEED™ buildings show high variability, with some buildings rated very positively and others having modest comfort and satisfaction levels (Elzeyadi 2012; Hwang et al. 2009).

A detailed IEQ assessment procedure that assesses visual, thermal, and spatial comfort in the building from the physical building performance perspective and the latent occupants' perception of comfort using questionnaire and interviews was implemented. Multi-comfort parameters and metrics with the thermal and visual environments were assessed and analyzed of the different spaces inside the building. Environmental sensors and data loggers measuring temperature, relative humidity, air velocity, and air movement stratified across the different floor levels of the buildings, were deployed over the winter, spring, and summer seasons respectively. In addition, infra-red (IR) imagery was taken over the course of sampled seasonal days for the occupants workstations to document surface temperature and mean radiant temperature indices over the study period.

16.6.1 Visual Comfort Analysis

To evaluate visual comfort inside multiple LEED™ certified work environments, it is essential to measure both spatial daylighting autonomy (sDA) distribution and useful daylight autonomy (uDA300-1000) as metrics of illumination effectiveness of the building. In addition, it is essential to assess glare and visual asymmetry in the field of vision for different seated and standing positions that represents the occupants' patterns of use for the building. Multiple glare analysis metrics should be calculated such as Daylight Glare Probability (DGP), Daylight Glare Index (DGI), and Visual Comfort Probability (VCP).

A comparative analysis reveals how different spatial configurations of the building impact both daylight level distribution and glare. Figure 16.5 displays visual analysis performance of two comparative spaces with similar exterior window to wall ratios of approximately 70% for exterior facing walls but slightly different space proportions and window geometries. Space proportions, orientation, and window geometry show substantial impacts on daylighting performance and glare management of both spaces. The 1:2 plan proportion of space 1, which is elongated in the North South axis and facing east with internal windows to the atrium provided better daylighting distribution.

16.6.2 Thermal Comfort Analysis

Discomfort with the thermal environment is one of the most reported dissatisfactions by occupants in work environments. In a recent national survey of 400 office workers, 73% of the surveyed occupants felt too cold at some point while 63% of the same occupants felt too warm (Elzeyadi 2012). It is important to both assess not only the occupants' perceived thermal comfort but also the actual physical indoor climate parameters according to ASHRAE-55 standards for thermal comfort. For this case study, various environmental sensors and data loggers were deployed

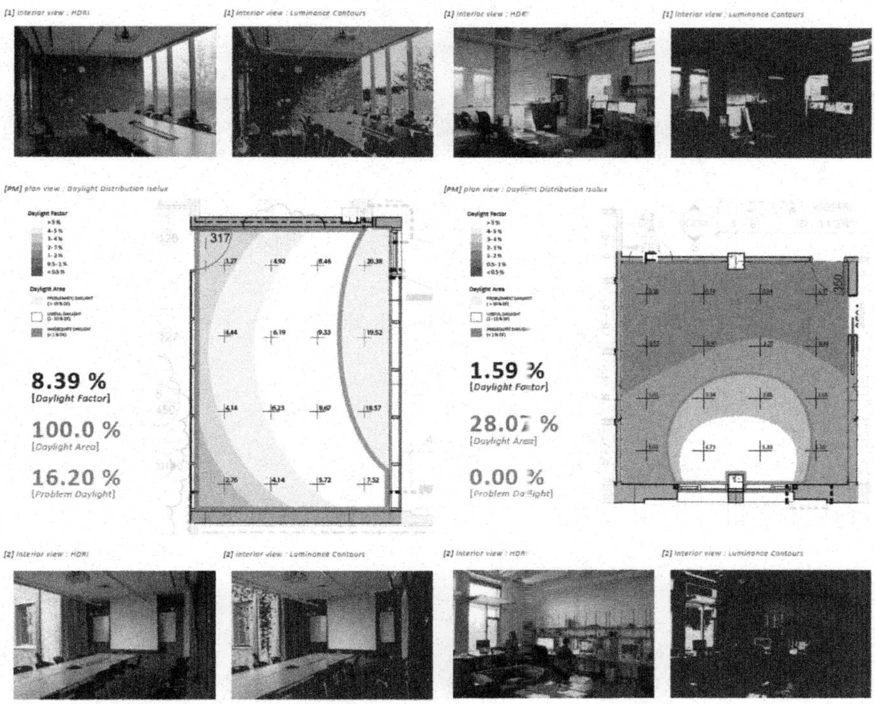

Fig. 16.5 Daylighting and glare analysis of work environments over four seasons. *Source* Author

throughout the building. These sensors recorded temperature, relative humidity, air velocity, and air movement stratified across the different floor levels of the buildings over fall, winter, spring, and summer seasons respectively.

A summary of the indoor thermal comfort indices for multiple spaces in the studied buildings reveal multiple discrepancies of thermal comfort and occupant satisfaction. Figure 16.6 shows a comparative analysis between two spaces; Space 1 (Bldg. 1 = wet labs) and Space 2 (Bldg. 2 = dry labs). While Space 1, maintained average indoor conditions within the ASHRAE-55 thermal comfort standards of the occupied hours, yet the percentage of time within the adaptive comfort zone tend to be more than Space 2. Further analysis performed by plotting thermal comfort parameters on the psychrometric chart reveals that although Space 2 was able to maintain the majority of hours within the green zone stipulated by ASHRAE-55 recommendations for thermal comfort. This resulted in indoor diversity of climatic conditions that offered occupants thermal options to accommodate for clothing and metabolic levels within the space.

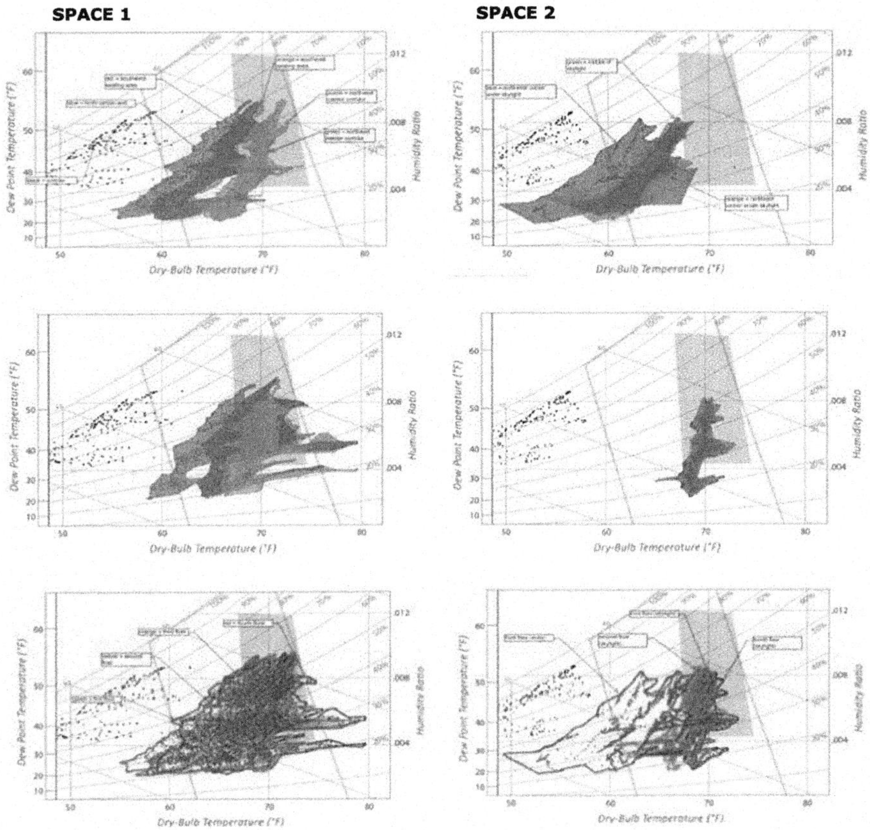

Fig. 16.6 Thermal comfort analysis example of spaces for the three different seasons (Winter, Spring, and Summer; note Spring and Fall data are similar). *Source* Author

16.6.3 Occupants Satisfaction Analysis

In addition to physical assessments and visualization of the multi-comfort metrics of the environment, an occupant Space Performance and Evaluation Questionnaire (SPEQ) was administered to solicit employees' satisfaction of multiple spaces within the building. Questions were added to address specific issues such as thermal and visual comfort of the various spaces. An average of 33% of the employees in the wet labs portion of the building (Building 1) and 42% of the employees in dry labs area of the building (Building 2) completed the questionnaire. Preliminary results of the survey show strong occupants satisfaction with the environmental agenda and LEED™ certification of both spaces. More than 75% of occupants in both buildings agree about the importance to work in a building that is environmentally conscious.

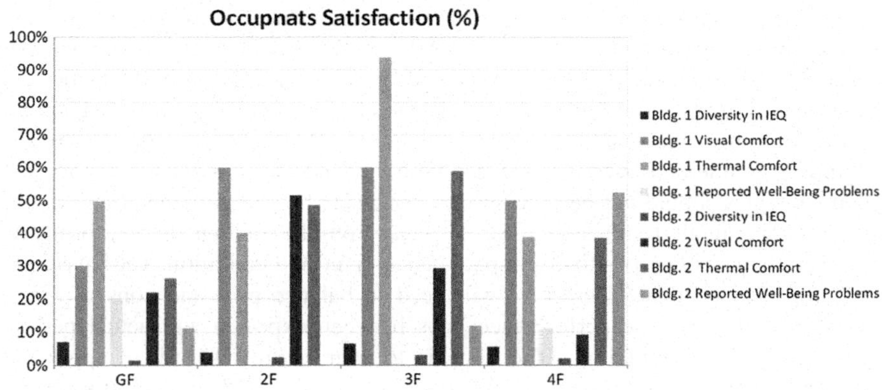

Fig. 16.7 Summary of occupants satisfaction with Indoor Environmental Quality for different floors – Bldg. 1 (Wet Labs) and Bldg. 2 (Dry Labs). *Source* Author

While both studied areas of the buildings exceeded occupants' expectations and were perceived to have reported positive impacts on their productivity, the perceptions do not confirm with the designers expectations of the building and LEED™ credits achieved. Figure 16.7 summarizes overall occupants' satisfaction with visual, thermal, and spatial comfort attributes of the spaces under study, as well as the perception of diverse climatic and lighting conditions within the comfort range that provided occupants with choice and engagement in their settings. In general, IEQ parameters for the spaces facing both the atrium and the exterior envelope, providing dual orientation, were positively perceived by the occupants whereas spaces with singular orientation or south facing orientations only were negatively perceived due to excess glare and thermal shifts in temperatures between the perimeter area and the internal areas of the same space.

16.7 Conclusions

To evaluate the effectiveness of IEQ parameters on multi-comfort impacts in LEED™ buildings, designers need to establish clear performance goals that acknowledge both the physical performance as well as the impacts of the selected green design strategies on occupants' well-being and comfort. An established process to ensure fine tuning of the systems and engaging the occupants in positive energy behavior that maximize indoor comfort and satisfaction is essential. Continuous post-occupancy performance evaluations (POPE) of both process expectations and product performance should be established in green and LEED™ buildings to provide necessary feedback loops for designers, building managers, and occupants. Such feedback loops help engage occupants in the building management and ensure that the building as designed performs to its predicted goals.

The main objective of this chapter is to provide detailed as well as context specific information for POPE of LEED™ buildings from a comprehensive approach. By establishing a detailed procedure of assessing the building performance against its predicted goals from both the physical metrics and the perceptual levels of comfort, the chapter provides an evidence-based guide to future designs that moves architecture towards evidence-based practice. It is important to note that green strategies should not be perceived as "one size fits all" in general and might not be suitable in all design situations. It is clear from the findings that performance of some strategies can positively impact behavior in one condition, yet have negative impacts on others. Designers will need to balance pros and cons of green systems as they manage performance gaps between expected and actual performance of LEED™ buildings. The hope is to spur future research to apply the proposed methods and contribute to a better understanding of the nature of POPE in LEED™ buildings beyond the fascination of the plaques they achieve.

References

Abbaszadeh, S., Zagreus, L., Lehrer, D., & Huizenga, C. (2006). Occupant satisfaction with indoor environmental quality in green buildings. In E. de Oliveira, M. da Silva, & J. Rosado Pinto (Eds.), *Healthy buildings 2006: Creating a healthy indoor environment for people* (Vol. III, pp. 365–370). Lisbon: Universidade do Porto, Portugal.

Altomonte, S., & Schiavon, S. (2013). Occupant satisfaction in LEED and non-LEED certified buildings. *Building and Environment, 68,* 66–76.

Elzeyadi, I. (2003). Environmental quality - shaping places for people: A systemic framework for conceptualizing people and their workplaces. In J. Robinson, K. Harder, H. Pick, & V. Singh (Eds.), *EDRA 34th proceedings: Shaping places for people* (pp. 71–79). St. Paul, MN: Environmental Design Research Association.

Elzeyadi, I. (2012). Post-occupancy evaluation: A design, operations and performance assessment of a LEED Platinum building. *World Health Design Journal*, 60–69.

Elzeyadi, I. (2015a). A tribute to performance arrows: Designing for better indoor environmental quality. In C. Pyke et al. (Eds.), *Thought and leadership in green buildings research. Greenbuild 2015: Monumental Green* (pp. 4–11).Washington, DC: USGBC Press, accessed on August 15th, 2016. http://gocms.hanleywoodexhibitions.com/GAL/Shows%5CGreenbuild%5C2015%5CPDF/GRB15_RESEARCH%20111115.pdf.

Elzeyadi, I. (2015b). A green lesson: Measuring the impacts of LEED certification credits on people, planet and profit of K-12 schools. In N. Fernando & G. Baker (Eds.), *EDRA 2015 – brainSTORM: Dynamic interactions of environment-behavior and neuroscience* (pp. 48–59). St. Paul, MN: Environmental Design Research Association.

Elzeyadi, I. (in press). Multi-performance spaces: Assessing the impacts of an innovative LEED platinum educational complex on the triple bottom-line. In W. Preiser, A. Hardy, & J. Wilhelm (Eds.), *Adaptive architecture: Changing parameters and practice.* London: Routledge.

Hwang, R., Lin, T., Liang, H., Yang, K., & Yeh, T. (2009). Additive model for thermal comfort generated by matrix experiment using orthogonal array. *Building and Environment, 44,* 1730–1739.

Newsham, G. R., Birt, B. J., Arsenault, C., Thompson, A. J. L., Veitch, J. A., Mancini, S., et al. (2013). Do 'green' buildings have better indoor environments? New evidence. *Building Research and Information, 41*(4), 415–434.

Newsham, G. R., Mancini, S., & Birt, B. J. (2009). Do LEED-certified buildings save energy? Yes, but…. *Energy and Buildings, 41*(8), 897–905.
Preiser, W. F. E. (2009). The triad of programming, post-occupancy evaluation (POE) and universal design: Toward continuous quality improvement. In A. Taylor (Ed.), *Linking architecture and education: sustainable design of learning environments*. Albuquerque, NM: University of New Mexico Press.
Preiser, W. F. E., Rabinowitz, H. Z., & White, E. T. (1988). *Post-occupancy evaluation*. New York: Van Nostrand Reinhold. (republished in 2015, London: Routledge).
Preiser, W. F. E., & Wang, X. (2012). Quantitative (GIS) and qualitative (BPE) assessments of library performance. In S. Mallory-Hill, W. F. E. Preiser, & C. Watson (Eds.), *Enhancing building performance* (pp. 179–190). Oxford: Wiley-Blackwell.
Rapoport, A. (1988). Levels of meaning in the built environment. In F. Poyatos (Ed.), *Perspectives on non verbal communication* (pp. 317–336). Toronto: C.J. Hogrefe.
Rapoport, A. (1990). Environmental quality and environmental quality profiles. In N. Wilkinson (Ed.), *Quality in the built environment* (pp. 75–83). New Castle: Urban International Press.
Schiavon, S., & Altomonte, S. (2014). Influence of factors unrelated to environmental quality on occupant satisfaction in LEED and non-LEED certified buildings. *Building and Environment, 77,* 148–159.
Schiavon, S., & Altomonte, S. (2013). Occupant satisfaction in LEED and non-LEED certified buildings. *Buildings and Environment, 68,* 66–76.
Scofield, J. (2009). Do LEED-certified buildings save energy? Not really. *Energy and Buildings, 41,* 1386–1390.
Scofield, J. (2013). Efficacy of LEED-certification in reducing energy consumption and greenhouse gas emission for large New York City ofice buildings. *Energy and Buildings, 67,* 517–524.
Stoppel, C. M., & Leite, F. (2013). Evaluating building energy model performance of LEED buildings: Identifying potential sources of error through aggregate analysis. *Energy and Buildings, 65,* 185–196.
Turner, C. & Frankel, M. (2008). *Energy performance of LEED for new construction buildings*. White Salmon, WA: New Buildings Institute, Electronic Report http://www.usgbc.org/Docs/Archive/General/Docs3930.pdf. Accessed on August 15, 2016.
Zimring, G., & Peatross, D. (1997). Cultural aspects of workplace organization and space. In G. Moore & R. Marans (Eds.), *Advances in environment, behavior, and design*. New York: Plenum Press.

Author Biography

Ihab M. K. Elzeyadi is a professor of architecture and the director of the High Performance Environments laboratory (HiPE) and Façade Innovative Technologies (FIT) testing facility at the School of Architecture & Allied Arts, University of Oregon. Dr. Elzeyadi has been engaged in the design, construction, and research of high-performance buildings for more than 25 years. He has conducted post-occupancy evaluations, building performance assessments, and field studies of more than 75 buildings, 30 of them are LEED™ rated. Professor Elzeyadi teaches design studios and lectures in building performance and metrics; integrated design and delivery; indoor environmental quality; as well as environmental impacts of buildings on human performance and health.

Dr. Elzeyadi has conducted grant-supported research on the relationship between people and buildings including impacts of the physical environment on health, productivity, and well-being as they relate to sustainable design strategies in commercial and educational environments. His studies produced evidence-based design guidelines and design-assistance services on various commercial projects with an emphasis on energy and resource effective design. He just completed a number of research projects investigating cost and financial benefits of green and LEED™ educational environments, livable communities' physical infra-structure, and the Green Classroom Toolbox Project for energy retrofits of existing schools. He is published in journals and international conference proceedings as well as being a featured speaker at various schools of architecture and international conferences around the world.

Chapter 17
Post-Occupancy Evaluation: Research Paradigm or Diagnostic Tool

Franklin Becker

17.1 Introduction

Post-Occupancy Evaluation (POE), now incorporated into Building Performance Evaluation (BPE), is a form of systematic inquiry intended to discover and document how a building, product, or service has worked for its intended use. It is an invaluable tool for increasing the likelihood that time, money, and effort invested in such endeavors achieves anticipated benefits for targeted users. This chapter argues that the role of POE in improving building performance has been inadvertently limited by trying to make POE an academically acceptable form of rigorous evaluation research. Of more value is developing POE as a diagnostic tool, essentially a clinical technique, conducted and used within organizations for their own continuous improvement, and as an important management tool for improving the business value of capital improvements.

Post-occupancy evaluation (POE), considered as less extensive and rigorous than what Preiser and Vischer (2005) call building performance evaluation (BPE), developed over the past forty years as a technique by which design practitioners could learn which aspects of their design worked well, and which did not, with the lessons learned fed forward into subsequent projects (Preiser and Vischer 2005; Zimmerman and Martin 2001). By doing so, presumably the cost of design development could be lessened; occupant satisfaction, comfort, and performance could be enhanced; and organizations could get better value for money from their facilities. Members of both the design professions and social scientists (Sanoff 2000; Cooper 1975; Marcus and Barnes 1999; Sommer 1969; Altman 1976) participated in studies assessing environmental performance, and in proposing alternative design approaches which more seriously took into consideration what Zeisel

F. Becker (✉)
Cornell University, Martha Van Rensselaer Hall, 116 Reservoir Ave, Ithaca, NY 14850, USA
e-mail: fdb2@cornell.edu

W.F.E. Preiser et al. (eds.), *Building Performance Evaluation*,
DOI 10.1007/978-3-319-56862-1_17

Table 17.1 Characteristics of POE and environment-behavior research. *Source* Author

POE	Environment and behavior
Semi-systematic	Systematic
Sampling casual	Sampling rigorous
Minimal research design	Explicit research design
Broad focus	Narrow focus
Satisfaction key measure	Attitude and behavior
Serendipitous with time and place	Deliberate selection of time and place
Minimal data analysis	Elaborate data analysis
Single-site case study	Comparative analysis

(1974) called the "non-paying client," as in the case of people such as prisoners, students, and employees. That is, the person who occupies a facility without necessarily owning or paying for it.

The POEs that emerged contained a number of characteristics (see Table 17.1). Many tended to be serendipitous. They were selected more because of convenience and proximity, than because they provided the opportunity to test some specific idea or hypothesis. Many relied on surveys of occupants, with the exception of some academic-oriented studies, but rigorous population sampling procedures were rarely followed, and statistical analysis of results was often minimal or non-existent. Clearly formulated research designs, particularly those involving comparisons among groups experiencing different environmental constraints and opportunities, were rare. Single-site studies were the rule, but unlike more traditional case studies, they rarely provided in-depth analyses of a case. Little attempt was made to understand the case as a complex eco-system shaped by the interplay of design, social, technological, organizational, and cultural factors. Contextual factors were rarely explicated, such as how long the occupants had been using the space, the circumstances under which they came to use it, e.g., compulsory or voluntary, or the nature of their relations with the "paying clients". More typically, studies reported users' self-reported satisfaction with dozens of aspects of the built environment, ranging from site planning and design to the interior arrangements and size of rooms, lighting, signage, storage, and privacy (Becker 1974; Brill 1984; Marans and Spreckelmeyer 1982; Shepley et al. 2009).

17.2 Environment-Behavior Research

Today, these research endeavors would be considered part of the Evidence-Based Design (EBD) movement. EBD, although widely associated with healthcare design, is applicable to all settings. It has been defined as "a process for the conscientious, explicit, and judicious use of current best evidence from research and practice in making critical decisions, together with an informed client (Stichler and Hamilton 2008, p. 3). The fundamental premise is straightforward: better designed facilities, ones that are more likely to support valued outcomes, will result from using the evidence generated by high quality, formalized and rigorous research.

EBD and environment-behavior research (EBR) tend to have a different set of characteristics (see Table 17.1) than those associated with POEs. In general, they are more scientifically rigorous. Most have clearly formulated research designs and follow accepted canons of population sampling. Statistical analyses of data are used to test for differences among groups or treatments, and to test specific research questions and hypotheses. The research focuses on one or two variables selected for theoretical or other deliberate reasons. Behavioral measures, e.g., actual seating locations, communication patterns, avoidance behavior, as well as interviews and questionnaires, are used to assess the effect of environmental variables (Reeves and Lewin 2004; Sundstrom et al. 1980; Ulrich 1984).

Early environment-behavior studies, Sommer (1969) for example, *observed* the relationship between environmental factors like seating distance and interaction *behavior,* i.e., what people did in actual environments. Over time that early work, based largely on direct observation, evolved into studies that relied more extensively on "users" *subjective* self-reports using surveys, to evaluate the facility design. That seemingly subtle shift in what is measured is, in fact, extremely important. For in many settings, such as the workplace, user satisfaction may not be as highly valued as specific behaviors. That is especially true in the healthcare context, where outcomes such as medical errors, patient falls, and hospital-acquired infection carry more weight than satisfaction ratings. However, with changes driven by Obamacare, ratings of satisfaction have become more important because they affect federal reimbursement rates.

17.3 Why Distinguish Between Practice-Based and Academic-Based Research?

EBR intended for peer-reviewed publication generally takes from 1 to 3 years from initiation to publication. Therefore, it is of little value in the context of the much shorter timeframe in which decisions are made for a specific project. With its emphasis on rigorous scientific methodology, it discourages the practitioner from becoming involved in briefer, simpler, and less scientific, but still useful, efforts that can influence design decisions for a specific project. Preiser, Rabinowitz, and White (1988) call this form of facility evaluation the "diagnostic" POE. The author calls it "practice-based" research, in contrast to "academic-based" research. Both are part of the EBD movement, but they are different.

The conundrum for practitioners is that the more rigorous environment-behavior published evidence is never complete. In the conclusion of every academic study the need for further research is explicitly stated. It is also always difficult to interpret findings in the context of one's own specific project and organization, in part because little information about the context in which the data was collected is provided. Then what does the practitioner who wants to base decisions on more than personal experience do? One answer is to supplement the published academic research with their own small, fast *project-specific* diagnostic studies that provide timely insight grounded in empirical data.

The key difference between "academic" and "practice-based" research is not where each is done, or who does it. The difference is in the scale and sophistication of research design, data collection, and analysis. Practice-based research draws on the same data collection techniques—surveys, observation, interviews, and institutional data– used in longer term and more rigorous academic-based research. And it may involve experimentation, like collecting data before and after some small-scale intervention in the same setting where the results will be interpreted and applied. The evidence and data produced in these practice-based research projects is not sufficiently extensive to warrant publication in a peer-reviewed journal. But it is far better than just guessing or talking informally with a few people about how well a process or space is working.

17.4 Quantitative and Qualitative Data

Quantitative data are viewed as more rigorous, and therefore, more credible than qualitative evidence. Because it is fast and easy to collect, especially in the form of surveys, ethnographic approaches that use structured observational methods are used to a limited extent in both EBD and POE research (Sandelowski 2000). That is unfortunate. In combination, each provides a unique form of evidence that provides a more holistic view of a situation than each does alone.

Nate Silver (2012), whose work with data analytics was the basis for the movie *Moneyball*, revolutionized baseball scouting by exploiting detailed analyses of baseball's voluminous statistics to predict which players are likely to be the best prospects. Scouts, traditional baseball players and fans worried that the nerdy "statheads" mining baseball's seemingly inexhaustible supply of data would render them baseball's dinosaurs. Not so, Silver wrote that "The fuel of any ranking system is information—and being able to look at both scouting and statistical information means that you have more fuel" (p. 91). He went on to say that rigor and discipline is achieved by the way in which "the organization processes the information it collects, and not in declaring certain types of information off-limits. [...] The key to making a good forecast [...] is not in limiting yourself to quantitative information" (p. 93).

17.5 Ethnography

In the context of healthcare design, Rick Iedema's (Iedema 2003; Long et al. 2007; Carroll et al. 2008) and other healthcare ethnographers' (Coiera and Tombs 1998; Johnson and Barach 2008) work around communication issues in hospitals has shown just how valuable qualitative, ethnographic data can be in understanding the complexity of any eco-*system*. Using ethnographic methods and video observation, Debbie Long (Long et al. 2007) one of Iedema's students, discovered that communication between doctors and nurses occurred primarily in hospital corridors. Understanding communication patterns is especially important in healthcare because communication problems are a leading cause of medical errors and incidents. Where communication occurs has not, however, been of much interest to healthcare researchers.

Medical researchers have typically seen corridor conversations as a waste of time, because they are often social and involve gossip; unprofessional, because they are associated with end-of-life and palliative care conversations with patients in semi-public areas with minimal privacy; and the locus of inappropriate "hallway medicine," where doctors and nurses ask each other for health advice, including asking for prescriptions. In fact, such conversations are essential to providing quality care. Long concluded that "The corridor, it turns out, proved to be the site par excellence for enacting emerging roles and responsibilities for oversight and coordination [...]. It functioned as a 'meta-space' whose lack of interactive-communicative prescription and ritual definition enabled clinicians to adopt a bird's-eye view on their tasks and on patients' care trajectories, while relaxing their status and formal power habits in favour of attentiveness to the logic of the work" (p. 189).

None of these kinds of insights could be gleaned from surveys, even in combination with interviews. They depend on detailed descriptions of situated practice grounded in detailed and direct observation, which is then used to guide deep interviews. For designers such situated observations are extraordinarily valuable in helping them understand the social and organizational *context* in which they are formulating design strategies. The research by Long and other medical anthropologists takes complexity as its starting point. The starting point is not about design per se. In fact, design may have a limited effect on desired organizational outcomes. It is all about understanding the interplay between an organization's complex social, organizational, design, and technological factors. Fritz Steele and the author called this more holistic approach to understanding environment-behavior relationships the study of "organizational ecology" (Becker and Steele 1995). It considers not just physical design factors, but also how such factors interact with information technologies, organizational culture, work processes, and user demographics.

17.6 Facility Management and POE

Initially, the design community was the primary audience for POE. Over the past twenty years, facility planning and management (FM) evolved as an additional practice-based client-centered function responsible for coordinating all aspects of building planning, design, and operation (Becker 1990; Cotts et al. 2010; Rondeau et al. 2012; Teicholz 2001). For facility managers, the concept of *building-in-use* is central to their professional responsibilities in a way that has never been true for the architecture and design community. Fundamentally, the FM function starts before and ends after the responsibility of the architectural and design professions does. It is deeper as well as broader. It embraces development and implementation of planning processes, design and construction, and management policy concerning allocation and use of space after occupancy. In this way FM incorporates the broader term BPE (building performance evaluation), with its more systemic and holistic focus than POE, which focuses on evaluation *after* occupancy. The FM "audience" is an internal one, namely, end-users who occupy and use the space, and management. For these audiences POE is extremely useful as a diagnostic tool for pinpointing specific aspects of the facility that need to be improved, or should be preserved in renovation and new design projects. It helps justify budget requests by demonstrating how design initiatives contribute to the business enterprise. Done well, interviews, observations, focus groups and other data collection techniques become a part of the process itself, and a form of organizational development.

Considering POE as a diagnostic tool intended largely for internal use by practitioners is likely to result in this form of structured inquiry being used more often because it is faster and less expensive than traditional academic research. While the data is unlikely to meet scientific canons of research, such an approach is far better than no evaluation or haphazard and deliberately skewed "guesstimates" about how well a facility is working for occupants. The internal organizational focus also avoids the reluctance of many organizations and design firms to formally document and *make public* aspects of plans and designs that did not work as expected.

17.7 Conclusion

Evaluation must become a staple of organizational practice, given that the purpose of POE is to increase the likelihood that how a facility is planned and designed supports organizational goals and an individual's and team's ability to work safely, comfortably, and effectively. For this to happen, it must be simple and quick to conduct, and its results must be immediately meaningful to the client. To be useful, it does not have to meet academic canons of scientific respectability. It simply must be conducted in a way that provides the client with information grounded in empirical data that stimulate insight.

In the long term, a more formalized database is critical to the development of a credible research tradition similar to that found in engineering and medicine. It is in large part this research tradition that has distinguished the medical and engineering professions from architecture and given them their far wider social influence and respectability. Practice-based research is akin to the medical model in which the clinician reports on interesting observations from work with patients, with those observations stimulating more focused and rigorous study by academic medical researchers. Diagnostic POEs, unlike academic EBR, place greater emphasis on influencing the course of events in the particular situation in which data is collected. The efforts of the practitioner and the academic build on and expand each other, and at their best, science and practice become flip sides of the same professional coin.

References

Altman, I. (1976). Privacy: A conceptual analysis. *Environment and Behavior, 8,* 7–29.

Becker, F. (1974). *Design for living: The residents' view of multi-family housing*. Ithaca, New York: Cornell University Center for Urban Development and Research.

Becker, F. (1990). *The total workplace: Facilities management and the elastic organization*. New York: Van Nostrand Reinhold.

Becker, F., & Steele, F. (1995). *Workplace by design: Creating the high performance workscape*. San Francisco: Jossey-Bass.

Brill, M. (1984). *Using office design to increase productivity*. Buffalo, New York: Workplace Design and Productivity.

Carroll, K., Iedema, R., & Kerridge, R. (2008). Reshaping ICU ward round practices using video reflexive ethnography. *Qualitative Health Research, 18,* 380–390.

Coiera, E., & Tombs, V. (1998). Communication behaviours in a hospital setting: An observational study. *British Medical Journal, 316,* 673–676.

Cooper, C. (1975). *Easter hill village*. New York: Basic Books.

Cotts, D. G., Roper, R. O., & Payant, R. P. (2010). *The facilities management handbook*. New York: American Management Association.

Iedema, R. (2003). *Discourses of post-bureaucratic organization* (Vol. 5). London: John Benjamin's Publishing.

Johnson, J. J., & Barach, P. (2008). Role of qualitative methods in designing health care organizations. *Environment and Behavior, 40*(2), 191–204.

Long, D., Iedema, R., & Lee, B. (2007). Corridor conversations: Clinical communication in casual spaces. In R. Iedema (Ed.), *The discourse of hospital communication* (pp. 182–200). London: Palgrave Macmillan.

Marans, R. W., & Spreckelmeyer, K. F. (1982). Evaluating open and conventional office design. *Environment and Behavior, 14,* 333–351.

Marcus, C. C., & Barnes, M. (1999). *Healing gardens: Therapeutic benefits and design recommendations*. New York: Wiley.

Preiser, W. F. E., & Vischer, J. (Eds.) (2005). *Assessing building performance*. Oxford, UK: Elsevier. (republished in 2015, London: Routledge).

Preiser, W. F. E., Rabinowitz, H. Z., & White, E. T. (1988). *Post-occupancy evaluation*. New York: Van Nostrand Reinhold. (republished in 2015, London: Routledge).

Reeves, S., & Lewin, S. (2004). Interprofessional collaboration in the hospital: Strategies and meanings. *Journal of Health Services Research & Policy, 9*(4), 218–225.

Rondeau, E. P., Brown, R. K., & Lapides, P. D. (2012). *Facility management*. New York: Wiley.

Sandelowski, M. (2000). Whatever happened to qualitative description? *Research in Nursing & Health, 23,* 334–340.
Sanoff, H. (2000). *Community participation methods in design and planning.* NY: Wiley.
Shepley, M. M., Zimmerman, K. N., Boggess, M. M., & Lee, Y. J. (2009). Architectural office post-occupancy evaluation. *Journal of Interior Design, 34,* 17–29.
Silver, N. (2012). *The signal and the noise: Why so many predictions fail-but some don't.* New York: Penguin.
Sommer, R. (1969). *Personal space, The behavioral basis of design.* Englewood Cliffs, NJ: Prentice-Hall.
Stichler, J. F., & Hamilton, D. K. (2008). Rigor and replication. *HERD: Health Environments Research & Design Journal, 2,* 3–5.
Sundstrom, E., Burt, R., & Kamp, D. (1980). Privacy at work, architectural correlates of job satisfaction and job performance. *Academy of Management Journal, 23,* 101–117.
Teicholz, E. (2001). *Facility design and management handbook.* New York: McGraw-Hill.
Ulrich, R. S. (1984). View through a window may influence recovery from surgery. *Science, 224,* 420–421.
Zeisel, J. (1974). Fundamental values in planning with the non-paying client. In J. Lang, C. Burnette, W. Moleski, & D. Vachon (Eds.), *Designing for human behavior* (pp. 293–301). Stroudsburg, PA: Dowden Hutchinson & Ross.
Zimmerman, A., & Martin, M. (2001). Post-occupancy evaluation: Benefits and barriers. *Building Research & Information, 29*(2), 168–174.

Author Biography

Franklin Becker Ph.D., is Professor Emeritus in the Department of Design and Environmental Analysis in the College of Human Ecology, Cornell University. He co-founded Cornell's *Facility Planning and Management Program* in 1980, the first university program of its kind in the world; and he founded the *Cornell International Workplace Studies Program* (*IWSP*) in 1989. The IWSP is internationally recognized for its pioneering research on innovative workplace strategies. Professor Becker's work is characterized by an ecological systems approach to understanding why facilities take the shape they do, how they work in practice, and their effects on individuals, teams, and the organization.

Professor Becker has conducted research, lectured, and consulted in England, Canada, Europe, Japan, Australia, New Zealand and China, as well as the United States. He is a Fellow of the American Psychological Association, and has served as the Academic Affairs member of the Board of Directors of The International Facility Management Association, from which he has received the Chairman's Award, Outstanding Educator Award and the Distinguished Author Award. Professor Becker is the author of eight books, including *Workplace By Design*: *Mapping the High Performance Workscape,* and *Offices at Work*: *Uncommon Workspace Strategies that Add Value and Improve Performance.* He is a co-author of *The Practitioner's Guide to Evidence-Based Design.*

Professor Becker is President, IDEAworks LLC, a consultancy firm specializing in the planning, design, and management of innovative workplace strategies.

Chapter 18
Towards Wellbeing: Hospital Evaluation Using the Problem-Seeking Method

Steven Parshall and Sofia Fonseca

18.1 Introduction

A post-occupancy evaluation (POE) feeds data back into the design process as a measurement of the gap between planned and actual performance of a building. The most common application is to evaluate the performance of a facility once it is occupied. The army calls this post evaluation "ground truth," an assessment of what happened in the field differently from how the strategy was planned. A POE aims at improving the quality and performance of the design process and its final product (see Fig. 18.1). One definition of excellence is the quality of experiences the project brings to all the categories of stakeholders: clients, design team, builders, users, the public, and the profession. A POE is thus part of the building performance evaluation (BPE) process and provides the benefits of a holistic view on the building life cycle of a project, including life cycle phases from strategic planning and programming, through design, and to post occupancy—with the related review loops, offering feedback loop as adjustments to existing building forms and feeding forward into design standards and guidelines (Dodson 2011).

Evaluating the final product of a health care building program is an important, but often overlooked, step. As part of an on-going process that looks back with its benefits in the future, a POE helps the healthcare administrator, the architectural team, and ultimately the patient. This type of research documentation of results gives rise to a new body of research called "evidence-based design" that feeds wellbeing studies (Merkel 2003).

S. Parshall (✉)
HOK, Suite 900, 3200 Southwest Freeway, Houston, TX, USA
e-mail: steve.parshall@hok.com

S. Fonseca
KYO Consulting Inc., 14006 Swiss Hill Drive, Houston, TX, USA
e-mail: sofia.fonseca@kyoconsulting.com

W.F.E. Preiser et al. (eds.), *Building Performance Evaluation*,
DOI 10.1007/978-3-319-56862-1_18

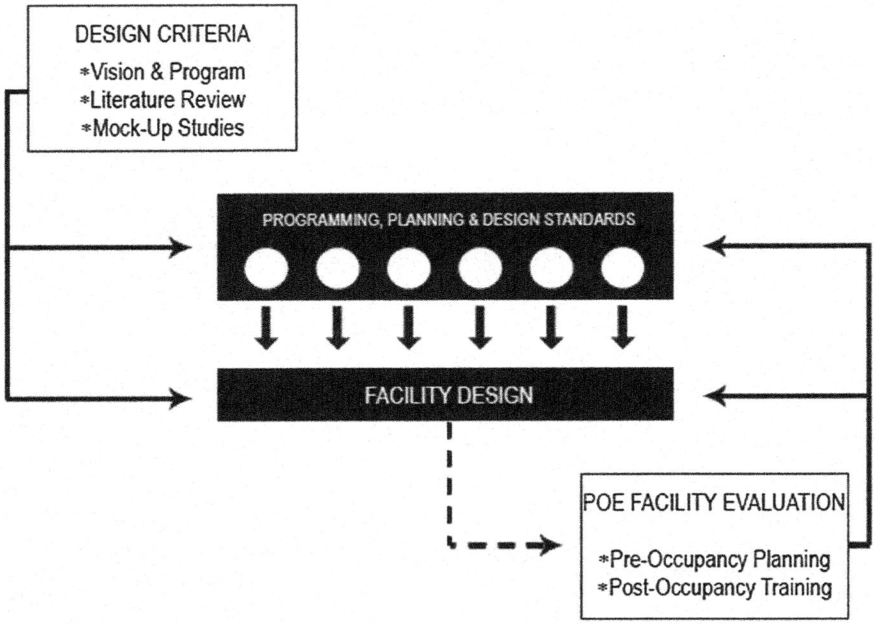

Fig. 18.1 Two loops for organizational learning. *Source* Authors, HOK

Here are some common intentions why an evaluation might be undertaken (Pena and Parshall 2011):

- To justify actions and expenditures: Accountability in institutional building programs like a hospital especially if public funds are involved (Fig. 18.2).
- To measure design quality, i.e., conformance to requirements of the end product and implementation of a vision for the institution, e.g., policies, actions and expenditures by management that the architect materializes (Fig. 18.3).
- To fine-tune a facility: A medical facility is a sophisticated building type that may require adjusting to ensure effective functionality (Fig. 18.4).
- To adjust a repetitive program, or to prepare for a future building program, renovation, or expansion: The results of a POE are most useful when they contribute to the program and design of a subsequent step in accomplishing a masterplan, a new unit or upgrading existing facilities. An evaluation can also help iron out the snags encountered the first time around in phases like additional wings, or free standing elements (Fig. 18.5).
- To research human-environment relationships: A medical facility can be a stressful place for its patients and staff. Information from a POE may improve other facilities, or the performance of the evaluated facility (Fig. 18.6).
- To test the application of new ideas and assess risks these imply before further application is made.

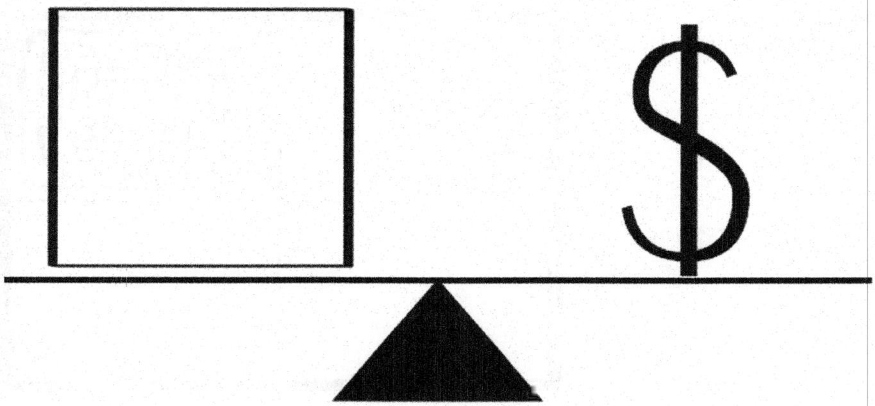

Fig. 18.2 Balancing money and building efforts. *Source* Authors, HOK

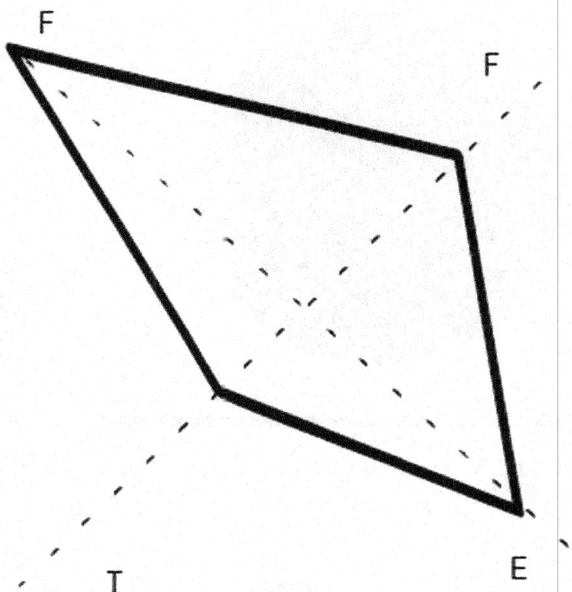

Fig. 18.3 Rose chart for decision making. *Source* Authors, HOK

- To educate past and future participants, i.e., client, architect, and user, through the POE process itself as well as through the lessons learned.

This chapter describes a comprehensive, practical, and feasible method that addresses scope, budget, and time commitments of a majority of Hellmuth, Obata + Kassabaum (HOK) and KYO clients. An important consideration in the development of this method is economy of effort while achieving the greatest value for the clients.

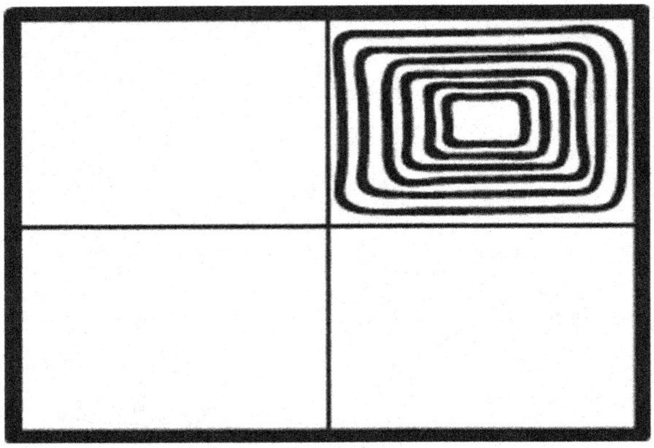

Fig. 18.4 Whole versus detail. *Source* Authors, HOK

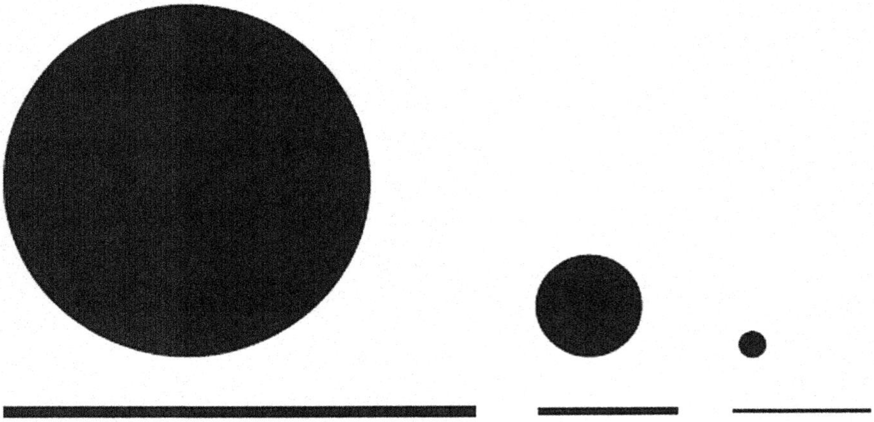

Fig. 18.5 Adjusting repetitive elements. *Source* Authors, HOK

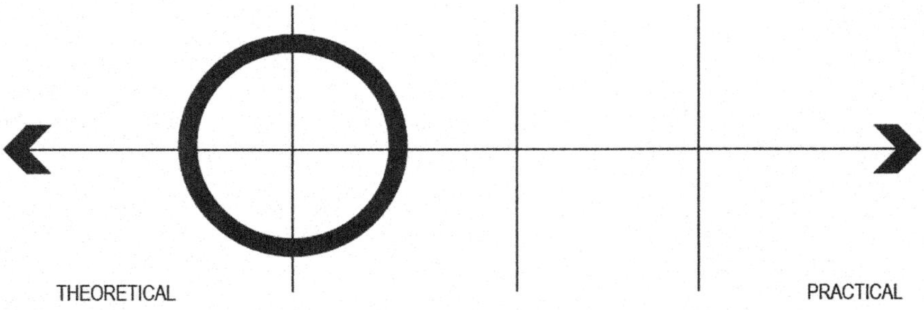

Fig. 18.6 Performance spectrum. *Source* Authors, HOK

18.2 Method

A POE considers responses from the best information source to understand how well a building performs (Gonchar 2008): facility users. It should take place between six months and two years after occupancy after solving the shakedown problems, and after the novelty has worn off.

The evaluation team should represent different fields and perspectives. Collecting from a diverse group contributes to a more objective evaluation. The size of the team will vary between three and seven, with one team leader. A team might encompass the following roles:

- Administrator
- Facilities Manager
- User Representative
- Programmer
- Designer
- Project Manager

Different data gathering techniques may be used by the evaluation team ranging from qualitative self-reports to quantitative analyses (Preiser and Nasar 2008). Hospital staff, physicians, and patients are the prime source of information. Collecting data most usefully involves a triangulated approach: (1) anonymous surveys, (2) observations, (3) interviews. Comprehensive surveys reach a greater number of people resulting in graphs and charts that can feed the hospital's quality improvement process. Space utilization studies or cultural anthropology involve detailed observation including touring of the facility. Interviews, either formal or informal, allow a deeper investigation to vet the other two sources of reported versus observed data. This data can be coded back to the Program and the POE.

18.3 Five Steps and Four Considerations

Hospital facilities are easily evaluated when a pragmatic, comprehensive, yet simplified enough method such as this one is applied. The design of a hospital must respond to unique demands and requirements The process to gather those requirements has five steps:

1. Establish the PURPOSE;
2. Collect and analyze QUANTITATIVE information;
3. Identify and examine QUALITATIVE information;
4. Make an ASSESSMENT;
5. State the LESSONS LEARNED; and four major considerations:

   FUNCTION, FORM, ECONOMY, and TIME (Fig. 18.7)

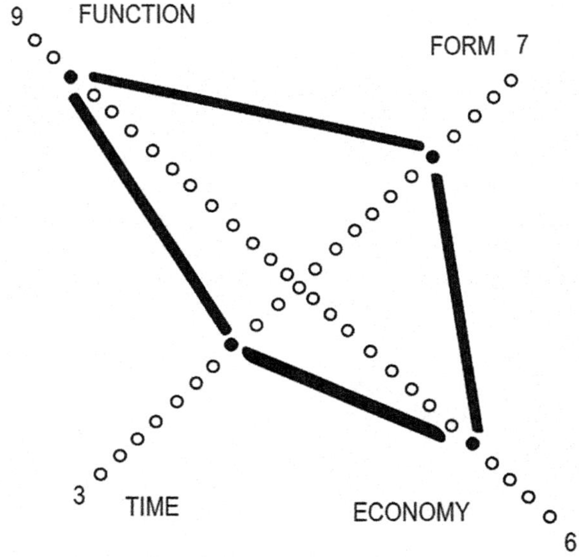

Fig. 18.7 Four considerations. *Source* Authors, HOK

Organization of an evaluation, i.e., feedback, should correspond to the framework used for programming, i.e., feedforward. The similarity of organization, content, and format will increase the usefulness of the POE results for programming and design. Evaluating and programming both involve an organized process of inquiry.

Successful evaluation depends on cooperation. All participants—owners, users, architect—should agree on what they hope to gain. The purposes will affect the type of information collected. An expenditure justification might differ from a human-environment relationship evaluation. Users or participants on the team may be different.

18.3.1 Basis of Design

The first step involves establishing the purpose or basis of design during the programming process. This initial picture of future performance involves data gathering from client input and is compiled into a statement of requirements containing goals, facts, concepts, and needs. A record of the initial programming requirements and design intentions is essential for a rigorous evaluation of the result.

18.3.2 Quantitative Description

The second step, preparing a quantitative description, includes collecting factual data on the building as designed; for example, the floor plans. Analyzing parametric data provides a basis for comparing this facility with similar ones.

18.3.3 Qualitative Description

A qualitative description (see Table 18.1) includes examining the client's goals for the facility, the programmatic, and design concepts for achieving those goals, and the statements representing design problems that the designer intended to solve. This step also includes identifying changes that have taken place since occupancy, and unresolved issues.

First, goals are gathered to convey the client's stated intentions. Sometimes clients express great aspirations that are not fully achievable in the end, due to budgetary, or operational constraints. Concepts are also collected, i.e., ideas for realizing goals

Table 18.1 Definitions of adequacy, quality and satisfaction. *Source* Authors, HOK

Functional adequacy	Space adequacy	Construction quality	Technical adequacy	Energy performance	User Satisfaction
Measures the amount of area per the facility's primary unit of capacity. It might also compare the capacity for planned procedures with the actual operations performed[a]	Gross area of a building is the sum of the net assigned area and the unassignable area. The ratio of net assignable area to the unassigned area measures the building efficiency	The unit cost associated with the quality level of the building measured as the building cost per gross square feet (including unique building systems which minimized cost. and unusual constraints, such as codes or site location which increased costs)	The cost of fixed and special equipment (such as renal dialysis or laser surgery devices). Measured as a percentage of the building cost, though it is also possible to represent it as a unit cost	A measure of the amount of energy per gross square foot consumed for the standard operation of a building	From data collectors. survey and interview data collected 011 how satisfied users are with the facility

[a]Examples: GSF/hospital bed for inpatient facilities, # of patient visits for ambulatory facilities

which represent abstract relationships and functional arrangements. Design concepts are physical responses that provide a unifying theme to the building solution. These are summarized in the end as Problem statements representing a recognition of the critical project conditions, and a direction for the design effort. Collected changes are indicators since occupancy of new requirements, or inadequacies as well as actions taken to alleviate undesirable conditions. Issues are tracked as unsettled and controversial decisions in dispute. Occupants or owner of the facility pose them during the evaluation, or the evaluation team raises them during the POE.

18.3.4 Assessment

The evaluation criteria are standard questions reflecting important values. See "Example Question Set for Facilities Evaluations" Table 18.2. The evaluation team

Table 18.2 Example question set. *Source* Authors, HOK

EXAMPLE QUESTION SET FOR FACILITIES EVALUATIONS

Instructions

1 Use this form for rating a facility

2 Review the criteria for each consideration and agree on the meaning for the particular building type being evaluated

3 Enter score from 1 to 10 for each criterion:

1	Complete Failure	6	Good
2	Critically Bad	7	Very Good
3	Far Below Acceptable	8	Excellent
4	Poor	9	Superior
5	Acceptable	10	Perfect

4 Add the scores and divide by 6 to get an average score for each major consideration

FUNCTION

		Score
A.	THE OVERALL ORGANIZATIONAL IDEA (the big functional concept)	______
B.	EFFECTIVE ARRANGEMENT OF SPACES (activities and functional relationships)	______
C.	WELL-PLANNED CIRCULATION (entry, orientation, flow)	______
D.	ADEQUATE SPACE ALLOCATIONS/PARKING (net assignable/unassigned area, parking)	______
E.	RESPONSE TO USER PHYSICAL NEEDS (comfort, safety, convenience)	______
F.	RESPONSE TO USER SOCIAL NEEDS (privacy, interaction, sense of community)	______
	SUM TOTAL	______
	DIVIDE BY SIX	6
	AVERAGE FUNCTION SCORE	______

should review the question set to understand the meaning of the criteria. Each evaluator forms a subjective response about the degree of excellence attained by the facility. A comprehensive evaluation concerns the equilibrium of all the forces that shaped the project. Quantification is useful even if quality is a subjective value judgment that varies with every individual.

First, rating provides a mechanism for identifying the differences in perception of a building by the various evaluators. Better understanding is possible when the evaluation team discusses these differences. Second, rating provides an explicit pattern of how the parts contribute to the whole assessment. Clearer knowledge of the strengths and weaknesses is possible when the evaluators compare these patterns and discuss them.

18.3.5 Lessons Learned

Lessons learned are conclusions about strengths and/or weaknesses. Rarely should an evaluation conclude with more than twelve statements. At a minimum, four statements will cover each of the major considerations: function, form, economy, and time.

- Function: The original program provides an immediate focus on the important client decisions that influenced the design, when evaluating functional performance. Refer to the original goals and concepts of the program.
- Form: The evaluation must include aesthetic standards to determine the physical design excellence of the building. This is the most difficult part of the evaluation, since aesthetic standards are subjective.
- Economy: A consideration of the original quality goals for the facility commensurate with the initial budget. It is unrealistic to expect grand quality results if the original budget allowed for no more than an economical level.
- Time: Because two or three years may elapse between programming and occupancy, the initial users may be different from those involved in the original planning. A certain amount of user satisfaction, therefore, depends on periodic interior design, or on the degree that partition and utility service changes are possible within the basic structure.

With a trained evaluation team, it is possible to complete the evaluation procedure within two to four weeks. Elaborate user satisfaction surveys (see Table 18.2) may extend the duration of the preparation phase.

18.4 Evaluation Activities

The typical sequence of activities (see Table 18.3):

Table 18.3 Evaluation Activities. *Source* Authors, HOK

1 Initiation	2 Preparation	3 Tour	4 Assessment	5 Summation, presentation and documentation
Establish the purpose of the evaluation	Research background	Make a visual inspection of the facility	Each evaluator makes a judgement as to the facility's success by assigning a score	Review the quantitative and qualitative descriptions, along with the assessment ratings, and prepare a statement of the lessons learned
Identify the background data requirements	Prepare the quantitative and qualitative descriptions. It may also entail the user surveys to record their perceptions	Possibly, undertake random interviews with users and probe for responses about performance	The ratings are recorded oil a special graph, which illustrates the pattern of each assessment	The team leader presents the conclusions. The team leader prepares a report of the evaluation process
		Meet to discuss observations		

The surveys employed should encompass four major areas: (1) description of the users, (2) their satisfaction with the facility, (3) importance of program elements, and (4) evaluation of particular features of the health care facility design.

18.5 Case Study

In 2009, Methodist Sugar Land Hospital (MLSH) Expansion, the 442,000 square-foot (41,063 square-meter) 6-story facility, designed by HOK, was developed by The Methodist Hospital System (TMH) to provide increased bed capacity, and to support new and expanded service lines including Cardiology, Neurology, and Orthopedics. The building consists of 168 new Acute Care Patient Rooms (56 of which are shelled), 20 Intensive Care Units (ICU), Patient Rooms, 12 Procedure Rooms, and supporting facilities.

18.5.1 Case Study Evaluation Process

The POE collected quantitative planning metrics to be compared across several facilities over time, and qualitative information (see Fig. 18.8) about the users' thoughts about the facility to list lessons learned for subsequent projects. Specific objectives were:

1. To document the final space allocations of departments, key rooms and support areas, for the entire building, and
2. To find out from the users' viewpoint what worked, what did not work and what could be improved, including mechanical/electrical/plumbing (MEP) functional elements.

A team of twelve members, each representing a different role consisted of TMH Core Team: HOK, Core Project Team of TMH Project management together with the MSLH Leadership Lead Team reviewed the POE goals, scope, project related policies and procedures, confirmed roles for collecting information and decided approval process. HOK presented survey question categories to receive feedback from the TMH/MSLH Core Project Team. Based on that feedback, HOK developed questions to pilot them in interviews with key leadership of departments thus identifying specific areas of concern or interest to be addressed in a hospital-wide staff survey.

The hospital-wide staff survey was administered 6 months after the kick-off. HOK collected, organized and analyzed the data from the survey during a 4-month period.

The HOK team conducted semi-structured interviews with a Director and/or Manager of each department listed below, to learn the top issues for the department and staff. The interviews were based on a set of predetermined questions previously

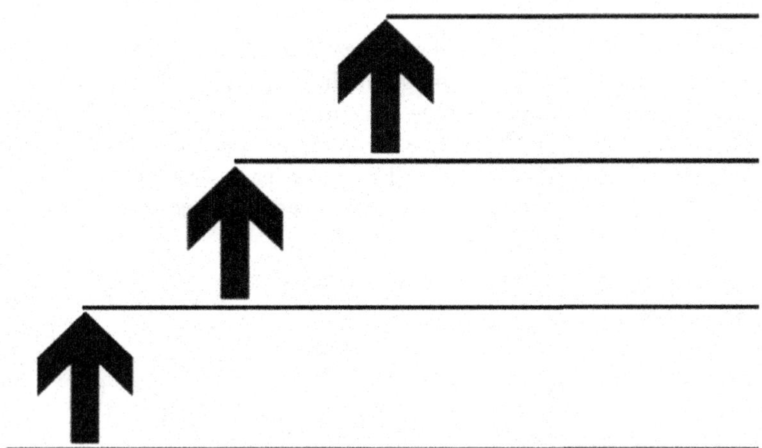

Fig. 18.8 Incremental steps for quality improvement. *Source* Authors, HOK

vetted with hospital leadership. After each interview the HOK team walked the department with the interviewee(s) and took photos of areas discussed.

Departments interviewed

- Acute Care Unit
- Admit Observe Discharge (AOD)
- Diagnostic Services
- Emergency
- Endoscopy
- Intensive Care
- Post Anesthesia Care Unit (PACU)
- Pharmacy
- Surgery

Each interview, digitally recorded with permission of the interviewee(s) and transcribed by HOK staff, identified significant issues tabulated for each department.

Content Analysis by HOK staff identified common issues to be explored in a facility-wide survey of staff. HOK's recommendations for the survey were presented to the Hospital Leadership and reduced to a core set that were of interest and value to the Hospital. These focused on the interface with the built environment with regard to four topics:

- Space
- Equipment and Supplies
- Quality and Environment
- Patient Care Delivery

18.5.2 Lessons Learned from the Case Study

The fifth and final step in the evaluation is the compilation of lessons learned. These were insights gained through the programming, design, and construction of the facility and brought to light through the evaluation process providing the benefits of a holistic view on the building life cycle through the review loops of thorough building performance evaluations (BPE). Consistent themes across several departments that had a relationship with staff operations and care delivery turned "Report Cards" for the results:

1. Centralized and decentralized charting
2. Work surfaces
3. Equipment, supply, location, and management
4. Electric outlet and data port location
5. Ergonomic and ambient conditions, e.g., noise, lighting, nursing station seating, privacy

A sample Scorecard for the Admit-Observe-Discharge (AOD) area survey is provided revealing basic needs met in areas: noise and distance to the primary nursing station. The open comments captured spatial and functional needs: adequate room space and family waiting area, blinds a good feature, doors can be wider for beds to easily move through, reduce noise from adjacent staff areas, provide adequate storage solutions (see Table 18.4).

This POE suggested further investigations on the following topics (see Table 18.5).

18.6 Synopsis of the Problem Seeking Method

HOK programmed its first project in 1951 using the Problem-Seeking method. The first publication about the method appeared in 1969; and in 2011, Wiley published the fifth edition of Problem-Seeking with a revised a chapter on "Building Evaluation."

Evaluating facilities in a time of big data is gaining acceptance as a meaningful practice. Development of new electronic tools has improved its implementation. New input mechanisms include:

Table 18.4 Sample POE results for a Hospital Unit. *Source* Authors, HOK

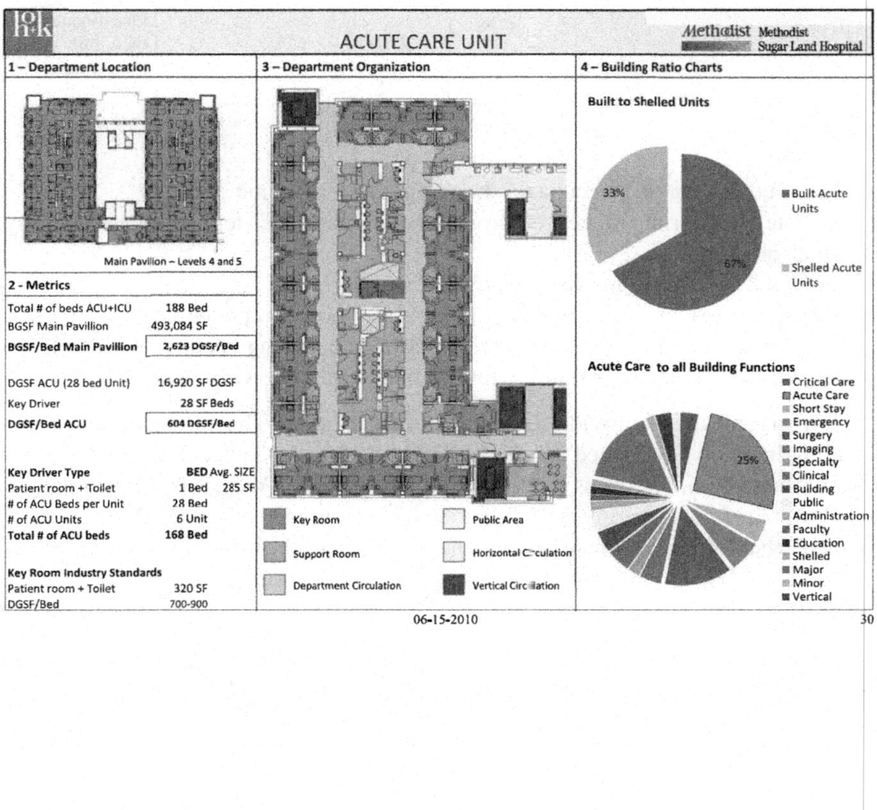

Table 18.5 Sample recommendation chart from POE. *Source* Authors, HOK

Accessibility to patients	Patient room size	Light motion sensors and lighting	Counter top dimensions	Ergonomic and ambient conditions	Centralized and decentralized charting
Practice emergency scenarios. Also, standardize equipment storage and location during emergency scenarios and daily operations	Closer examination of the patient room size and its impact on key care delivery tasks (e.g., decentralized charting) should be conducted to determine the optimal room size)	Optimal light levels during work-related tasks may need additional investigation (set a standard and monitor it)	A closer examination of countertop dimensions is warranted	A closer examination of noise, its sources, its prevalence, and its impact on performance is needed	A detailed study of centralized and decentralized nursing strategies on the nursing unit is needed and should involve
			A closer examination of countertop dimensions and their relationship to light levels and performance is warranted	A closer examination of seating at the main nursing station is desirable	Time*motion studies of work flow for Patients and staff
					Questionnaires mcludmg performance metrics

- Space utilization and occupancy tracking with electronic devices.
- Remote monitoring of key environmental metrics, i.e. temperature, humidity, water quality.
- Occupant wearables measuring physical comfort.
- Laboratory testing, i.e. measuring brain and blood components.
- Observing outcomes with individual controls, i.e. giving occupants control over some aspect of their workplace environment and observing what they do.

New organizations provide new ways of measuring. The International WELL Building Institute has created the WELL Building Standard providing Guidelines to promote health and wellness in the design of buildings. WELL is a performance-based system for measuring, certifying, and monitoring features of the built environment that impact human health and wellbeing, through seven concepts of wellness: air, water, nourishment, light, fitness, comfort, and mind. HOK now also uses the Leesman Index (independent database of workplace effectiveness) across global locations and market sectors to collect POE information and measure the performance of office environments.

The drive for wider application of evaluation comes from client demand. POEs achieve value for clients and users of facilities and demand comparable to that of programming. Evaluation methods vary, each suited to a particular application. Some are rigorous and strive for objectivity; others provide expedient answers, and are more subjective. Problem Seeking is pragmatic—comprehensive, yet simplified enough for practice. The process is general enough to be suitable for many types of facilities. The content makes the evaluation specific. The office can use the framework to consolidate new data points whether from a WELL guideline or wearable technology. The five step process and four considerations used both in pre and post design remain significant and useful.

18.7 Conclusions

Two loops for organizational learning are a feed-forward during programming and feed-back during a POE. This method is practical for evaluating most health care facilities. POEs provide a competitive advantage (Fairley 2015).

The five-step process suits many purposes. The criteria are comprehensive, encompassing four major considerations: function, form, economy, and time.

The method acknowledges user satisfaction, though the final assessment requires the judgements of an evaluation team. Six months to two years after initial occupancy is the best period for conducting a POE. As an aid to the programming process, an appropriate time for conducting an evaluation is prior to initiating a new building program, remodeling, or discontinuing the use of a facility. Strive for the format and organization of the evaluation results to be compatible with those informational formats used in programming and design. The procedure allows for quick starts and for timely completion.

Hospital projects are highly complex in the diversity of considerations to be resolved. Necessary trade-offs made along the way from inception to final completion in a process, which may last from three to five years, involves a multi-participant team of client and architect. An evaluation offers an excellent opportunity to visit the implemented plan and to reconfirm the priorities and concepts set forth in the program and design.

Has the investment achieved the ideas set out to achieve? An evaluation strengthens the healthcare administrator's plans to follow through with the operational activities necessary to achieve the full value of the asset. It is also an opportunity to recognize the minor adjustments that will realize the fullest performance of the facility in place.

Acknowledgements Thanks are owed to members of the project team for their professionalism and dedication: Sid Sanders, Kate Rose (Core), Mike Smith (Core); MLSH Lead Team: Chris Siebenaler (Lead), Lowell Stanton (Lead), Janet Leatherwood (Lead), Becky Chalupa (Lead), Sean Sevy (Lead), and HOK: Ron Smith, Sara Bullington, Ricardo Socorro, Nick Watkins.

Bibliography

Dodson, M. (2011). A dossier on post occupancy evaluation: Post occupancy. *Architecture Australia, 100,* 5.

Fairley, P. (2015). Feedback loop: With a frequent gap between predicted and actual performance, post-occupancy evaluations begin to catch on. *Architectural Record, 203*(8), 114–120.

Gonchar, J. (2008). Looking back and moving forward. *Architectural Record, 196*(2), 160–169.

Merkel, J. (2003). Healthcare architecture now. *Architectural Record, 191*(3), 11–17.

Pena, W. M., & Parshall, S. A. (2011). *Problem Seeking: An architectural programming primer.* Washington, DC: Wiley.

Preiser, W. F. E., & Nasar, J. (2008). Assessing building performance: Its evolution from post-occupancy evaluation. *ArchNet-IJAR, International Journal of Architectural Research, 2*(1), 84–99.

Author Biographies

Steven Parshall is Senior Vice President and Regional Leader of Consulting for HOK in Houston, Texas. He is responsible for developing real estate and facility plans to further the vision and goals of organizations to make a business case for facility change—both long-term strategies and immediate implementation. He manages projects for major corporations, educational institutions and government agencies which define planning models that tie organizational mission, population forecasting, work process improvement, workplace requirements, capital and occupancy cost into a comprehensive facilities plan.

Steve has published more than 23 articles and research reports and co-authored five books, including the recently released 5th edition of Problem Seeking: An Architectural Programming Primer, used by colleges and universities as the basic text on programming. Steve is a frequent lecturer and an established teacher. He has led over 80 training workshops and lectures at professional meetings and educational campuses, where he has shared new workplace ideas and innovations in analysis and all aspects of research as applied in the design and facilities industry with clients, students and other professionals. He is the leading expert on the Problem Seeking® process, having taught more than 40 external workshops and over 30 workshops within HOK. He held the William Peña Visiting Professorship at Texas A&M University.

Sofia Y. Fonseca is a Workplace Strategy and Programming Specialist. She is delights in using design thinking to innovate generating business strategies and visions for the built environment. Her 20 years of experience in pre-design services providing facilitation, programming, workplace innovation, post occupancy evaluations, people engagement, strategic planning and master planning services include institutional and corporate clients with an emphasis in corporate, high-tech and educational facilities. She bridges corporate practice and academia conducting research and benchmarking at KYO Consulting while teaching at the University of Houston, Gerald D. Hines College of Architecture and Design Masters Thesis Program, and Methods of Research and Human Factors. She is the editor of the fourth edition of Problem Seeking, the architectural primer in programming. A frequent lecturer at Universities and Professional Forums, she holds a Master of Architecture from *Harvard University* and a Bachelor of Science, Environmental Design from the Honors College at *University of Houston*.

Chapter 19 Evaluating Changes in Sustainability Culture: A Model for Universities and Other Organizations

Robert W. Marans and John Callewaert

19.1 Introduction

Approaches to evaluating built environments are adaptable to environmentally-related programs. This chapter presents an overview of a multi-year project designed to monitor programs aimed at moving toward a culture of sustainability on the University of Michigan (U-M) campus. Culture of sustainability is meant to reflect a set of attitudes, behaviors, levels of understanding and commitment, degrees of engagement, and dispositions among a population within a single entity- a university, a corporation, a city, or a building. The evaluation is intended to inform and provide feedback to U-M administrators, faculty, and others responsible for day-to-day operations of the University including its academic programs. It also is used to assess experiments/trials of new sustainability initiatives. Finally, the evaluation program serves as a model demonstrating how behavioral research can be used to address critical environmental issues within universities and in other settings.

For nearly a half century, there has been a plethora of books, book chapters, and reports about the systematic evaluation of different building types and their performance. Methods used in conducting these evaluations are equally varied and typically involve the collection of information about the buildings and their occupants. The former focuses on hard measures such as light and noise levels. Information about building occupants often includes employment records, performance data, and their perceptions, assessments, and behaviors.

R.W. Marans (✉)
Institute for Social Research, University of Michigan, Ann Arbor, MI 48106, USA
e-mail: marans@umich.edu

J. Callewaert
Graham Sustainability Institute, University of Michigan, Ann Arbor, MI 48106, USA
e-mail: jcallew@umich.edu

W.F.E. Preiser et al. (eds.), *Building Performance Evaluation*,
DOI 10.1007/978-3-319-56862-1_19

Evaluators often use a combination of qualitative and more structured quantitative approaches to gather information about the perceptions, assessments, and behaviors of building occupants. Qualitative approaches include informal interviews, focus groups, walk-throughs, and behavioral observations whereas the quantitative approaches include systematic people counts at key locations within buildings, behavioral mapping, i.e., space syntax, and survey research. The latter involves the use of primary data collected at the individual level using surveys of all or a representative sample of building occupants.

Approaches to conducting systematic evaluations of buildings can also be used to evaluate environmentally-related programs. Examples of environmental-related programs evaluated systematically deal with housing (Francescato et al. 1979; Peroff et al. 1979), environmental education (Marans et al. 1972), and office productivity (Brill et al. 1984). In recent years, the University of Michigan (U-M) has used a similar strategy to evaluate a range of programs designed to make its Ann Arbor campus more environmentally sustainable.

19.2 Sustainability at the University of Michigan

For more than a decade, U-M has embarked on a program of creating sustainable buildings. Included in this initiative are building renovation projects and the building new buildings reflecting Leadership in Energy and Environmental Design (LEED) standards. The construction of green buildings is a key mechanism for addressing the University's sustainability goals, particularly with regard to reducing greenhouse gas emissions and preventing waste, UM's sustainability goals however go beyond buildings. They also address programs and activities involving transportation, natural environment protection, and a movement toward more sustainable foods. These goals were established as part of a presidential initiative in 2011 under the themes of Climate Action, Waste Prevention, Healthy Environments, and Community Awareness (Callewaert and Marans 2017).

Whereas climate action, waste prevention, and creating healthy environments are prevalent at other universities, the focus on raising community awareness and dealing with the behavioral aspects of campus sustainability is viewed as critical and innovative. The articulation of the fourth theme of community awareness and its goal of moving toward a campus-wide culture of sustainability reflects U-M's belief that institutes of higher education have a critical role to play in bringing about a societal shift toward a more sustainable future (Marans and Shriberg 2012).

Community awareness involves multiple actions to educate and engage members of the university community with the intent of creating a culture of sustainability on campus. A culture of sustainability is "a culture in which individuals are aware of major environmental (and social/economic) challenges, are behaving in sustainable ways, and are committed to a sustainable lifestyle for both the present and future" (Callewaert and Marans 2017).

Mechanisms for bringing about a cultural shift within universities and colleges are varied, complex, and often not well articulated. U-M efforts involve programs to expand recycling and reduce energy use in buildings, encourage alternative modes of travel to/from campus, promote the use of foods from sustainable sources, and introduce the concept of sustainability in university coursework (Marans et al. 2014).

19.3 The Sustainability Cultural Indicators Program

Whereas the above initiatives are essential to creating a more sustainable campus culture, the Sustainability Cultural Indicators Program (SCIP) represents a critical and complementary component of the Community Awareness theme. That is, SCIP is the mechanism for measuring, evaluating, and tracking progress in the movement toward a sustainable campus culture. SCIP is also used to evaluate experimental or trial sustainability programs that could be tested in one or a cluster of buildings before being implemented throughout all university buildings.

U-M cultural change initiatives stem from the principles outlined under theme of Community Awareness. They indicate that U-M will "pursue evaluation strategies toward a campus-wide ethic of sustainability and scientifically measure and report progress and behavior as a community" (Coleman 2011).

The evaluation strategies involve a groundbreaking program for monitoring the U-M's progress in moving toward a culture of sustainability. Progress is measured using annual web surveys of students, faculty, and staff regarding sustainability awareness and behavior along with tracking changes over time.

19.3.1 Questionnaires

Two questionnaires, similar in content are used. One is sent to a student sample and the other to the sample of staff and faculty. Since SCIP builds on the U-M's sustainability goals, questionnaire modules contain questions focusing on transportation, waste prevention, conservation, the natural environment, food, and U-M sustainability efforts. The questionnaire is designed to be completed in 15 min.

19.3.2 Population and Sample

In order to ensure proportional representation from all segments of U-M's population and from all geographic parts of the campus, the sample design aims to obtain large numbers from the entire student body and from the population of staff and faculty. Specifically, a stratified sample is selected each year so as to yield 1000 respondents from the freshmen class, 350 respondents from each of the sophomore, junior and senior classes, and 400 graduate students. Another sample is selected that targets 750 staff and 750 faculty members annually. Since the inception of the surveys in 2012, targets have been reached or exceeded. Annual response rates over the 4 years ranged from 44 to 22%.

19.3.3 Findings

In order to summarize findings covering key concepts reflecting the culture of sustainability, indicators are created that combined responses to closely related questions or items about a common idea, concept, or action. Items used to create the indices are shown in Table 19.1.

As shown in Table 19.2, several key findings are gleaned from indicator scores covering the 4 years. First, there is considerable room for improvement with regard to pro-environment behaviors, levels of awareness, degrees of engagement, and expressed commitment to sustainability. Second, the travel behavior of students is more in line with the goal of greenhouse gas reduction than travel to and from campus by the staff and faculty. Not surprisingly, students are most likely to walk, bike, or bus to campus. Similarly, students are likely to know more about available transportation options available and are more engaged than either staff or faculty in sustainability activities on campus.

Third, compared to students and staff, faculty tend to report acting in a more sustainable manner with respect to conserving energy, preventing waste, purchasing food, and more generally, engaging in pro-environmental activities outside the University. Faculty members also express a higher level of commitment to sustainability than staff or students. Fourth, students tend to be less knowledgeable than staff or faculty about protecting the natural environment, preventing waste, and sustainable foods. But they know more about sustainable transportation options than staff and faculty and are equally knowledgeable about sustainability at the University. Finally, the indicator scores over the 4-year period show that in general, awareness or knowledge about sustainability has increased among members of the University community. In some instances, indicator scores from 2015 are significantly higher than 2012 scores and/or higher than the 2014 scores. In the case of sustainable foods, significantly positive changes between the 2015 scores and the earlier scores reflect a growing understanding of sustainable foods over the 4 years.

The relatively large numbers of student, faculty and staff respondents each year enable the calculation of index scores for each of Ann Arbor's campuses, regions, and sub-regions (see Fig. 19.1). Differences in indicator scores by campus, region, and sub-region help U-M operational units determine where their outreach programs are successful and where they need to be enhanced. The SCIP data covering behavioral responses are also being examined relative to hard measures covering energy use and waste for buildings, regions, and sub-regions.

Table 19.1 Items used for creating sustainability indicators. *Source* Authors

Themes and indicators	Name of items	Number of items
Primary indicators		
Climate action		
Conservation behaviour	Turn off lights, use computer power-saver, turn off computer, use motion sensor	4
Travel behaviour	Most often mode of travel to campus since fall semester	1
Waste prevention		
Waste prevention behaviour	Print double-sided, recycle paper, etc., use reusable cups, etc., use property disposition	4
Healthy environments		
Sustainable food purchases	Buy sustainable food, organic, locally-grown	3
Protecting the natural environment	Use fertilizer, herbicides, water lawn	3
Community awareness		
Sustainable travel and transportation	Ann Arbor area transportation authority, UM buses, biking, Zipcar rental	4
Waste prevention	Recycle glass, plastic, paper, electrical waste, property disposition	5
Natural environment protection	Dispose hazardous waste, recognize invasive species, residential property, protect Huron River	4
Sustainable foods	Locally grown, organic, fair trade, humanely-treated, hormones-free, grass-fed, sustainable fish	7
U-M sustainability initiatives	Sharing, recycling, sustainable food, reduce greenhouse gas, maintain grounds, protect Huron River	8
Secondary indicators		
Sustainability engagement at U-M	Participate in sustainability organization, Earthfest, took a sustainability course (not for staff/faculty)	3
Sustainability engagement generally	Give money, voting, volunteering, serving as officer	4
Sustainability commitment	How committed to sustainability	1
Sustainability disposition	Willingness to pay for expanded waste prevention, alternative transportation, and greenhouse gas reduction efforts at UM	3
Rating U-M sustainability initiatives	Save energy, encourage bus or bike, promote ride sharing, recycling sustainable food, reduce greenhouse gas, maintain grounds, protect Huron River	8

Finally, the different geographic areas present opportunities to conduct experiments or trial programs in some places and not in others in order to test the effects, if any of new initiatives. That is, experiments or trials can take place in clusters of buildings as well as in individual buildings.

Table 19.2 Change in sustainability cultural indicators for students, Staff & Faculty—2012, 2013, 2014, 2015 (means scores). *Source* Marans et al. (2016)

INDICES	STUDENTS				STAFF				FACULTY			
	2012	2013	2014	2015	2012	2013	2014	2015	2012	2013	2014	2015
Primary Indicators												
Climate Action												
Conservation Behavior (4)*	6.1	6.1	6.1	6.1	6.6	6.6	6.5	6.5	6.9	6.9	7.0	7.0
Travel Behavior (1)	7.6	7.5	7.4	7.6	1.6	1.6	1.7	1.5	2.2	2.0	1.8↓	2.3
Waste Prevention												
Waste Prevention Behavior (4)	6.6	6.6	6.7↑	6.9↑	7.0	7.0	7.0	7.1↑	7.3	7.3	7.4↑	7.6↑
Healthy Environments												
Sustainable Food Purchases (3)	5.5	5.3	5.6	5.5	5.7	5.8	5.8↑	5.9↑	6.3	6.2	6.3	6.4
Natural Environment Behavior (3)	8.6	8.9↑	8.8	8.8	6.5	6.4	6.6	6.7	6.1	6.1	6.4	6.6
Community Awareness												
Sustainable Travel Awareness (4)	4.4	4.3	4.2↓	4.1↓	3.0	3.0	3.2	3.1	3.4	3.3	3.4	3.5
Waste Prevention Awareness (5)	4.0	4.2↑	4.2↑	4.1	5.0	5.1	5.0	4.9	5.1	5.4↑	5.5↑	5.3
Natural Environment Awareness (4)	3.1	3.3↑	3.4↑	3.4↑	4.1	4.3↑	4.3↑	4.2	4.3	4.6↑	4.6↑	4.5
Sustainable Food Awareness (7)	4.3	4.5↑	4.8↑	4.7↑	4.7	5.1↑	5.0↑	5.2↑	5.6	5.7	5.7	6.0↑
U-M Sustainability Intiatives (8)	5.1	5.1	5.0	5.1	5.4	5.6	5.3	5.3	4.9	5.1	5.1	5.1
Secondary Indicators												
U-M Sustainability Engagement	1.3	1.4	1.6↑	1.9↑	0.9	0.7	0.7	1.1↑	0.7	0.7	0.7	1.2↑
Sustainability Engagement generally	1.9	1.8	2.0↑	2.0	1.9	1.9.	1.8	1.8	3.0	2.9	3.0	2.7↓
Sustainability Commitment (1)	6.3	6.3	6.3	6.5	6.3	6.4	6.4	6.4	7.0	7.2↑	7.1	7.2
Sustainability Disposition (3)	3.5	3.3	3.4	3.4	2.9	2.6↓	2.6↓	2.5↓	5.3	4.8↓	4.9↓	4.8↓
Rating UM Sustainability Intiatives (8)	6.6	6.4	6.5	6.7	6.7	6.8	6.6	6.7	6.4	6.5	6.4	6.5

* Numbers in parentheses are the number of items used to create the index. For U-M sustainability engagement, 3 items & 2 items are used for students, and faculty/staff, respectively
Significant changes reflect the differences between each year's mean score and the 2012 mean scores.

↑ significant change (p<.001)
↑ significant change (p<.01)
↑ significant change (p<.05)

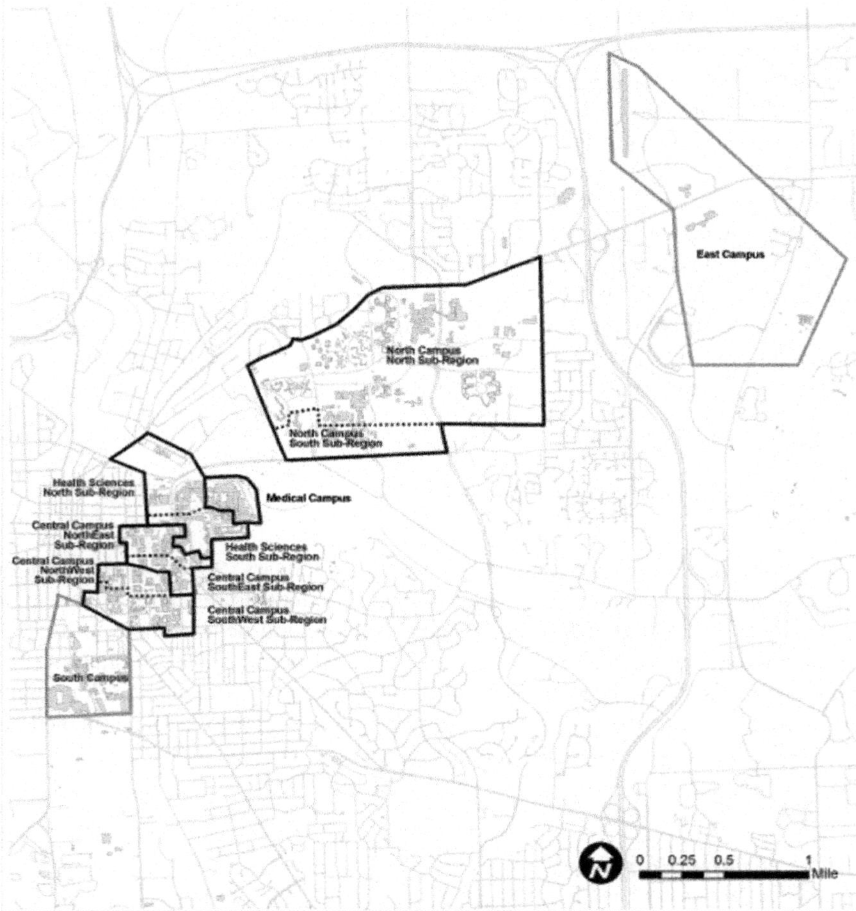

Fig. 19.1 Campus regions and sub-regions. *Source* Authors

19.4 Using SCIP to Test New Initiatives

SCIP data are being used to assess the impacts of proposed new sustainability initiatives that are set up as experiments or trials. Data collected before and after a particular intervention in a selected region or building can reveal if and by how much awareness and behavioral change has occurred as a result of new sustainability initiatives.

In 2015, several new initiatives were recommended as part of a series of sustainability reports to U-M's new president. While some initiatives are inexpensive and easy to implement, others will require substantial planning and financial resources. For cases where initiatives are relative easy and inexpensive to implement, annual SCIP data can demonstrate to decision makers whether the initiative accomplished what it was intended to do. If it does not, the initiative will be discontinued with relatively little cost to U-M. However, in the case where recommended new initiatives require substantial resources in terms, it is prudent to set up trials or experiments in one part of campus or in one or two buildings, evaluate their impacts, and determine whether it should be extended to other parts of the campus, modified, or discontinued. SCIP data will be instrumental in making that assessment. One such initiative being tested on a trial basis covers composting.

Currently, only a small amount of the University's compostable waste is diverted from landfills. Much of that waste is food scraps coming from dining facilities in residence halls. In efforts to expand composting beyond the dining halls into other parts of the students' living-learning environment, a pilot or trial program was initiated in Bursley, one of the University's largest residence halls.

19.4.1 Composting Experiment

The pilot program was planned and implemented in January, 2016 and ran through the end of the academic year. It recruited Bursley residents to regularly collect their individual food scraps or other compostable material and deposit them in a composting container in the nearest waste closet. The residence hall custodians would collect the compost material daily and take it to building's waste center (along with recyclables and other trash) where it is weighed weekly prior to its being picked up by a private industrial composting company.

19.4.2 Evaluation Plan

As shown in Fig. 19.2, there are two models for evaluating the impact of the Bursley composting experiment. The first model, Residence Hall Change, considers the impact of the experiment on residence halls whereas the second model, Individual Student Change, examines the impact of the experiment on individual students over time.

In the Residence Hall Change model, the plan shows that prior to the experiment, SCIP data covering survey participants in Bursley Hall are compared to survey participants in other University residence halls. Specific data to be compared cover students' general understanding of composing, their composting behavior, and their aware of U-M's efforts to promote composting. It is hypothesized that for there would be no significant differences between Bursley residents and those living in all other residence halls in both the 2014 and 2015 surveys.

During the 3-month intervention period, surveys are administered to all Bursley residents at two points in time. The initial survey, administered a few weeks after

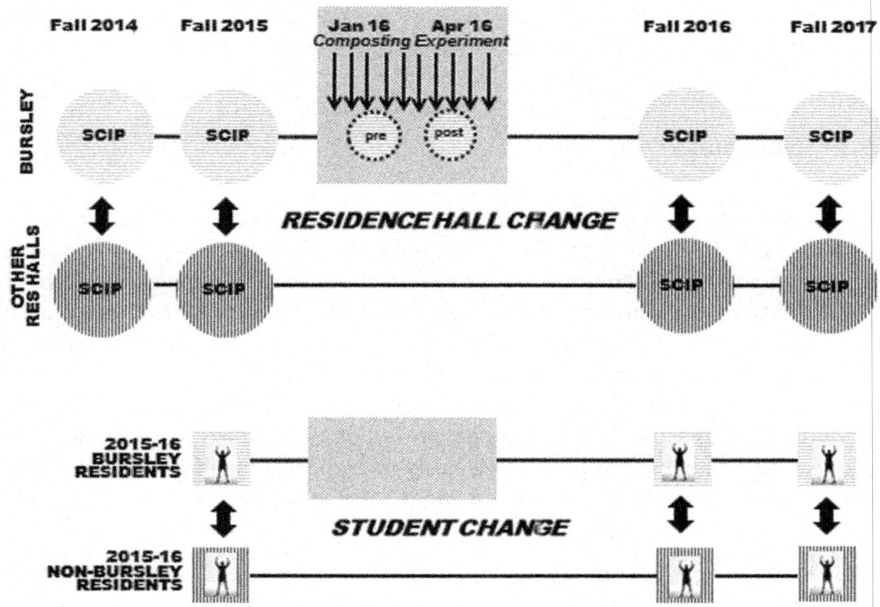

Fig. 19.2 Composting evaluation models using SCIP data. *Source* Authors

the pilot composting program was launched, would determine whether or not students had volunteered to compost and for those who had, the difficulties they were having in doing so. Students who had not participated in the composting experiment were asked why they had not done so. Both participants and non-participants were asked the SCIP questions covering their overall understanding of composting, their composting behavior, and their awareness of U-M's composting efforts. A similar follow-up survey would be administered to students prior to their leaving the residence hall at the end of the semester in late April.

The Residence Hall Change model shows that SCIP data collected in the fall 2016 and fall 2017 will compare Bursley Hall residents with residents living in other residence halls. The expectation is that as a result of the experiment, Bursley residents would be most likely to (a) know more about composting, (b) engage in composting, and (c) be more aware of what U-M was doing to promote composting on campus.

In the Individual Student Change model, the emphasis is in tracking 2015–2016 Bursley residents over the next 2 years. Some of these students are expected to be participants in the composting experiment while many others will not have participated. There is also the possibility that some will remain in Bursley as sophomores in the 2016–2017 year.

In the 2016 and 2017 SCIP surveys, students would be asked where they had previously lived while at U-M. It is hypothesized that because of the experiment, those who lived in Bursley during the 2015–2016 academic year would be more aware composting at U-M and more likely to engage in composting than students who had not exposed to the Bursley program and experience.

19.5 Summary

This chapter has posited that systematic approaches to conducting evaluations of buildings are applicable to environmentally-related programs. An example of evaluating and monitoring of a set of programs aimed at changing the culture of sustainability on the University of Michigan campus was presented. The approach used involved annual surveys of students, faculty, and staff in order to determine the extent to which levels of awareness and pro-environmental behaviors have changed over time. The annual data are also used to evaluate a specific sustainability initiative, i.e., composting, indicating the power of evaluative tools for decision making. The evaluative approach could easily be replicated in other universities and organizations where efforts are being made to change the culture of sustainability.

References

Brill, M., Margulus, S. T., & Konar, E. (1984). *Using office design to increase productivity* (Vol. 1). Buffalo, NY: Workplace Design and Productivity Inc.

Callewaert, J., & Marans, R. W. (2017). Measuring progress over time: The sustainability cultural indicators program at the University of Michigan. In F. W. Leal (Ed.), *Handbook of theory and practice of sustainable development in higher education* (Vol. 2). Berlin: Springer.

Coleman, M. S. (2011). http://sustainability.umich.edu/news/going-green-staying-blue-sustainability-michigan. Accessed 9 Dec 2016

Francescato, G., Weidemann, S., Anderson, J., & Chenoweth, R. (1979). *Residents' satisfaction in HUD-assisted housing: Design and management factors.* Washington, D.C.: U.S. Department of Housing and Urban Development, Office of Policy Development and Research.

Marans, R. W., Callewaert, J., & Shriberg, M. (2014). Enhancing and at the University of Michigan. In F. W. Leal (Ed.), *Implementing campus green initiatives: Approaches, methods and perspectives* (pp. 165–181). Frankfurt: Peter Lang Scientific Publishers.

Marans, R. W., Callewaert, J., & Webster, N. (2016). *Monitoring the culture of sustainability at the University of Michigan: Fall 2015.* http://graham.umich.edu/media/files/SCIP-Year-4-Report.pdf.

Marans, R. W., Driver, B. L., & Scott, J. C. (1972). *Youth and the environment: An evaluation of the 1972 Youth Conservation Corps.* Ann Arbor: Institute for Social Research, University of Michigan.

Marans, R. W., & Shriberg, M. (2012). Creating and assessing a campus culture of sustainability: The University of Michigan Experience. In F. W. Leal (Ed.), *Sustainable development at Universities: New horizons* (pp. 557–568). Frankfurt: Peter Lang Scientific Publishers.

Peroff, K. A., Davis, C., & Jones, R. (1979). *Gautreaux housing demonstration: An evaluation of its impact on participating households.* Washington, D.C.: U.S. Department of Housing and Urban Development, Office of Policy Development and Research.

Author Biographies

Robert W. Marans, Ph.D. is Research Professor at the University of Michigan's Institute for Social Research and Professor Emeritus of Architecture and Urban Planning in the university's Taubman College of Architecture and Urban Planning. For more than four decades, he has conducted evaluative studies and research dealing with various aspects of communities, neighborhoods, buildings, and parks as well as environmentally-related programs. His research has focused on user requirements and the manner in which attributes of the physical and sociocultural environments and environmental programs influence individual and group behavior and the quality of community life. Dr. Marans' most recent book, *Investigating Quality of Urban Life: Theory, Methods, and Empirical Research* was published by Springer (2011). He is the recipient of the 2012 Environmental Design Research Association (EDRA) Career Award and is a Fellow in the American Institute of Certified Planners (FAICP) His current research considers the impact of the built and natural environments on quality of life, the long-term benefits of new town living, and issues of sustainability and energy conservation in buildings and institutional settings.

John Callewaert, Ph.D. is the Integrated Assessment Director at the Graham Sustainability Institute and a Lecturer in the College of Literature, Science, and the Arts at the University of Michigan. As Integrated Assessment Program Director of the Graham Institute, Dr. Callewaert is responsible for designing, implementing and overseeing the day-to-day operations for the Integrated Assessment (IA) Program—the Graham Institute's engaged problem solving/translational research initiative. Dr. Callewaert came to the Graham Institute in 2009 after serving for two years as the Director of the University of Michigan-Flint Office of Research. From 2000–2007, Callewaert was the founding director of the Institute for Community and Environment at Colby–Sawyer College and an Associate Professor of Social Sciences and Education. Between 1984 and 1987 he served as an agricultural volunteer with Peace Corps Nepal working on improving seed supplies in remote hill villages. Dr. Callewaert serves as an associate editor for the Journal of Environmental Studies and Sciences and as an advisory board member for The Integrated Assessment Society.

Chapter 20
Mind the Gap: Studying Actual Versus Predicted Performance of Green Buildings in Canada

Shauna Mallory-Hill and Mark Gorgolewski

20.1 Introduction

As the building industry moves towards embracing new sustainable practice, outcomes are less assured than in the past. "Gaps" are reported in predicted and actual outcomes, indicating that "green" buildings are not performing as expected (e.g. Fedoruk et al. 2015; Newsham et al. 2012; Bartlett et al. 2014). Are owners and occupants of such buildings getting what they paid for? Bordass et al. (2004) suggest that "credibility gaps" are not necessarily due to mistakes but rather assumptions made at the design stage are not well enough informed by monitoring buildings in-use. Akerstream et al. (2012) point out that most green building rating systems such as Leadership in Energy and Environmental Design (LEED), Green Globes, and Building Research Establishment Environmental Assessment Method (BREEAM) have focused on predicted performance at the design stage. Delivered performance is rarely voluntarily verified by building owners. Thus, measurement and monitoring of in-use performance of "green" buildings is crucial to advance confidence and innovation in sustainable design.

This chapter examines the gaps between anticipated and achieved performance of nine "green" buildings located across Canada. The evaluation covers key performance indicators (KPI) for buildings under five major categories. Data collection involved both qualitative and quantitative techniques. Additional information about the findings highlighted here can be found at http //iisbecanada.ca/sb-14/.

S. Mallory-Hill (✉)
Faculty of Architecture, University of Manitoba, Winnipeg, MB R3T 2N2, Canada
e-mail: S.Mallory-hill@umanitoba.ca

M. Gorgolewski
Interim Associate Chair Graduate Studies, Building Science,
Department of Architectural Science, Ryerson University, Toronto, ON, Canada
e-mail: mgorgo@ryerson.ca

W.F.E. Preiser et al. (eds.), *Building Performance Evaluation*,
DOI 10.1007/978-3-319-56862-1_20

20.2 Nine Green Buildings

This study involved the in-depth building performance evaluation (BPE) of 128,164 m^2 (1,379,000sf) of high performance "green" buildings, including academic, office, and community buildings, housing more than 6000 occupants (see Table 20.1). The buildings are located in five different bio-regions across Canada. Most of the buildings are certified by one or more green building rating systems. Each building has been in operation for at least two years, some for considerably longer. The buildings use a typical selection of sustainable building technologies and strategies employed in green building in Canada. Many focused on natural daylighting and natural or mixed-mode ventilation strategies, passive solar strategies

Table 20.1 Details of study buildings. *Source* Authors

Building	Location	Type	Net floor area (m^2)	ASHRAE climate zone	Construction cost (Can. \$/m^2)	Type
MMM Group office	Kitchener Ontario	Small Office	1900	6	$2900	New build
Manitoba Hydro Place	Winnipeg Manitoba	Large Office	64,590	7	$3550	New build
Surrey District Education Centre	Surrey, BC	Medium Office	11,420	5	$2500	New build
Canal Building	Ottawa, Ontario	Medium Academic	7310	6	$4160	New build
Ron Joyce Center	Burlington, Ontario	Medium Academic	9340	5	$1980	New build
Roblin Centre	Winnipeg Manitoba	Large Academic	19,210	7	$1950	Re-use & new build
Jim Pattison Centre of Sust. Building	Okanagan, BC	Medium Academic	6780	5	$4150	New build
Centre for Interactive Research on Sust.	Vancouver, BC	Medium Academic	5500	5	$5611	New build
Alice Turner Library	Saskatoon, Saskatchewan	Small Community	2070	7	$3200	New build & addition

including improved thermal insulation and the use of thermal mass. A wide range of HVAC systems from simple to complex were used. Four buildings included renewable energy systems and three used water collection or recycling systems.

20.3 Project Methodology

The multidisciplinary team consisted of researchers from three Canadian Universities: University of British Columbia, University of Manitoba, and Ryerson University. The project, initiated by iiSBE Canada (International Initiative for a Sustainable Built Environment), was supported by the industry and public funds. The BPE methodology compared three types of building performance: (1) design-stage "predicted", (2) in-use "actual" and (3) reference standards. The protocol focused on measuring KPI with readily deployable collection instruments. The categories of KPI studied include: occupancy factors, energy and emissions, water use, economic factors, and indoor environmental quality (IEQ) (see Table 20.2).

The study used qualitative and quantitative data collection and analysis methods based on an extensive literature review of previous studies (including: Leaman et al. 2010; Newsham et al. 2009; ASHRAE 2010; Baird 2010; Mallory-Hill et al. 2012) and further refined by the research team. Three levels of increasingly in-depth investigation were identified (see Table 20.3). A Level 2 POE was carried out at almost all locations. The data collection took place during the spring and summer of 2014.

The operational performance of each building is compared to its design goals or targets and to relevant benchmarks. A difference between actual and expected/benchmark outcomes is noted as a "gap". Individual building results are not compared to each other, because of differences in size, type, and location. Some general indicative trends and lessons however, were drawn from the overall performance across the buildings and are mentioned later in this chapter.

Data was taken from utility bills, design documentation, rating system submissions, semi-structured interviews with members of the design and operations teams, an occupant survey, and spot IEQ measurements. Qualitative and quantitative data collection methods allow for a robust study of the buildings, with the comparison of measured energy, water, and IEQ data to the experience of inhabitants and operations personnel. The period of data collection was relatively short and limited to spring-summer. On-line survey response rates varied from 16 to 90% of occupants, some low response rates are likely due to vacations. Results are therefore considered indicative rather than conclusive. The following sub-sections highlight some findings from the KPI categories studied.

Table 20.2 Excerpt of data collection and analysis spreadsheet for performance comparison. *Source* Authors (Bartlett et al. 2014)

	No.	Performance indicator	Reference standard	Predicted performance	Actual operational performance	Difference between actual and reference (%)	Difference between actual and predicted (%)	Units
E: Energy and emissions								
Required	E1	Total delivered electricity		81.0	81.7		1	kWh/m^2 *yr
Required	E7	Building energy use intensity for all operating uses (E1 + E2 + E3 + E4 + E5 − E6)	372	81.0	81.7	−78	1	kWh/m^2 *yr
Required	E8	Net delivered energy use intensity for all operating end uses (E1 + E2 + E3 − E6)	372	71.3	81.7	−78	15	kWh/m^2 *yr
Required	E9	Greenhouse gas from delivered energy for all operating end uses	55	8.0	10.0	−82	25	Kg GHG/m^2 * yr
W: Water								
Required	W1	Delivered water per occupant to the building	7.3	1.3	2.2	−70	70	m^3 water delivered per occupant per year
Required	W2	Recycled or captured water used in the project (If available)		1.91	1.41		−26	m^3 per occupant per year
Required	W3	Gross water use per occupants (W1 + W2)	7.3	3.2	3.6	−51	12	m^3 gross water use per year
Required	W4	Water use intensity (W1 + W2) per m^2 of conditioned area	0.327	0.142	0.147	−55	3	m^3 gross water use per m^2 per year
$: Economic factors								
Required	$1	Construction cost	Not available		2899	0		Cost per m^2 of net floor area
Required	$2	Commissioning cost	Not available		Not available	0		Cost per m^2 of net floor area
Required	$3	Annual operating water cost	Not available	0.40	0.41	0	3	Cost per m^2 of net floor area per year
Required	$4	Annual operating energy cost	Not available	13.22	13.33	0	1	Cost per m^2 of net floor area per year

Table 20.3 BPE levels of investigation (as proposed by authors). *Source* Authors

	Activity
Level 1	Discussion with building owner/operator
	Walk through observation of building
	Document analysis: energy consumption and demand
	Document analysis: water consumption
	Discussion with owner/operator
Level 2	Interview design team/owner
	Occupant satisfaction survey
	Observation study (technical)
	Field measurement (IEQ)
	Report findings
Level 3	Field measurement (H_2O, Energy, Occupants)
	Re-calibration of energy model
	Building envelope investigations
	HVAC system investigations
	In-depth study of specific key performance indicator
	Observation study (occupant)
	Occupant interview/focus groups
	Recommending actions
	Reviewing outcomes (validity, bias)
	Determining Design Guidelines

20.4 Anticipated and Actual Building Performance

20.4.1 Occupancy

Occupancy affects many aspects of building performance. Table 20.4 and Fig. 20.1 show design stage predictions, actual daily average number of occupants, and the typical weekly operating hours. The buildings are used considerably differently than expected. Daily occupant numbers vary by a range of −57 to +20% and building operating hours up to +82% longer (see Fig. 20.1). Only three buildings were within 20% of their predicted usage rate. Two were over double the predicted rate. The Roblin Centre has 20% more people for 75% more hours per week. So in total it has 185,430 occupant hours per week (no. of occupants × average hours), compared to an expected 88,000 ohr/w. Such occupancy "gaps" impact water and energy use.

Accurately predicting and measuring occupant loads is challenging. For office buildings occupant numbers can be based on employee numbers and workstations. However, in educational buildings full and part-time staff and students change hourly throughout the year. For two buildings in this study it was not possible to establish accurate actual occupant numbers.

Table 20.4 Comparison of predicted and actual occupancy levels. *Source* Authors

Building	Typical daily occupancy during operating hours (persons/h)			Typical weekly operating hours (no. of h)		
	Predicted	Actual	Difference (%)	Predicted	Actual	Difference (%)
MMM Group office	85	78	−8	57	57	0
Manitoba Hydro Place	2000	1800	−10	50	60	20
Surrey District Education	396	465	17	45	82	82
Roblin Centre	2200	2649	20	40	70	75
Jim Pattison Centre	250	137	−45	45	69	53
CIRS	378	252	−33	45	57	27
Alice Turner Library	42	32	−24	66	57	−14

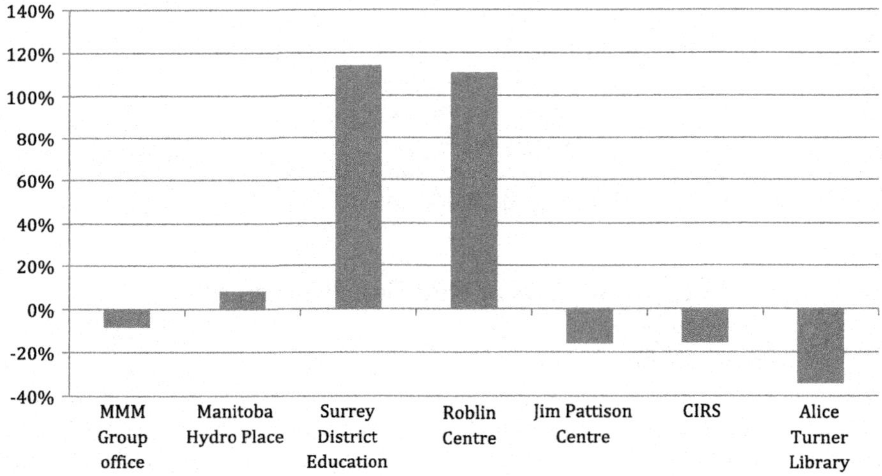

Fig. 20.1 Variation of building weekly occupant hours (average no. of occupants/h × h of operation per week) from predictions. *Source* Authors

20.4.2 Energy

Figure 20.2 shows comparisons of predicted, i.e., modelled, measured and typical reference building energy use intensity (EUI) values for the nine buildings. All but one building are performing considerably better than benchmarks. Five are more than 50% more efficient. The gap between measured and predicted performance varies significantly. MMM office and Manitoba Hydro Place perform close to predicted energy use, however it took several years of refinement and tuning to achieve this.

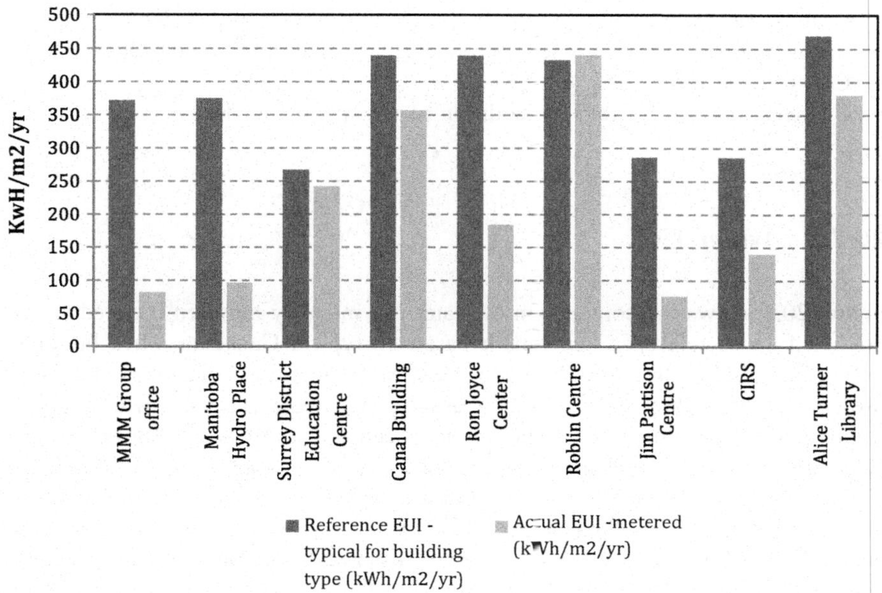

Fig. 20.2 Comparison of building EUI (kWh/m^2/yr) predicted, actual and reference for 9 buildings. *Source* Authors (Bartlett et al. 2014)

Three buildings failed to meet their predicted performance by a significant margin, but were not seen as "failures". The Surrey District Education Centre consumes 9% less energy than reference, but the actual EUI (242 kWh/m^2/yr) is significantly higher than predicted. The evaluation process identified technical issues related to mechanical systems that can now be fixed. Also, the building is used for longer hours (82%) and by more occupants (17%) than predicted (see Table 20.4). Similarly, the Roblin Centre uses 69% more energy than predicted, but the building is used for longer hours (75%) and by more occupants (20%). Both buildings use more energy because they are used more intensely, which in turn eliminates the energy and environmental costs of constructing another building. The Jim Pattison Centre performs 22% better than predicted partly because occupancy is 16% lower than expected.

It was found that energy models using standardized conditions were generally poor predictors of actual building performance in the nine buildings due to:

- Occupancy numbers and patterns
- Non-standard hours of operation
- Occupant behaviour
- Unregulated loads, e.g., IT, office equipment, coffee makers etc., and external lighting
- Technical problems with HVAC equipment and controls

- Lack of thorough and on-going commissioning of building systems
- Insufficient resources/knowledge to manage/operate the building efficiently
- Construction problems
- Changes in use, e.g., tenants, addition of unexpected loads

20.4.3 Water Use

Figure 20.3 shows the comparison of water use intensity between metered actual, predicted and a reference benchmark based on typical Canadian office buildings (REALPac 2012). All the buildings use less water than the benchmark, but some use considerably more their predicted usage. The predictions vary widely and in some cases seem imprecise. The higher occupancy levels of Surrey District Education Centre likely led to higher water use. The Roblin Centre also had a higher occupancy rate than predicted, but a lower water use due to a major retrofit of water fixtures. The Alice Turner Library uses very little water, most likely because most visitors stay for only a short time and only staff regularly use the washrooms. The MMM Group office and Jim Pattison Centre reduced municipal water consumption to below a very low 0.15 $m^3/m^2/yr$ by including rainwater

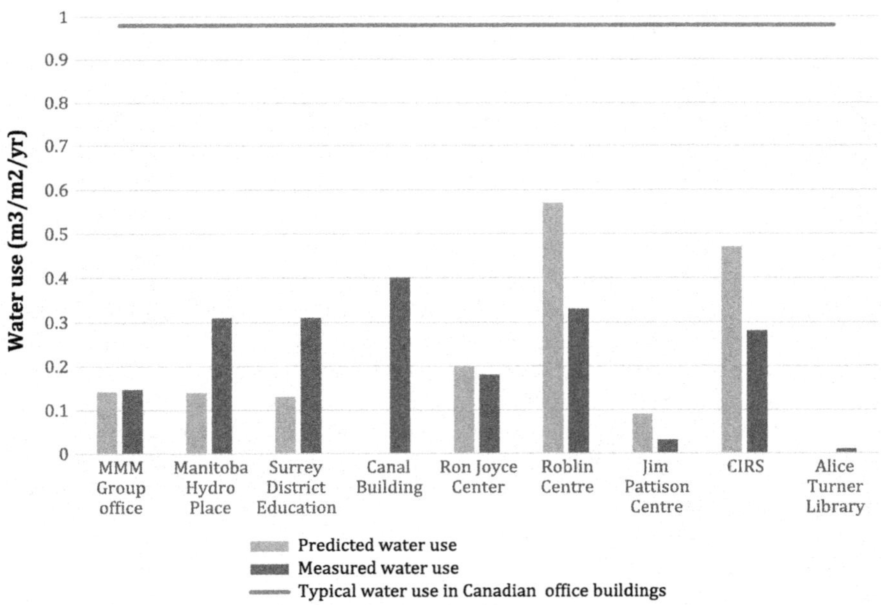

Fig. 20.3 Comparison of water use for 9 buildings—predicted versus actual versus reference. *Source* Authors (Bartlett et al. 2014)

collection/reuse systems. Other buildings achieved low water use by careful specification of fixtures and with minimizing landscape irrigation.

20.4.4 Indoor Environmental Quality (IEQ)

Figure 20.4 compares occupant survey scores for five IEQ categories. Occupants scored performance on scale of 1–7 (see Newsham et al. 2012 for the methodology). The range of responses varied widely from building to building, and within each building. Respondents generally expressed satisfaction with lighting conditions and dissatisfaction with speech privacy. Thermal comfort scored well, although some buildings had significant variations in occupant satisfaction between winter and summer. Occupants wanted more personal control over windows/lighting, ventilation, temperature, and noise. Many comments centered on specific technologies and local problems.

The percentage of time that workspace lighting met the recommendations from the Illuminating Engineering Society of North America is compared with occupant satisfaction in Fig. 20.5. Generally, occupant satisfaction is high even when lighting conditions do not meet recommended levels. In the CIRS building only 34.5% of the measurements were within the recommended artificial lighting range of 300–750 lx; most were higher due to daylight contribution. However, surveyed occupants were satisfied to very satisfied with lighting levels, with a mean score of 5.7 out of 7. Findings suggest that occupants are more tolerant of higher illumination levels in daylit spaces.

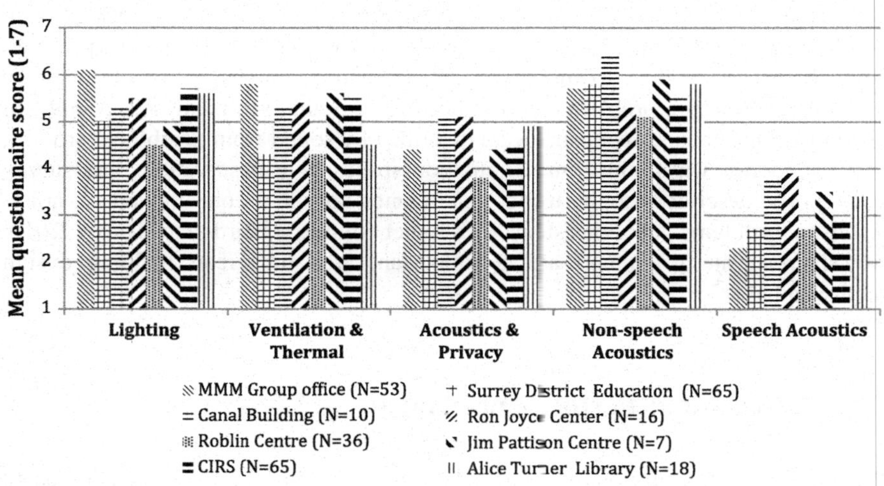

Fig. 20.4 Questionnaire mean scores for eight buildings compared (using a 7 point scale with 1 = dissatisfaction and 7 = satisfaction). *Source* Authors

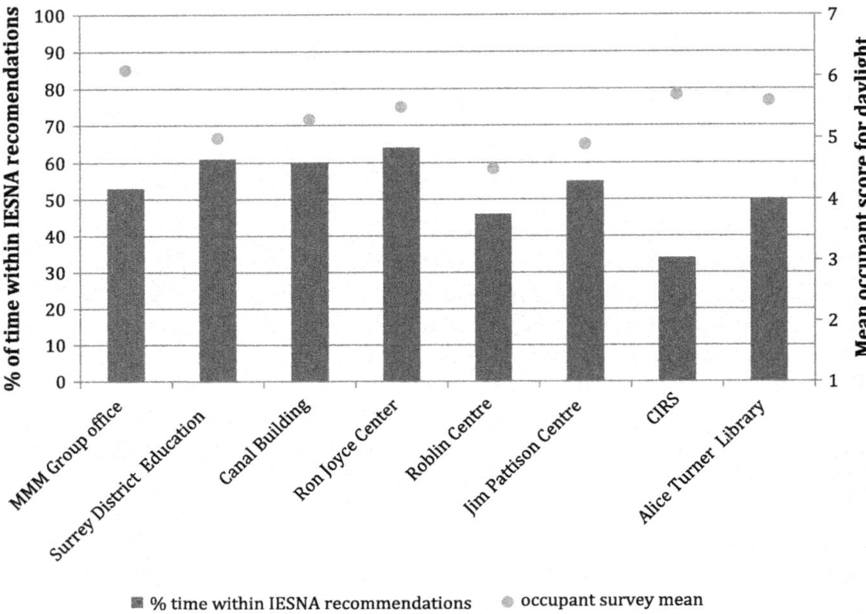

Fig. 20.5 Lighting measurements and occupant scores compared. *Source* Authors

Compared to all other aspects of the indoor environment, building occupants showed the lowest levels of satisfaction with acoustics and privacy, and were especially dissatisfied with speech conditions (see Fig. 20.6). Sound pressure levels were measured on site in A-weighted decibels, then converted into Noise Criterion Balanced (NCB) values and compared to reference standards. Only 45% of open plan spaces and 55% of private offices were below the recommended maximum NCB values. According to the occupants distracting conversations and acoustic privacy were often a problem (scoring between 2.3 and 3.9, mean 3.2, $N = 9$). In open plan office areas, occupants were disturbed by sound from nearby classrooms, meeting rooms, washrooms, and circulation spaces. Site observations and survey self-reports describe people using headphones to block out unwanted noise. Acoustic problems are consistent with other studies (Baird and Dykes 2010; Newsham et al. 2012) and are an important area for green building design improvement.

20.5 Discussion: Performance Gaps

What do gaps in performance mean? Are buildings that do not meet anticipated performance targets "failures"? The answers are not straightforward. A walkthrough observation of adaptive behaviours like headphone use can suggest

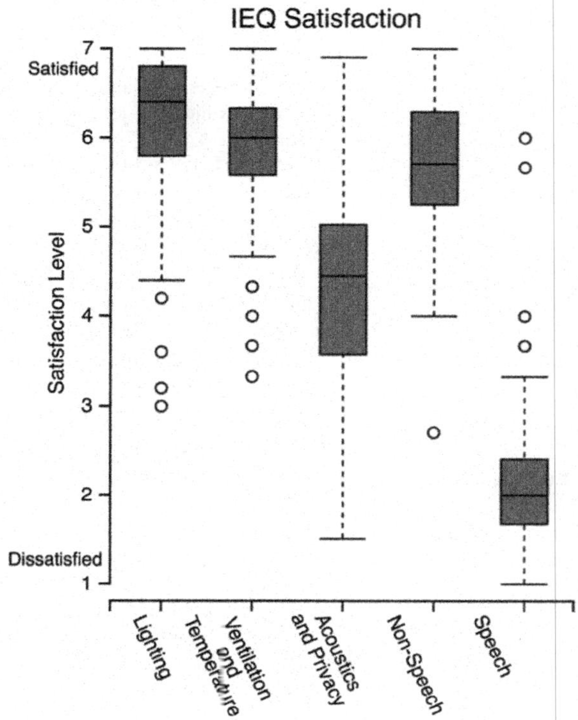

Mean	6.1	5.8	4.4	5.7	2.3
Standard Deviation	0.97	0.81	1.09	0.87	1.01

Fig. 20.6 Survey results for the key IEQ factors for the MMM building showing the median and four quartiles into which that data is divided. *Source* Authors (Bartlett et al. 2014)

acoustic issues even if the area appears quiet. Spot measures of lighting levels may not comply with recommended reference standards, but occupants are satisfied because they are more tolerant of a natural source. Design teams highlight original goals while building managers are aware of changes to building use, maintenance, and operational procedures over time that explain unexpected outcomes. A "failure" in worse than predicted energy performance, can still be "win" overall for the environment in an intensively used building. As Bordass et al. (2004) indicate, it is not so much that the predictive techniques are wrong, but "gaps" occur because the assumptions used were not well enough informed by what really happens in buildings once they are in occupied. Understanding and assessing the performance of the buildings in-use, requires consideration in the context of each individual building's "story", from design, through construction to occupancy and into the future.

This study gave providers a good picture of in-use performance of each building compared to predicted and reference standards. A summary of lessons-learned from the nine study buildings is provided in Table 20.5. The scope of this study did not

Table 20.5 Lessons learned. *Source* Authors

	Observation
Occupancy	• Not well understood and often varies considerably from design prediction which can lead to operational issues (e.g. energy and water use) • Actual numbers hard to determine
Energy Models	• Design stage energy models are better suited to comparison of design options than accurate prediction of actual final performance • Recalibrate energy model based on actual building use and occupancy
Water use	• Linked highly to occupancy, this variable needs to be better understood both at the prediction stage and during building use
Benchmarks	• With the exception of energy, finding relevant, accurate, local reference KPI benchmarks for comparison can be difficult
Acoustics	• Of variables associated with IEQ, green buildings consistently score worst on acoustics (speech privacy, other occupants)
Lighting	• Occupants appear to be more accepting of high lighting levels from a daylight source, than from an artificial source
Operations and management	• Continuous commissioning and operation and management of systems are crucial to successful performance, yet often overlooked by owners • Exemplary performance in several buildings linked to actions by building management and operational staff
Sub-metering	• Lack of sub-metering hinders both system evaluation and opportunity for building operator to improve performance

allow researchers to gauge the size of the performance gaps, or look "within" each "gap" to determine causes and/or suggest extensive remedial action. This would require reconciling design stage performance projections with the actual building use and occupancy. Energy and water use analysis would need a greater degree of granularity, e.g., specific end-uses. The lack of sub-metering and/or data availability is a critical limitation for investigators, as it is for building operators trying to monitor, maintain, and improve the performance of their buildings.

Assessing new green building technologies by conventional building benchmarks and standards may be inappropriate. The predictive limitations of energy models and satisfaction of occupants with broader ranges of daylighting compared to artificial lighting levels were noted. Leaman and Bordass' 2007 study found occupants tend to be more "tolerant" of green buildings, ranking them higher for overall IEQ comfort than conventional buildings. This sometimes obscures flaws on individual IEQ variables, e.g., acoustics that consistently rank poorly in green buildings. Metrics need to be adapted for buildings employing newer, green technologies with naturally sourced lighting, ventilation, and conditioning.

20.6 Conclusion

This chapter has provided an overview of a BPE study of nine "green" buildings in Canada. The performance "gaps" between predicted, actual, and benchmark performance in the areas of occupancy, energy use, water use, and IEQ were presented to explore the implications "failing" or "succeeding" to achieve targets set at the design stage. An overarching observation is that closing the gap to achieve high levels of building performance in "green" building has far more to do with communication and feedback mechanisms from buildings in use to the design stage, than it does with technical issues. No building provider or designer is anxious to disclose the "failures" of their buildings; however, it is increasingly clear that advancing innovative systems requires understanding them in use and focusing on lessons learned. The adoption of green building rating systems and municipal energy disclosure bylaws provides an opportunity to begin sharing the results of the measurement and verification of buildings in use, including making building energy performance publically available. The challenge is to determine how to encourage the adoption of feedback mechanisms and communication between building researchers, design teams, building providers, operators, and occupants to inform the design of buildings in the future.

References

Akerstream, T., Knirsch, A., & Pauls, M. (2012). Manitoba hydro place: Energy efficiency 2.0. Dallas, TX. *ASHRAE Transactions*, *119*(1), Special section p 1.

ASHRAE. (2010). *Performance measurement protocols for commercial buildings*. Atlanta, GA: ASHRAE.

Baird, G. (2010). *Sustainable buildings in practice: What the users think*. New York: Routledge.

Baird, G., & Dykes, C. (2010). Acoustic conditions in sustainable buildings—Results of a worldwide survey of users' perceptions. *Journal of Building Acoustics, 17,* 291–304.

Bartlett, K., Brown, C., Chu, A. M., Ebrahimi, G., Gorgolewski, M., Hodgson, M., Mallory-Hill, S., Ouf, M., Scannell, L., & Turcato, A. (2014). Do our green buildings perform as intended? In *Paper presented at World Sustainable Building Conference (SBE2014)*, Barcelona, Spain. Paper retrieved from: http://iisbecanada.ca/umedia/cms_files/Conference_Paper_1.pdf. Accessed 18 Nov 2016.

Fedoruk, L. E., Cole, R. J., Robinson, J. B., & Cayuela, A. (2015). Learning from failure: Understanding the anticipated–achieved building energy performance gap. *Building Research & Information, 43*(6), 750–763.

Bordass W., Cohen R., & Field, J. (2004). Energy performance of non-domestic buildings: Closing the credibility gap. In *Paper presented at Building Performance Congress*, Frankfurt. Paper retrieved from: http://www.usablebuildings.co.uk/Pages/Unprotected/EnPerfND Buildings.pdf. Accessed 18 Nov 2016.

Leaman, A., & Bordass, B. (2007). Are users more tolerant of "green" buildings? *Building Research & Information., 35*(6), 662–673.

Leaman, A., Stevenson, F., & Bordass, B. (2010). Building evaluation: Practice and principles. *Building Research & Information, 38*(5), 564–577.

Mallory-Hill, S., Preiser, W. F. E., & Watson, C. (Eds.). (2012). *Enhancing building performance*. Oxford: Wiley-Blackwell.

Newsham, G., Birt, B., Arsenault, C., Thompson, L., Veitch, J., Mancini, S., Galasiu, A., Gover, B., Macdonald, I., & Burns, G. (2012). *Do green buildings out-perform conventional buildings?* Indoor environment and energy performance in North American offices, Research Report RR-329: National Research Council Canada.

Newsham, G., Mancini, S., & Birt, B. (2009). Do LEED-certified buildings save energy? Yes, but…. *Energy and Buildings, 41,* 897–905.

REALPac. (2012).*Water benchmarking pilot report: Performance in the Canadian Office Sector.* Retrieved from: http://c.ymcdn.com/sites/www.realpac.ca/resource/resmgr/ industry_sustainability_-_water_benchmarking/rp_water_report_05_hr_final.pdf. Accessed 18 Nov 2016.

Author Biographies

Shauna Mallory-Hill is an Assistant Professor in the Faculty of Architecture, University of Manitoba. She has a PhD in Architecture from the Eindhoven University of Technology (TU/e), in The Netherlands, and a Masters of Architecture (M. Arch) from the University of Manitoba. A LEED BD + C Accredited Professional, she has over 25 years' experience in teaching and research in building systems, universal & sustainable design, and building performance evaluation.

From 1992 to 1996, Mallory-Hill was Executive Director of the Canadian Institute for Barrier-Free Design. She served on the Canadian Commission on Building and Fire Codes (CCBFD) Standing Committee for *Sect. 3.8 Barrier Free Design* and several municipal and provincial level Universal Design committees. She has held executive positions on the Boards of Directors of the Manitoba Chapter of the Canada Green Building Council (Director of Education and Training) and Environmental Design Research Association (EDRA–Chair and Treasurer).

Her first book, ACCESS (Lane & Hill, CPA, 1989) was followed by a chapter on UD Codes in Canada co-authored with B. Everton in *Universal Design Handbook* (Preiser et al., McGraw-Hill, 2010). For the past two decades her research work has concentrated on exploring how sustainably designed environments impact on occupant health and productivity. In addition to publication in books and journals, Mallory-Hill is regular presenter of her green building evaluation work at EDRA conferences along with recent presentations at the World Sustainable Building Conference (Barcelona, WSB14), CaGBC Summit (Vancouver, 2015) and SBE16 (Toronto, 2016). Her latest book, co-edited with Wolfgang Preiser and Chris Watson, is *Enhancing Building Performance* (Wiley-Blackwell, 2012).

Dr. Mark Gorgolewski is a Professor in the Department of Architectural Science at Ryerson University in Toronto. He has worked for many years as an architect, researcher and sustainable building consultant in Canada and the UK. He has been a director of the Canada Green Building Council and chair of the Association for Environment Conscious Building and is a LEED Accredited Professional. Mark has written many papers and books on the subject of sustainable built environments. Currently areas of research include building performance, reuse of components and materials in buildings, and design for urban agriculture. He was co-curator of the exhibition "Carrot City: Designing for Urban Agriculture," which has travelled around the world, and is co-author of a Carrot City book and web site. Mark has participated in various sustainable building projects, including a winning design for the CMHC Equilibrium (net zero energy) Housing Competition and is also co-recipient of the 2007–2008 ACSA/AIA Housing Design Education Award, and recipient of the 2012 H.A. Krentz Research Award from the CISC and the CMHC 2013 Excellence in Education Award.

Chapter 21
The College and University Campus: Facility Assessments for Long Term Decision Making

Brodie Bain

21.1 Introduction

Post-Occupancy Evaluations (POEs) have been defined as "the process of evaluating buildings in a systematic and rigorous manner after they have been built and occupied for some time" (Preiser et al. 1988, p. 3). One important objective of POEs is to glean lessons-learned from the review of completed buildings to improve design strategies for future projects. For Vischer (2001, p. 23), it is about "learning how a building performs once it is built, including if and how well it has met expectations." This is a 'rear-view' approach to POEs, in order to feed forward future design decisions, and has served as an important tool for designing better spaces. Over the many years since POEs were first developed, there has been concern by researchers over the utilization of results by designers (Duffy 2005; Karim and Crozier 2009). In fact, scores of studies and presentations have touched on this issue over the years. The Environmental Design Research Association (EDRA) has published some of these studies and/or hosted presentations as evidenced by conference proceedings since the late 1960s. A common conclusion is that POEs are most useful when results are applied to ongoing building programs where designs are replicated over and over, for example, the U.S. Postal Service's 'Store of the Future' prototypes of the early '90's, or where findings can be generalized to specific types of settings, such as studies on senior housing or hospitals. Generalizing has its own limitations however, with the challenge of control for the many variables that affect person-environment relations.

B. Bain (✉)
AICP, LEED AP, Perkins & Will, Suite 2300, 1301 Fifth Avenue,
Seattle, WA 98101, USA
e-mail: Brodie.Bain@perkinswill.com

W.F.E. Preiser et al. (eds.), *Building Performance Evaluation*,
DOI 10.1007/978-3-319-56862-1_21

Preiser discusses the usefulness and value of POEs for designers to reflect on and review results of their work (Preiser et al. 1988). It has been stated that design practitioners ought to bear the cost of POEs to learn and improve on their design approaches for future projects (Bordass and Leaman 2009). At Perkins + Will and other firms, POEs have been applied through systematic review of building performance such as energy and water use and technical concerns, as well as user and owner feedback. In the world of practice however, challenges remain related to building function and behavioral findings much as in the world of academic research, in particular, resources required (time and funding) to produce generalizable results that apply to every functional dimension of a specific building or space.

21.2 Building Evaluations and Future Decision-Making

Another application of POEs that is addressed less frequently is one of its most powerful, a version of the Building Performance Evaluation (BPE) concept. In campus environments where multiple buildings are owned and managed by one entity, the campus land and facilities must continually be considered a portfolio of assets with buildings regularly assessed for their value and life expectancy. The cost of upgrades to extend a building's life through renovation, along with maintenance costs over time, are compared against the price of replacement. With this approach, life expectancy criteria typically focus on the condition and capacity of a building's physical systems such as structural, mechanical/electrical/plumbing systems (MEP), building envelope, extent of hazardous materials, and compliance with current codes. These are the common drivers for decision making about whether to retain or replace buildings. Often buildings are also assessed for their physical performance, as in energy and water use.

But what about the appropriateness of the building's design for meeting institutional goals and supporting needed activities? Functional criteria that address user needs are not commonly used. This is partly due to the challenge of capturing and evaluating more qualitative data without the use of robust research techniques, such as surveys designed to elicit responses that allow for analysis with scientifically valid results, requiring significant time and cost.

Yet, the role of facilities to serve a need, that is, facilitate a set of activities, is their primary purpose. Understanding how well facilities support user goals and activities through user input and observations can be hugely valuable: Which existing buildings are worth keeping and can continue to support an institution's mission and vision? Which buildings can be adapted at a lower cost, and/or more effectively, than the cost of constructing a new building? Campus environments in particular can benefit from this 'forward-view' approach.

Campus representatives continually face decisions about the future of campus buildings—how to move ahead with the buildings and environments they steward. The steadfast goal must be to support the institution's mission, which typically encompasses some combination of education, research, and service. Moreover these activities are quickly evolving and changing. For example, new advances in neuroscience have significantly changed our understanding of the way people learn, resulting in direct impacts to curriculum and learning environments (Tokuhama-Espinosa 2011). The active-learning model where students are directly engaged in problem-solving and team-based work has proven to be highly effective (Freeman et al. 2014). Classrooms that support this new pedagogy must be open, flexible, and fully digital, yet highly accessible with direct visual connections to and among all participants - resulting in more square footage per seat.

Similarly, many pressing challenges of our day are being studied through interdisciplinary research such as Bioengineering and Environmental Sciences. The result is that with research activities are becoming more integrated across disciplines. Direct space implications are that academic environments must encourage, more than ever, interaction, interdisciplinary exchanges and informal, serendipitous encounters.

Such evolutions, including an evolving character of openness and cross-fertilization, are flourishing and extending beyond the walls of higher education. In addition, the Academy's sense of responsibility to its service role and relationships to external communities is becoming increasingly important, resulting in campuses that are much more porous and integrated with their surrounding neighborhoods. Among the many examples are the University of Washington,

Fig. 21.1 University Crossing, UMass Lowell: Student center with ground level retail along the campus edge. *Source* Edward Dudley, Perkins + Will

South Seattle College, The Ohio State University and the University of Massachusetts Lowell (see Fig. 21.1).

While such evolutions make sense now in the 21st Century (C.), many of the buildings on campuses throughout North America are out of date and do not support new ways of thinking, acting, and operating as an institution. How well do existing 19th and 20th C. buildings support current and future changes in higher education? This is where a functional evaluation that examines an existing building's seminal qualities against the activities that must be supported can be extremely useful. While this approach is not formally called POE in the author's practice, it is very similar to the Indicative POEs described by Preiser et al. (1988).

21.3 Functional Assessments

In campus planning work, the primary focus should always be on the mission and vision of an institution, and how the campus might help these to be achieved through a supportive physical environment. This requires an evaluation of the existing conditions, including campus buildings, by considering their likely future effectiveness and life expectancy to determine whether it makes sense to plan with or without them.

Typically a good amount of time is spent with stakeholders to understand the institution's vision for the future and resulting total space needed based on the institution's culture, expected growth, and anticipated changes in ways of doing things. Depending upon the institution's needs and the focus of the plan, the level of detail studied ranges from a high level campus-wide assessment of space needs to a design-ready, room-by-room program for a specific building.

Space needs are then evaluated against the existing building stock for capacity and condition, answering questions such as: Does the existing space accommodate the need? Will the buildings last as long as they need to? Often the long-term plan incorporates a 10-year capital plan where specific projects are identified. Projects may be 'Growth' projects, to support a new program or increases in enrollment, 'Replacement' projects, to address uncorrectable issues of a specific building, or 'Major Renovations,' where a facility can be corrected and its life extended.

Renovation compared to Replacement decisions require an understanding of building performance, often without the resources and most critically time, to perform a robust POE. So, the most common approach to understanding the value of an existing building focuses on building systems through a Physical Evaluation without understanding function. Yet, a Functional Evaluation can provide critical information to the decision making process.

21.4 Functional Assessments—Data Gathering

Functional Assessments are much like an "Investigative POE", as they reach out to users, observe users, and develop evaluation criteria. Several social science data-gathering techniques are especially effective including Walkthrough Tours with department representatives and facilities staff; Focus Groups with functional units; and Observations. Observations can be augmented with Geo-based User input via Web Surveys; these can be very effective and powerful.

- *Walkthrough Tours*: This is an initial tour of the facilities, hearing from user representatives and facilities staff about what works and what doesn't. The issues are kept at a high level so as avoid getting bogged down in the details such as casework fixes or temporary space configurations—high level decisions are what need to be made. It should be noted that, this can also be a time to listen for the opportunity to solve easy-to-solve problems quickly, even while the long term plan is the focus.
- *Focus Groups*: Focus groups allow the team to dive more deeply into understanding the needs and aspirations of each functional unit—both in terms of how they operate and their adjacency requirements. The level of detail varies depending upon the scope of the study, e.g., whether a campus-wide effort or precinct plan is needed, or the effort is centered on specific program functions, like housing, learning spaces, recreation, or workplace, for example. In every case, the discussion generally focuses on how the current facilities fulfill current and expected future needs, both in terms of growth and evolving ways of learning, teaching, research and other work.
- *Geo-based User Input via Web Surveys*: Web surveys have proven to be effective and valuable, particularly for understanding or confirming use patterns at the campus and building scale. Users drag icons on a map or floor plan to places where they typically engage in certain activities such as 'socializing', 'studying', 'eating', or where they see and experience the most problems with 'wayfinding' or 'traffic.' They can also trace their typical use patterns for walking on campus, driving, or biking. At the University of Oregon, within a couple weeks, the Physical Framework Vision Plan team of Perkins + Will was able to collect information from over 2000 participants with a graphic that compiled all responses to a particular question, real time, and illustrated geospatially (see Figs. 21.2 and 21.3).
- *Visual Observations*: Often the initial walk-through tour is followed by a more detailed team tour to survey each space and gain a sense of its current use, its potential for a new use, and an assessment to 'right size' its current function.
- *Documentation Review*: This includes the review of existing space data bases, floor plans, and structural drawings. Databases are organized by functional unit to gain a sense of current space. Floor plans are also annotated by functional unit and reviewed against adjacency needs. Structural drawings are reviewed to understand the building's capacity to support current or other functions.

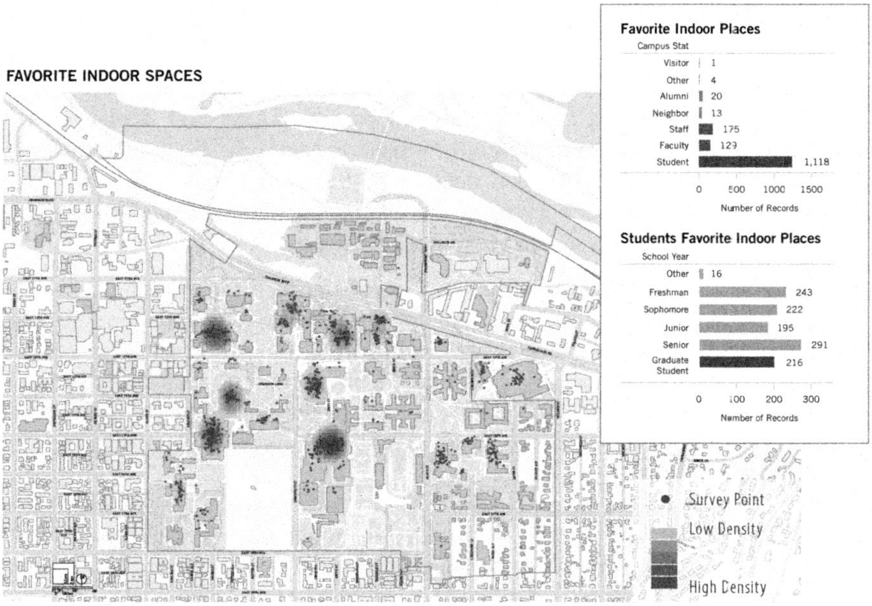

Fig. 21.2 University of Oregon My Campus Survey: Favorite Indoor Spaces. *Source* University of Oregon

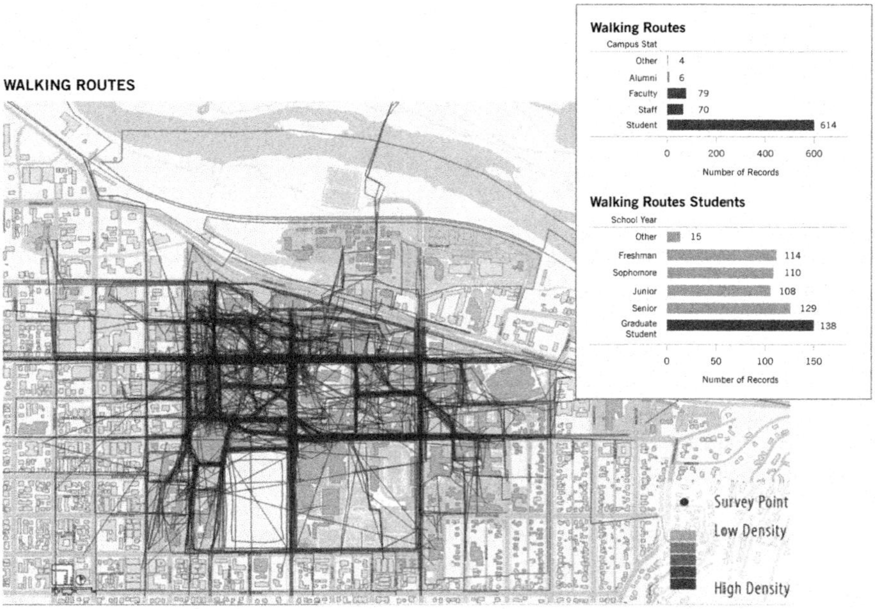

Fig. 21.3 University of Oregon My Campus Survey: Pedestrian Travel Patterns. *Source* University of Oregon

For example, buildings with large open bays are great for classroom space while more historic buildings might better support offices.

It should be noted that all of the data assembled as described above is paired with physical condition data gathered through tours with facilities staff and document reviews, as well as building performance data such as energy and water use.

21.5 Functional Assessment—Analysis

The Functional Assessment is the result of qualitative analysis by the consultant team, based on the data gathering techniques described above. A list of functional criteria are assessed that relate to permanent characteristics building which supports, or does not support and the overall functions and activities expected to occur in the present and the future are assessed. Each building is evaluated for its ability to support the following without a major overhaul to building structure, envelope, or configuration:

- *Accessibility*: Users' general ease of access, particularly for the mobility-impaired, beyond meeting ADA requirements. See, for example, the *Universal Design Handbook* (Preiser and Smith 2011).
- *Comfort*: A user's general level of comfort in relation to the physical space. This could include floor-to-floor heights, convenient travel throughout, access to daylight. It should be noted that this criterion is not related to the condition or capacity of the MEP System, as evaluated in the Physical Assessment.
- *Image/Identity*: The image and identity of the university, or uses within, expressed by the building. Questions asked include: Is it positive? Aesthetically pleasing? Does it reflect the image and identity of the institution?
- *Flexibility of Uses*: A building's ability to house a range of uses and be converted easily from one to another. For example, from lab to office to classroom, etc.
- *Mission, Vision, and Overarching Goals*: Typically the planning project of which the functional condition is a part, identifies important goals that must be achieved, in addition to support of the institution mission and vision (see Fig. 21.4). Goals may include issues related to the following, among many others:

  – Collaboration
  – Community-Building
  – New and Evolving Pedagogies
  – Growth in Research

- *Program Fit*: A building's ability to efficiently and effectively support its current or needed future use.
- *Wayfinding*: Users' ability to remain oriented within and outside the building

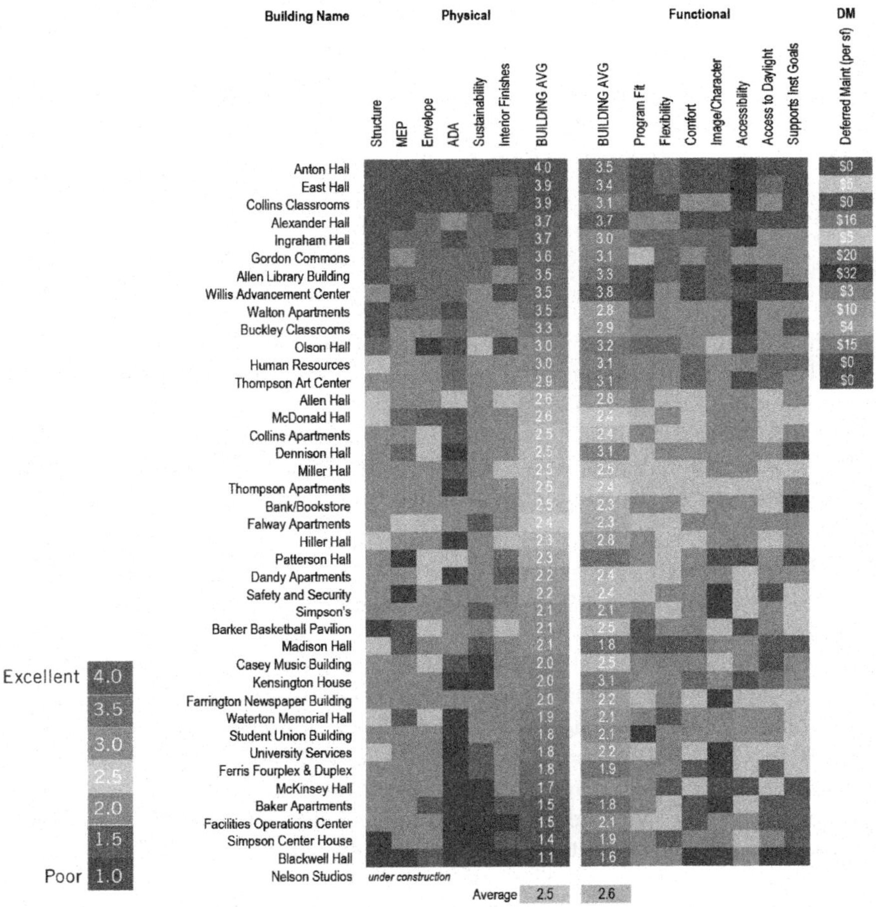

Fig. 21.4 Building Assessment: Physical and Functional Conditions. *Source* Brodie Bain, Perkins + Will

Given the subjective nature of the Functional Assessment method, results of the analysis are reviewed and edited in close collaboration with the users and the core project team. Scores on a 3-point or 5-point scale can be useful with criteria weighted based on issues most important to the institution and users and most supportive of the institution and project goals. In this way, priorities are clearly articulated and discussed, and decisions made with transparency and clarity on institution priorities.

21.6 Conclusion

While POEs are useful to understand the effectiveness of newly completed facilities, their utility can be challenged by realities such as cost, timing, and generalizability. This approach is rear-view-focused, with the results of past work meant to inform future work—critically important, but not the only application of POEs. Another valuable use of POEs is as a tool to help building owners make decisions about the future of existing assets, expanding beyond the physical capacity and condition lens. Functional Assessments, with the use of a wide range of data gathering techniques much like Indicative POEs, are effective in helping owners with a 'forward-view' as they determine the long term future value of specific buildings. Campuses can greatly benefit from Functional Assessents. In an era where our understanding of how people learn is undergoing major evolutions along with research and work processes constantly changing, facilities supporting these activities also need to change. Depending upon the goals and desired product, Functional Assessments are tailored to address current and future needs, life cycle cost and timing, and augment the Physical Assessment. The use of comprehensive, efficient, and effective evaluations as part of the campus planning process is important and valuable as institutions seek to determine how to solve current issues and move into their future. This version of Building Performance Evaluation (BPE) helps owners by incorporating evolving goals and offering a new dimension to their decision making process.

References

Bordass, B., & Leaman, A. (2009). *An overview of post-occupancy evaluation*. Retrieved from www.usablebuildings.co.uk. Accessed 6 Aug 2016.

Duffy, F. (2005). *Research, practice and architectural knowledge*. Retrieved from https://www.architecture.com/Files/RIBAProfessionalServices/ResearchAndDevelopment/Symposium/FrankDuffy.pdf. Accessed 6 Aug 2016.

Freeman, S., Eddy, S., McDonough, M., Smith, M., Okoroafor, N., Jordt, H., & Wenderoth, M. (2014). Active learning increases student performance in science, engineering, and mathematics. *Proceedings of the National Academy of Sciences, 111*(23), 8410–8415. doi:10.1073/pnas.1319030111.

Karim, H., & Crozier, C. (2009). Post-occupancy evaluation: Purpose, benefits and barriers. *Facilities, 27*(1/2), 21–33.

Preiser, W. F. E., Rabinowitz, H. Z., & White, E. T. (1988), *Post-occupancy evaluation*. New York: Van Nostrand Reinhold (republished 2015, London: Routledge).

Preiser, W. F. E., & Smith, K. H. (2011). *Universal design handbook*. London: Routledge.

Tokuhama-Espinosa, T. (2011). *Mind, brain, and education science: A comprehensive guide to the new brain-based teaching*. New York: McGraw Hill.

Vischer, J. C. (2001). Post-occupancy evaluation: A multifaceted tool for building improvement. In Federal Facilities (Ed.), *Council, learning from our buildings: A state-of-the-practice summary of post-occupancy evaluation* (pp. 23–34). Washington, DC: National Academy Press.

Author Biography

Brodie Bain, FAIA, AICP, LEED AP is Principal and Campus Planning Director at Perkins + Will. She has a Bachelor of Science in Environmental Studies/Biology from Trinity College, Hartford, CT and a Master of Architecture with a focus on Design Behavior Research from the University of Illinois at Urban/Champaign. Throughout her 28 years in practice Brodie has focused on helping clients think strategically about their physical environment by aligning their mission, vision and goals with campus/community needs. She has worked with over 30 higher education institutions throughout the U.S. Brodie has recently advanced methods in sustainable campus planning using comprehensive approaches to capture the 'triple bottom line.'

Chapter 22
Emerging Trends in Performance Evaluation of Pediatric Intensive Care Units in Japanese Children's Hospitals

Akikazu Kato, Shiho Mori and Masayuki Kato

22.1 Introduction

Japan enjoys the longest life expectancy in the world. However, when compared in statistical figures among industrialized nations, the Japanese mortality rate of age one to four is quite high, showing a sharp contrast to other age groups. The Japanese Ministry of Health, Labour, and Welfare (MHLW) claims that the infant mortality rate of those one-month after the birth is 1.8 in every 1000 births, which is the lowest in the world (Tanaka 2004). However, the pediatric mortality rate of age one to four is 1.2 in every 1000 births, which is 21st ranking in the World. When compared to seven industrialized nations: France, USA, UK, Italy, Germany, Canada, and Japan, the USA is highest and Japan is the second highest at this age range.

Takei et al. (2008) claim that the lack of Pediatric Intensive Care Unit (PICU) beds is one of the major causes of the above phenomena. They note that the medical results are improved when seriously ill children patients are treated in the PICU. Moreover, a survey carried out by the Japanese Association for Acute Medicine revealed that the numbers of PICU beds and facilities are quite low when compared to others of industrialized nations. To implement the remedy for this situation, the MHLW revised the payment system for PICU service to further the development of

A. Kato (✉)
Mie University, Tsu, Japan
e-mail: kato@arch.mie-u.ac.jp

S. Mori
Nihon Fukushi University, Mihama, Japan
e-mail: shihom@n-fukushi.ac.jp

M. Kato
Naito Architects, Osaka, Japan
e-mail: katomasayuki.arch@gmail.com

W.F.E. Preiser et al. (eds.), *Building Performance Evaluation*,
DOI 10.1007/978-3-319-56862-1_22

pediatric emergency medicine in 2012, and even further in 2014. Given the statistics and the government's interest, the development of PICU's is an urgent matter in Japan. The planning and design of PICUs is the main issue discussed in this chapter.

Children patients are more easily affected by their surrounding environment, and family ties are stronger when compared to those of adult patients, the rights of children patients and their family are needed to enhance the Quality of Life (QOL).

Finally, when compared to adult care, Family-Centered Care (FCC) is considered essential to childcare. The core concepts of FCC are Dignity and Respect, Information Sharing, Participation, and Collaboration (AHA 2004).

22.2 Research Objectives and Methodology

There are three main objectives of this chapter: Firstly, it will clarify the status-quo of planning, design, and management of PICU in children's hospitals in Japan, along with comparing situations with those of the USA for further understanding. The second goal is to grasp the essential requirements for planning, design, and management of future healthcare facilities. And finally, it will illustrate how building performance evaluation (BPE) supports the planning, design, and management methodology in Japan's healthcare system.

To implement these goals, various surveys were carried out. In a survey conducted in 2011, a set of questions were sent to 27 institutions, all members of the Japanese Association of Children's Hospitals and Related Institutions (JACHRI), and from that group, 20 institutions replied. Among those who replied were four institutions that have Intensive Care Unit (ICU) beds not constituting an independent nursing unit, and four institutions had neither a PICU nor ICU. Finally, thirteen institutions replied to have an independent PICU and these became the basis for analysis. Among these institutions, eight were children's hospitals, whose bed capacity ranges from 150 beds to 561 beds, while others were general hospitals with a children's ward units. Two of the institutions were established in 1970's, three in 1980's, three in 1990's, four in 2000's, and the newest in 2010. Parallel to the survey, site-visits were carried out at PICU's of three hospitals and ICU of one hospital.

In 2013, to add to data collected in 2011, information regarding time duration of visiting hours of PICU's were retrieved from web-sites of 61 member institutions of American National Association of Children's Hospital and Related Institute (NACHRI). Finally, 30 blogs written by parents having children patients were watched to gain information regarding their hospital visits (Kato et al. 2014).

22.3 Process of Evidence-Based Design

The topic of evidence-based design (EBD) is becoming a great concern in the architectural design field (Brandt et al. 2010). Hamilton and Shepley (2010) define the goal of EBD as the use of reliable information to support more effective design decisions and the examination of the current state of that information as it applies to critical care. This view is based on the definition of evidence-based medicine by Sackett et al. (1996, p. 71) which reads, "evidence-based medicine is the conscientious, explicit and judicious use of current best evidence in making decisions about the care of individual patients." Research carried out by Roger S. Ulrich of Texas A&M University focused on the scientific foundation of EBD (Hamilton and Shepley 2010). Moreover, Shepley discusses the early impact of the Architectural Research Laboratory at the University of Michigan by referring to the full-scale mock-up studies on patient room at the University Michigan Hospital and others, and that Michigan doctoral program spawned above (Shepley 2014). Shepley's book illustrates the role of EBD in the design of PICU and Neonatal Intensive Care Unit (NICU) by referring to numerous articles on children's healthcare environments that reflect research, descriptive studies, guidelines, and literature reviews. And, the book finds that most salient topics in EBD are positive distraction in pursuit of stress reduction; environmental congruence; safety, i.e., infection control, fall prevention, medical error reduction; and the use of private rooms staff.

The main topic of this book may be restated as the advocate of BPE as an effective method to verify significant evidence for EBD. And, among variety of BPE methodology this chapter focuses on layouts of furniture and equipment to clarify the medical and nursing activities.

22.4 Building Performance Evaluation of PICU

22.4.1 *Number of Beds and Staff*

Among the thirteen institutions the average number of PICU beds was 7.3, and the largest units were 8 beds among four institutions. Because PICU is the most critical treatment facility requiring extensive resources, its bed capacity should reflect the potential number of patients in the target region of the institution. The average utilization ratio of beds was 79%. For institutions marked over 89%, this meant securing a bed for new arrival might requires extra efforts.

The average number of single rooms was 1.5, and 11 hospital units had a ratio of one to three single rooms. Overall, the average ratio of single rooms in relation to the total number of the entire unit beds was 20%. Therefore, the remaining majority of beds were placed in an open room, or using open bay design. For those able to obtain a single room, the average width was 3750 mm and depth was 5080 mm resulting in average floor area of 19.2 m^2.

The average number of pediatricians dedicated exclusively to the unit was 3.9. However, this differed greatly among the institutions ranging from one to fifteen. The average number of nurses exclusive to the unit was 28.8, and in most of institutions the number of nurses was over 20.

22.4.2 Use of Nursing Rooms

When medicine injections are carried out on a child patient, the control of quantities is carried out carefully to suit the patient's specific body weight. The preparation for injection of fluid requires sterile environment in a hospital pharmacy; however, in PICU prescription is frequently changed to meet the alternating sickness status. Three hospitals placed a pharmacist at the Pharmacy Ward to cope with this issue. However, seven institutions replied not to have this feature and the preparation of injection fluid was carried out in the staff station.

Waiting rooms for the families are provided in seven units, and rooms for consultation in ten units. These rooms were placed so that one room inside of PICU counted eight units, two rooms inside was one unit, and one inside and another outside was one unit.

22.4.3 Attributes of Patients

The unaltered average duration of a patient's stay in PICU is 9.2 days including one with a very long average stay of 22 days. When this extended stay is excluded, the average duration is 7.8 days. Principle cause of admission to PICU varies greatly depending on the speciality of hospital. However, post-operative care was chosen as a major causation for unit admissions in 10 hospitals.

22.5 Analysis on Medical and Nursing Activities Surrounding PICU Beds

22.5.1 Average Space Planning for PICU Bed

Because the majority of PICUs surveyed have open floor plans, case studies on medical and nursing activities were carried out focusing on the area surrounding PICU bed. To analyse space issues the distance between bed centers was measured. This distance is equivalent to the average frontage for one PICU bed, and the average size of surveyed units was 3430 mm. Pediatricians of six hospitals surveyed commented their area was too small. Figure 22.1 shows a 3D view of typical equipment layout surrounding PICU bed, and this figure may improve the understanding of following figures regarding equipment layout plans.

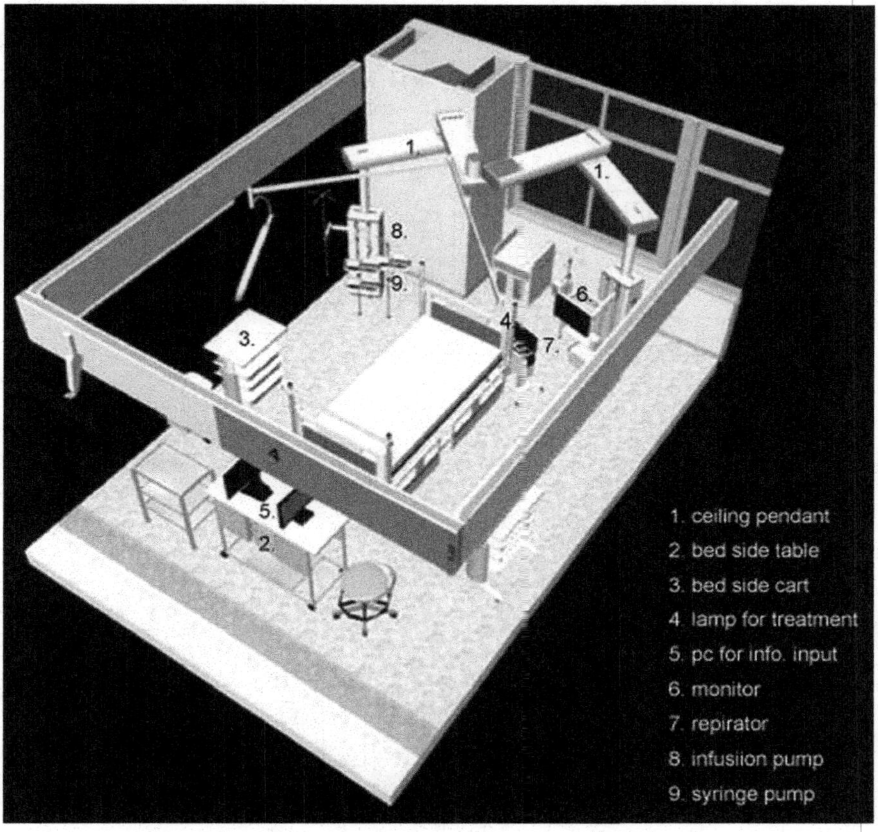

Fig. 22.1 Equipment layout surrounding PICU bed. *Source* Authors

22.5.2 Case Using ECMO

Two hospitals replied to the survey stating spatial problems when using artificial lungs and/or Extra-corporeal Membrane Oxygenation (ECMO) device. ECMO is a new type of artificial respirator used for infants who are not able to use conventional respirators. The device may not be used often, in case of one hospital it was only used approximately ten days in one year. However, all of thirteen surveyed hospitals possessed the device, and the issue requires detailed space planning.

Figure 22.2 shows the case of one hospital where between-bed distance is 3000 mm, lacking space for medical equipment, staff movement, and medical activities. Although the average bed utilization rate is about 70%, so that space of an empty bed may be used, there are times when all beds are occupied causing serious spatial problems.

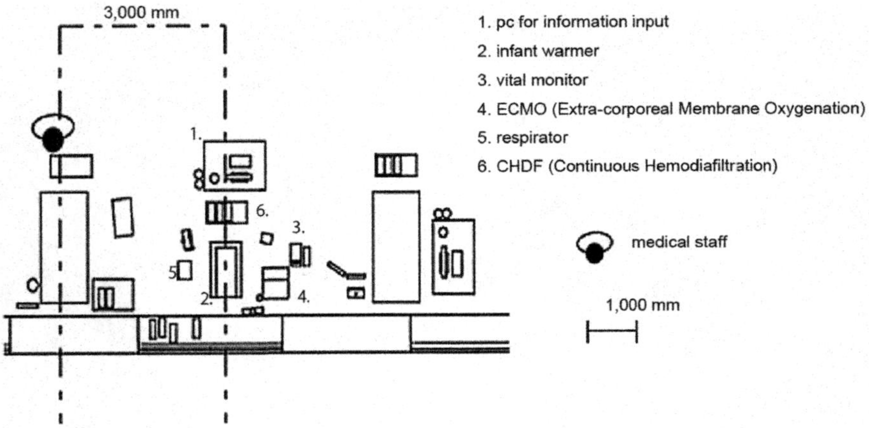

Fig. 22.2 Case using ECMO. *Source* Authors

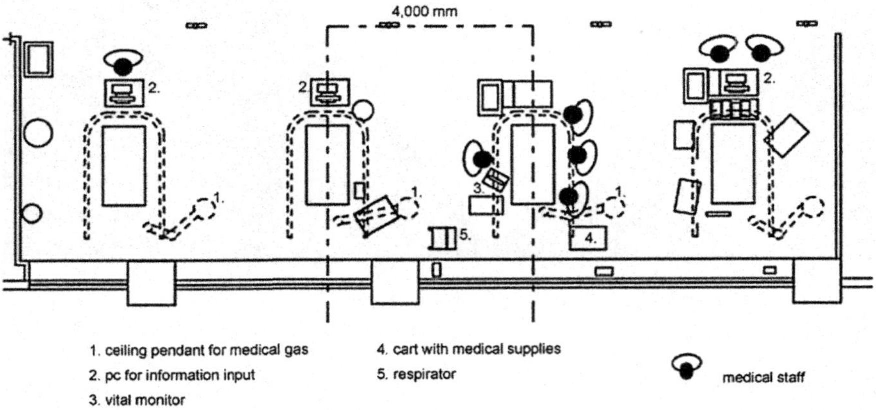

Fig. 22.3 Case with suitable space. *Source* Authors

22.5.3 Case with Suitable Space

At one newly constructed hospital built to function as a national facility, space issues are resolved to a certain level. The between bed distance is 4000–5000 mm. It is a 4-bed unit, and the average utilization ratio is 64%. It has an adjacent High Care Unit (HCU) and depending on situations patients may be transferred to the HCU (Fig. 22.3).

22.6 Healing Environment for Children Patients and Their Families

22.6.1 Resource to Remedy Patient Anxiety

Various measures are taken to remedy patient anxiety in the surveyed healthcare facilities. Six units name wall ornaments hand-made by nursing staff to help with patient anxiety, and eight units named the use of music and DVDs. In addition, one hospital introduced therapeutic dog.

The authors are in favour of all these measures. Although research on impact of ornaments and interior design is carried out in hospital wards (Suzuki and Okaniwa 2008), a rigorous survey is not yet carried out in the intensive care environment. Also, it should be noted that family support is quite significant.

22.6.2 Issues on Visitors

In all surveyed PICUs, visitors were restricted to patients' families. There are no age restrictions in three of the units, and four of them allow 24-h visiting time. Meanwhile, three units do not even provide curtains or roll blind to secure limited privacy during visits.

Thirty blogs were monitored for this survey, those written by parents whose child was admitted as a patient in PICU. Twenty-one blogs provided the information on duration of patient stay, which was noted to be an average of 8 days. Eleven cases mentioned about the process of admission, four from regular inpatient wards, five from operating departments, one from Neonatal Intensive Care Unit (NICU), and another from Coronary Care Unit (CCU). Oikawa (2000) reported that the average days of visits by patients' family per week was 6.7, and the trend was confirmed in surveyed units, where most of patients' family visited the unit almost every day of the week.

Thirteen cases referred to visitation restrictions by pointing out "It's sad to leave the unit" and "Continuous Attendance was not allowed." Mashimo et al. (2007) point out the merits of 24-h visiting system and its impact to reduce the anxiety of patient's parents. However, it should be noted that careful considerations should be taken not to disturb other patients in the present open bay environment.

22.6.3 Comparison with Visiting Situations

In the USA, the concept of Family-Centered Care (FFC) was developed in mid-1990's. Following the concept that family is considered as members of the medical team to maximize the impact of care toward the patient. Thus, 24-h visiting is carried out. In one of the previously mentioned blogs, a parent wrote, "(because 24-h visit was possible and information of disease was well taught,) I thought PICU in USA was well managed. For parents the possibility to visit the patient 24 h a day is very appealing" (source closed for privacy).

From web-sites of 239 member hospitals and institutions of NACHRI, 144 institutes possess PICUs, 61 institutes clearly described the visiting situations, and 52 institutes (85.2%) carried out 24-h visiting system. Moreover, 3 institutes mentioned the possibility to extend the visiting hours with prior notice. Padbury et al. (2013) clarified the merits of all single room NICU and PICU in line with the full implementation of FCC.

In the authors' survey, 4 hospitals responded that they provide 24 h visiting. However, a visitor is only given a small stool and as mentioned earlier the level of privacy is minimal. Thus, there is a long way to realize the concept of FCC.

22.7 Some Recent Development

Because Japan carries out a national insurance policy, the reimbursement system controlled by the MHLW acts as the implementation method of developments in healthcare services. And, as the result of recent revision in 2014, it is reported that only five institutions in Japan meet the requirements to receive the reimbursement as specialized PICU (Suga and Kawamura 2016). This is especially so due to the high standards of facility features and the performance achievement results are highly evaluated in the system. Here, facility features of newly completed three PICUs are compared to clarify the high standards of physical environment. In this facility, a suitable floor area is provided, and especially the number of single bed rooms provided to support the realization of FCC. It should also be noted that Building Information Modeling (BIM) was used in the planning, design, and construction stages of the project, and further use of BIM data in Facility Management stage is now being tested. It should be noted that 3D rendering by BIM system provides a perfect representation of designed environment. Thus, it realizes a visual simulation of design and a fine foundation of coming BPE stage.

22.8 Conclusions

This chapter clarifies the status-quo of physical facility and management issues of PICUs in Japan. The approach follows the principles of EBD. The quality of units varies greatly in view of physical settings, the number of beds, provisions of ancillary rooms, and in view of management issues, namely provisions of paediatricians. Secondly, the chapter pointed out the suitability of the area surrounding PICU beds. Even if Japan was to retain the open bay plan type, the space issue is critical to resolve. The recommended distance between the centers of beds is

4000 mm. All these items directly influence the concept of FCC and enable healing environments for children patients and their families, and thus, improvement of physical and managerial conditions for visiting is essential. Lastly, some recent developments are introduced to further clarify the emerging trends. The cyclic procedure of strategic planning, programming, design, management, and performance evaluation will be greatly enhanced by the practice of EBD supported by the technology of BIM.

Bibliography

American Hospital Association (AHA) Institute for Family-Centered Care (2004). *Strategies for leadership: Patient and family-centered care*. Retrieved from http://www.aha.org/advocacy-issues/quality/strategies-patientcentered.shtml. Accessed 10 Feb 2017.

Brandt, R. M., Chong, G. H., & Martin, W. M. (2010). *Design informed: Driving innovation with evidence-based design*. Hoboken, NJ: Wiley.

Hamilton, D. K., & Shepley, M. M. (2010). *Design for critical care, an evidence-based approach*. London: Routledge.

Kato, M., Kato, A., & Mori, S. (2014). A study on operational policies and architectural design and planning of PICU (Pediatric Intensive Care Unit) in children's hospital. *Journal of Architectural Planning*, Architectural Institute of Japan, 79(702), 1737–1743, (language: Japanese, English translation of this paper was handed out at the edra46 Los Angeles presentation, Emerging Trends in Planning and Design of PICU in Japanese Children's Hospitals, May 2015).

Mashimo, S., Ishiduka, C., & Yamakawa, S. (2007). Survey on consciousness of patient's family on introduction of 24-hour visiting system, assessment and future tasks. *Journal of Japanese Nursing Academy, 38,* 55–57. (language: Japanese).

Oikawa, I. (2000). Actual situations of visits to child inpatient, issues in the status-quo. *Journal of Japanese Society of Child Health Nursing, 23*(6), 719–722. (language: Japanese).

Padbury, J., Bender, J., & Taub, M. (2013). Millennium neonatology: A building for the future. In J. Wells & E. Pavlides (Eds.), *edra44 Providence Healthy + Healing Places [Proceedings of the 44th annual conference of the environmental design research association, EDRA]* (pp. 229–230). McLean, VA: EDRA.

Sackett, D. L., Rosenberg, W. M. C., Muir Gray, J. A., Haynes, R. B., & Richardson, W. S. (1996). Evidence based medicine: What it is and what it isn't. *British Medical Journal, 312,* 71–72.

Shepley, M. M. (2014). *Design for pediatric and neonatal critical care*. London: Routledge.

Suga, T., & Kawamura, M. (2016). Present conditions, patient demographics and skills required to staff in Japanese PICU. *Children Care, 10*(6), 2–9.

Suzuki, K., & Okaniwa, J. (2008). Study on interior design of health care environment in pediatric wards. *Journal of Architectural Planning, Architectural Institute of Japan, 625,* 511–518. (language: Japanese).

Takei, K., Shimizu, N., Matsumoto, H., Yagi, T., Ohara, S., Sakai, Y., & Mashiko, K. (2008). Patient intensification in PICU is needed in lifesaving of pediatric acute patients. *Journal of Japanese Association for Acute Medicine, 19,* 201–209 (language: Japanese).

Tanaka, T. (2004). *Present situation and future prospects of pediatric emergency medicine*. Tokyo: Shindan To Chiryo Sha (Diagnosis and Treatment Publishing). (language: Japanese).

Author Biographies

Akikazu Kato is Professor of Architecture at the Mie University Graduate School of Engineering in Japan. His appointment includes teaching and research responsibilities in the fields of architectural planning and facility management. Previous positions involved serving as faculty member at other national universities, and working as licensed architect at Kume Architects. He received his doctorate in engineering from Nagoya University, has published a number of books and refereed papers, and has presented at various international symposia. Also, he planned and designed a number of architectural projects, mostly in the health care field including those winning architectural prizes such as the 1991 Minister of Health Award for Hekinan City Hospital, and the 2001 Minister of Construction Award for the Asahi-honmachi Housing for Elderly.

Shiho Mori is Associate Professor of Architecture at Nihon Fukushi University. She consults on master plans for medical and welfare facilities, focusing on the relationship between management and space planning. She has presented at international congresses, and worked as a lecturer in training courses for facility directors of nursing homes. Her previous position was as the creator of universally designed housing and equipment, such as modular bathrooms and system kitchens. Using the experience of her previous position, she is participating in several projects, ranging from detached houses to a large-scale housing complex aimed at implementing truly accessible universal design. She received her doctorate in engineering from Nagoya University and worked at Mie University as an assistant professor.

Masayuki Kato is working at Toyota T&S Construction Co., Ltd., and has healthcare facility design experience working at Osaka Office of Naito Architects. He received his BEng and Master of Engineering in Architecture from Mie University. The research topic of his Graduate Thesis and Master's Thesis was Planning, Design and Management of PICU, Pediatric Intensive Care Unit. His first appearance at edra was edra44Providence in 2013, where he was deeply interested in Family-Centered Care in single-room-structured PICU and NICU. He is planning to enter the doctoral course at Mie U. as a working student.

Chapter 23
Feeding the Knowledge Forward: Advancements in Post-Occupancy Evaluation Application Through Collaboration

Lindsey L. Fay

23.1 Introduction

Post-occupancy evaluation (POE) plays an important role in the broader field of Building Performance Evaluation (BPE), offering critical feedback for improving that which has been evaluated. The full process of BPE should utilize POE to feed building information forward by measuring and comparing functional and technical criteria, reporting meaningful findings, applying research outcomes, and informing future designs, thus strengthening the overall design process. However, the final phase of a POE, "applying", is often disregarded in the evaluation process and instead ends with the step of reporting the findings rather than demonstrating a practical application of how this knowledge feeds forward. This speaks to the challenge of effectively translating research findings into concepts that are both accessible and usable to design practice. As noted by Hamilton (2007), "Members of the design community tend to be nervous about reading and understanding original research. They are unsure of their ability to understand academic language, much less to critically interpret the implications of research on their projects" (p. 29). This presents an opportunity to rethink how researchers are communicating building performance outcomes and develop new methods for integrating evidence into design practice. "Unless designers make optimal use of research evidence, both the research and design project are at risk of not reaching the desired objectives" (Fay et al. 2017, p. 2).

This chapter demonstrates the full cycle of a POE process by discussing meaningful methods of planning and conducting a POE, and actively reporting and applying outcomes through the use of a collaborative design charrette. The use of a design charrette presents an opportunity to engage with research that can inform

L.L. Fay (✉)
School of Interiors, University of Kentucky, 117 Pence Hall,
Lexington, KY 40506, USA
e-mail: lindsey.fay@uky.edu

W.F.E. Preiser et al. (eds.), *Building Performance Evaluation*,
DOI 10.1007/978-3-319-56862-1_23

future designs while also developing familiarity with the broader building performance framework, methodologies, and building occupants. Offering designers insight into the benefits of POE through collaboration has great implications for how this systematic analysis might be adopted throughout the full cycle of BPE.

23.2 Planning the POE

Researchers at the University of Kentucky (UK) were contacted by GBBN Architects of Cincinnati, OH to complete a post-occupancy evaluation of the newly completed emergency department (ED) within the University of Kentucky Chandler Medical Center. The new 40,000 square foot (3715 m^2) emergency department includes separate treatment areas for pediatric and adult patients and is organized around 5 pods: triage, trauma and imaging, critical care, acute care, and express care. The client's guiding principles for design included a focus on patient access and care, the academic mission, integration of clinical services, efficiency, flexibility, and image. This emergency department POE was initiated 18 months after the building's opening.

The POE research team, which included members from the university's College of Design, worked closely with the architectural firm and ED personnel to determine specific objectives of the POE. These included:

1. Assess the environmental quality of spaces;
2. Examine the design's impact on family, patients, and caregivers; and
3. Analyze the design layout and its impact on the delivery of efficient, dependable, and safe care.

It was determined that a multi-phased and multi-method diagnostic POE would be implemented due to the extent of information to be measured. The POE employed a timeline that allowed 18 months for completion with data collection occurring in two phases. Throughout the course of the POE, the research team met on a weekly basis to construct a strategic plan for examining focused objectives, developing and refining methodologies, and discussing research outcomes.

23.3 Conducting the POE

After the completion of the research planning phase, approval from the Institutional Review Board of the University of Kentucky was obtained and pilot testing occurred for the methodologies to be utilized. Data collection occurred in two phases. To ensure a complete understanding of the environment and culture of the ED, Phase 1 methodologies were primarily observational. A team of ten individuals worked in pairs to collect data in 4-h shifts. Instruments were distributed on a weekly basis to the research team. The instruments for the first phase of the POE included instructions for data collection, tables or plans for documentation, and an area to record notes, limitations, and additional observations. Each researcher was trained in the use of the instruments to ensure validity and reliability of the findings. Methods utilized for the observation process included measurements of wait times, acoustics, and occupancy counts. Data collection for Phase 1 took place over a 10-week period and involved over 200 h of recorded observations and physical measurements obtained over the course of all 24-h. The Phase 1 methodologies informed the development of Phase 2 data collection methods, which included questionnaires, focus groups, and interviews. Phase 2 began nine months after the completion of the first phase of data collection, which allowed time for data analysis, instrument development, pilot testing, and refinement.

23.4 Reporting POE Findings

While much of the evidence gathered from the research suggests that many attributes of the UK ED are contributing to a positive patient, visitor, and staff experience, other outcomes have revealed design attributes that should be further assessed by the designers to ensure a deeper understanding of environmental and operational factors shaping the ED experience. Specifically, the findings revealed that the entry/triage sequence was one area of the ED that required further design exploration (see Fig. 23.1). Findings related to the entry/triage sequence revealed

Fig. 23.1 University of Kentucky Emergency Department's entry. *Source* Scott Pease

four key issues that impacted patient, visitor, and staff satisfaction in this area. These included workflow, communication, privacy and confidentiality, and safety and security.

23.4.1 Workflow

Overall, UK staff reported that the design of the unit layout facilitated their ability to work efficiently, as a team, and as individuals. However, staff responses were relatively neutral when asked if the layout of the triage area worked efficiently, which reiterated the need to further examine this area. The outcomes from the staff focus groups helped to further understand this and indicated particular concerns regarding circulation patterns, including the absence of a staff corridor between registration and the centralized nursing station, wheelchair accessibility, room sizes, and walking distances.

23.4.2 Communication

Findings related to the issue of communication revealed that patients, visitors, and staff each had unique views. The staff survey outcomes indicated that the location of the consultation rooms and the overall open layout of the unit led to better face-to-face communication. When examining findings more specifically related to the entry/triage areas of the ED, both patients and visitors agreed that the registration desk was easy to locate and that it was easy to determine where to go to get staff help or information when needed. However, staff focus group outcomes revealed that staff had difficulties communicating with patients seated in the waiting area, which could be attributed to limited visibility into the waiting areas from the triage doors.

23.4.3 Privacy and Confidentiality

The issue of privacy and confidentiality largely impacts patient and visitor experiences in the ED. At UK, visitors of the ED were satisfied with the level of privacy during the check-in process and indicated that places existed where the doctor or nurse could talk to them alone. On the other hand, the majority of surveyed patients believed others could hear their private information while checking in. Outcomes from the focus groups further reiterated this as they expressed concern that patients could be overheard during check-in, and suggested the need for acoustical dividers. Staff outcomes indicated adequate space was provided to privately talk with patients' families. However, they expressed concern regarding the number of places they could talk confidentially to other staff members. Findings from the focus groups revealed that they had to seek back hallways and the enclosed medication areas to talk privately.

23.4.4 *Safety and Security*

Findings for the UK ED related to this issue of safety and security revealed that the majority of patient respondents indicated that they felt safe while in the emergency department; the staff responded in the same manner. However, staff comments from the focus groups revealed security concerns. The greatest security issue was the absence of a direct sightline from the registration area to the doors accessing the treatment area, which are used primarily by visitors. Building upon this, the security staff believed these doors stay open too long, resulting in the entry of numerous unobserved visitors. Security personnel also indicated that they were too far away from the triage and swing rooms to adequately monitor them, which is consistent with staff comments regarding feelings of isolation and insecurity while working in these areas.

23.5 Applying POE Outcomes

One of the primary challenges in presenting the results of POEs is to effectively transfer the information gained to the world of practice. Vischer (2002) states "most organizations have no established system for knowing how to process, direct, and act on the information they receive from a POE" (p. 30). Thus the design charrette served as an interactive method to assist design practitioners in both understanding and applying research findings.

During the week of the charrette, a pre-charrette document was shared with the participants that outlined summaries of previous research related to the four primary issues (workflow; communication; privacy and confidentiality; safety and security), key research outcomes, an existing floor plan annotated with research findings, and charrette objectives. The four primary objectives outlined for the charrette were:

- To present evidence of both the positive attributes of the UK ED and those areas that should be reconsidered based on the research.
- To redesign spaces in the UK ED utilizing research findings gathered through the POE process.
- To develop and test new strategies for reporting research findings and integrating evidence.
- To demonstrate the usefulness of academic and practitioner partnerships in generating evidence-based design solutions.

On the day of the charrette, two teams of design researchers, architects, and interior designers, worked over the course of 4 h to reconceive the entry/triage sequence of the ED through a series of exercises. Prior to beginning the charrette exercises, an overview of research findings was presented by the research team. Following the presentation, the two teams were asked to complete charrette exercises that included character profiles, an activity analysis, flow diagrams, a design forecast, and an annotated floor plan with proposed revisions to the space. The development of character profiles offered participants a chance to acquire new insights for the values and abilities of the building occupants. The activity analysis tasked participants with developing a list to describe all tasks, actions, and objects that were involved with the entry and triage processes. This exercise was completed to ensure consideration of the physical and psychological needs of various user groups so that future designs can more effectively respond. Additionally, each team created flow diagrams to represent the flow of information or activities taking place during the operational processes of the entry/triage sequence. The flow diagrams helped to ensure the developed design proposals would support a more efficient workflow and positive user experience. Teams were also asked to think outside the physical limitations of the extant space and develop scenarios that represented “big ideas with no limitations”. These were identified as “design forecasts” and considered the sociocultural and technological trends shaping the ED experience. Lastly, participants utilized these exercises to develop a revised floor plan of the entry/triage sequence that was annotated to communicate the application of research (see Fig. 23.2).

The charrette culminated with the presentation of exercise outcomes and resulting design recommendations. Each team identified what attributes of the design responded to the research, the guiding principles, and user experiences. The design charrette problem was approached differently by each team, creating two distinct ways of resolving the issues outlined in the POE findings regarding the entry/triage sequence. Following the presentation, a discussion took place that evaluated each designs’ effectiveness in responding to the four primary issues, user needs, and their impact on the entry/triage process.

One team kept the first encounter desk facing the entrances, but moved the security station to the opposite side of the entry lobby, maintaining views of the entire lobby and waiting areas. This scheme kept the triage rooms behind the first encounter desk, but reduced the number of rooms in this area to four. Observation of the doors to the treatment zones is now possible from the first encounter desk. The team suggested that security guards should be posted at each of these points to control traffic. Additionally, a zone was created for both adult and pediatric patients to mitigate inadvertent eavesdropping by those waiting in line.

The second team moved the first encounter desk to the opposite side of the entry lobby from where it presently sits. This location allows staff to see the traffic into the triage rooms and improves the observation of traffic into the treatment zones.

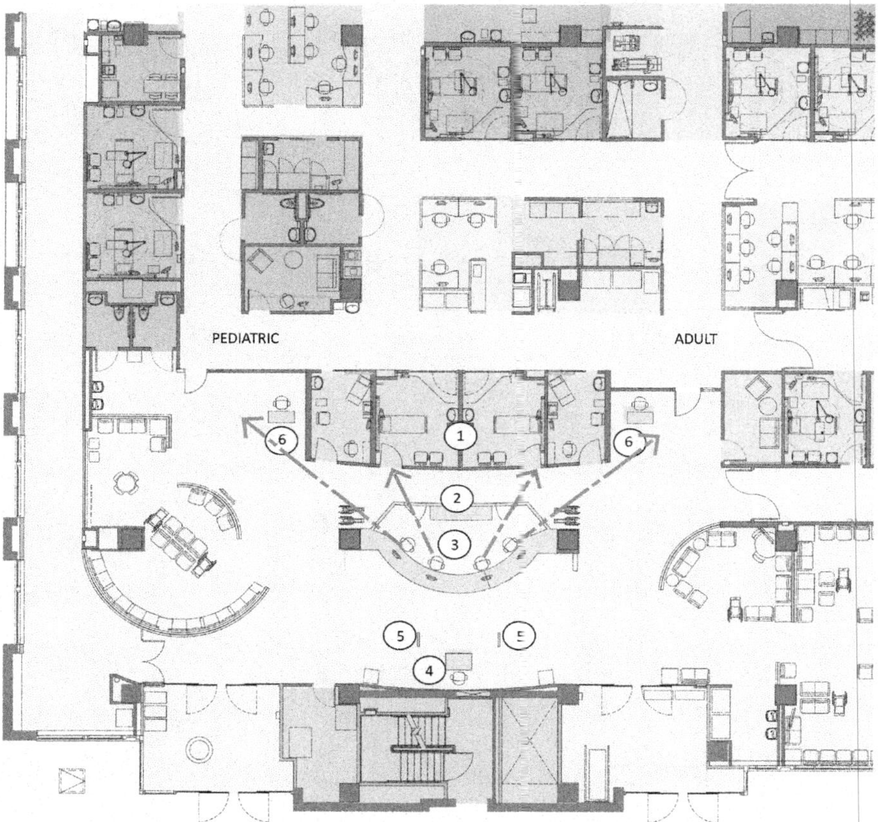

Fig. 23.2 Charrette outcomes illustrating two developed schemes for the entry/triage sequence. Description 1: *1* triage rooms, *2* first encounter desk wall, *3* first encounter desk, *4* security station, *5* queuing stations, *6* optimal security stations. Description 2: *1* triage rooms, *2* first encounter desk, *3* security station, *4* queuing stations, *5* optimal security stations. *Source* Jim Harrell

A security station was proposed opposite the first encounter desk, allowing the security staff to visually monitor both entrances. The proposed location of the security station allows for visibility to the entrances, but has reduced visualization of the waiting areas. This scheme offers enough space in front of each first encounter desk to form a queuing line and utilizes six triage rooms, three for each service. There is now a crossover corridor to permit access to all rooms when incoming volume may spike for either adult or pediatric patients.

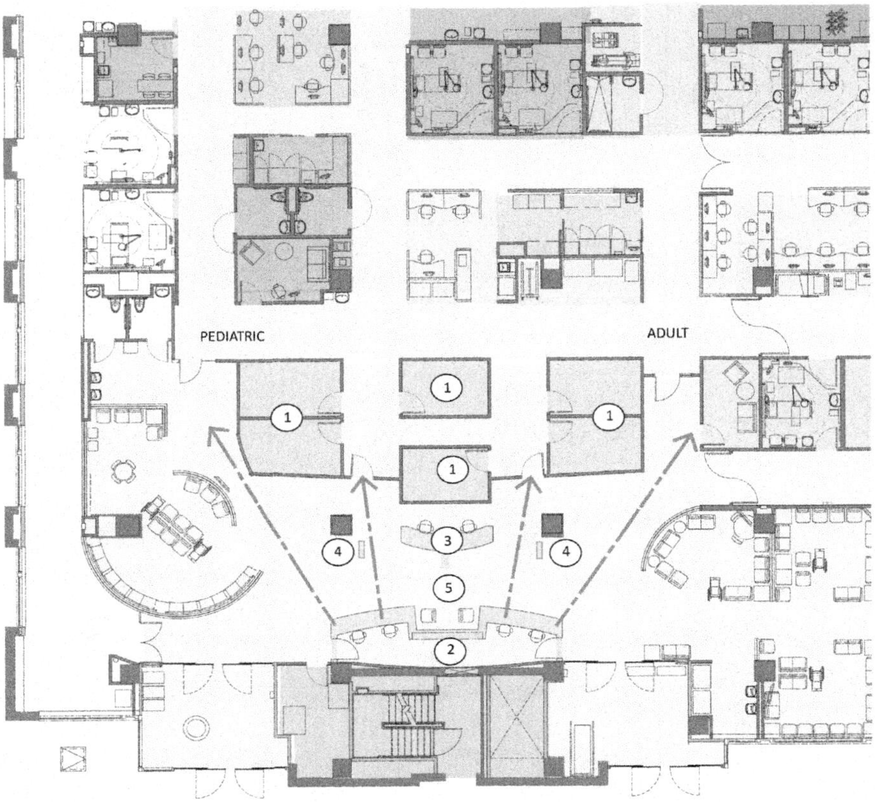

Fig. 23.2 (continued)

23.6 Discussion

Implementing the charrette process as a final stage of the POE can offer a positive mark of completion to the process and help participants align findings and design application. When considering the advancement of Building Performance Evaluations, it was determined that a POE charrette should be utilized and integrate four key standards to ensure effective outcomes (Fay et al. 2017). These include:

1. An all-inclusive, collaborative process
2. Easily interpreted evidence
3. Active participant engagement with the evidence
4. Feasible outcomes grounded in research

23.6.1 All-Inclusive, Collaborative Process

The use of POEs as a research tool can aid in better understanding the needs of building occupants and serves as the critical stage of BPE for initiating building feedback. However, the gaps between building design, occupancy, and evaluation raise concerns regarding the effectiveness of a design in meeting the unique demands of its users. Members of the design team, researchers, healthcare providers, healthcare administrators, and other end users such as patients and visitors should be included in the entirety of the BPE process. This engagement helps to bridge the gap between designers and users, and further exposes them to the importance of evidence-based decision making. Also, an all-inclusive, collaboration offers participants a voice in the design process, resulting in greater diversification in design ideation and increased confidence in design decisions.

23.6.2 Easily Interpreted Evidence

Practitioners are often unsure of how to effectively translate academic language into usable results for application. As design projects continue to grow in complexity, findings from BPEs should be presented in multimodal formats to ensure accurate interpretation. While visual communication is common to design practice, a variety of formats should be utilized to offer a more inclusive understanding of the evidence. This also ensures that research interpretation can align with a variety of learning styles, allowing people to better connect with evidence and in turn use it to impact future designs.

23.6.3 Active Participant Engagement with the Evidence

The more actively the design team is engaged in the BPE process, the greater the likelihood that research outcomes will be integrated into future designs and future POEs will be conducted. To ensure active participation in the POE charrette, a variety of engagement opportunities should be offered to participants. These might include the development and sharing of a pre-charrette document outlining objectives and findings, participation in a variety of design exercises, and opportunities to thoughtfully complete, reflect, and share outcomes from the experience.

23.6.4 Feasible Outcomes Grounded in Research

Ideally, outcomes from a POE should be used "to both improve the fit of the existing space and to be fed back into design research and programming of the next building, thus strengthening the entirety of the building performance evaluation process. Without a feedback loop, every building is, to some extent, a prototype" (Zimmerman and Martin 2001, p. 169). To ensure an optimal experience the charrette must produce feasible outcomes grounded in research. The development of action items for moving the design process forward can ensure both short term and long term outcomes from a POE can be acted upon.

23.7 Conclusion

While evidence-based design is increasingly utilized to guide decision-making, practitioners are still unclear on how to interpret evidence from a POE and within what context it should be applied. This suggests that more manageable criteria are needed for determining how research findings can be fed forward in a useable and accessible manner. The use of a collaborative design charrette involving researchers and practitioners offered a stimulating challenge in that the researchers had to present their findings in a meaningful and memorable manner, while the practitioners were challenged to critically think about the information and its implications for the built environment. The charrette exercises attempted to engage an all-inclusive participatory design process offering active participant engagement with the evidence. This process for feeding building information forward through a POE has greater implications for the entirety of the building performance evaluation process. Considering active participant and end-user engagement throughout the various phases of BPE will contribute to designs that are less experimental and more personal.

Acknowledgements Thank you to Allison Carll-White of the University of Kentucky and Jim Harrell of GBBN Architects for their contributions to this chapter and collaboration in the POE charrette.

References

Fay, L., Carll-White A., & Harrell J. (2017). Coming full cycle: Linking POE findings to design application. *HERD: Health Environments Research & Design Journal, 10*(3), 83–98.

Hamilton, D. K. (2007). Bridging design & research. *Health Environments Research & Design Journal, 1*(1), 29–30. http://www.herdjournal.com. Accessed 13 June 2014.

Vischer, J. C. (2002). Post-occupancy evaluation: A multi-faceted tool for building improvement. In Federal Facilities Council (Ed.), *Learning from our buildings: A state-of-the-practice summary of post-occupancy evaluation* (pp. 23–34) Washington, DC: National Academy Press.

Zimmerman, A., & Martin, M. (2001). Post-occupancy evaluation: Benefits and barriers. *Building Research & Information, 29*(2), 168–174.

Author Biography

Lindsey Fay an Assistant Professor in the School of Interiors at the University of Kentucky, is creating an extension of the academic learning environment for design professionals and students through advancements in the participatory design process and evaluations of extant environments. As an advocate of evidence-based design, her research has examined healthcare, learning, and community-based environments to assess the intercultural conditions and exchanges that exist within these spaces. Currently, her research utilizes post-occupancy evaluation to assess the design of healthcare spaces and their impact on care delivery. It further implements this methodology as a learning tool and immersive learning experience for interior design students. Fay has been published in a number of peer-reviewed journals and is a frequent presenter at national conferences.

Part V
Epilogue

Chapter 24
Epilogue: From Building Evaluation to Building Performance Evaluation and Beyond

Jacqueline C. Vischer

As this Epilogue was being written and this book being completed, our senior editor and team leader, Wolfgang Preiser, died. For most of the contributors to this book, and no doubt many of its readers, Wolf Preiser's work has blazed a trail that has influenced the practice of architecture and design as well as social science research into occupant (building user) behavior in countries all over the world. I myself have collaborated with Wolf since he published *Building Evaluation*—the precursor to the present volume—in 1989, thrilling me by giving me my first chance to see myself in print. Together we co-edited *Design Intervention: Toward A More Humane Architecture*, published in 1991, and some years later *Assessing Building Performance* (2005), to which a follow-up volume—*Enhancing Building Performance* (edited by S. Mallory-Hill, W. Preiser, and C. Watson)—was published in 2012.

It was one of my Ph.D. students who demonstrated to me how much my own work is influenced by Wolf Preiser. In her thesis, she compared Preiser's Habitability Framework (Preiser 1983) with my own Functional Comfort Pyramid (Vischer 1989). It seems that Wolf was already writing about the pyramid as a framework for evaluating environmental quality in the early 1980s, but until my student pointed this out I did not realize that my own idea had built on something Wolf had already thought up.

In view of the great loss to our field of such a prolific, effective, energetic and creative intellectual leader, this book has an important role in enabling all of us who wrote it—and all of you who read it—to gain a perspective not only on the future of Building Performance Evaluation, but also where it came from and what it has achieved. This Epilogue, therefore, first briefly summarizes the evolution of Wolf Preiser's ideas and his notion of Building Performance Evaluation among his other seminal contributions to our field of work and study. It is clear that Wolf's work laid the groundwork for the many teachers, students, researchers and practitioners who have built on and grown the powerful base he created for us. The second

J.C. Vischer (✉)
University of Montreal, 2453 rue Ryde, Montreal, QC H3K 1R5, Canada
e-mail: vischer.biu@gmail.com

W.F.E. Preiser et al. (eds.), *Building Performance Evaluation*,
DOI 10.1007/978-3-319-56862-1_24

section looks at the rich array of studies and ideas—of which select examples are included in this volume—illustrating how far building evaluation research has come in Wolf's lifetime. As the new and updated edition of a book first published in the late 1980s, some contributors have been given a chance to update and enrich our own and our colleagues' work with our intervening years of practical and research experience, while others are bringing a fresh new perspective not previously available in this volume. Finally, this Epilogue will turn to what Wolf felt an epilogue should do: focus on the future of the field and the advances that are moving our discipline forward, even though sadly this future is without him.

As a trained architect, Wolf insisted from early in his career on the importance of the user experience. This guided his research, his publications and his career decisions. His first three major publications came in quick succession and set in motion the now global movement that we call Building Performance Evaluation (BPE). Starting with *Programming the Built Environment* in 1985, this was followed by *Post-Occupancy Evaluation* in 1988: one of the most widely read and taught books in Architecture and Environmental Psychology curricula, giving a sound theoretical base to what was fast becoming a new field of knowledge and practice. *Building Evaluation*—the third in the series and predecessor of the present volume—followed in 1989, setting a pattern that characterized Wolf's long and rich career: working on more than one book at once and bringing them out at a rate that left most academics breathless. Wolf's custom of producing most of his books with co-editors means he has given a unique chance to numerous practitioners and researchers—both newcomers and 'old hands'—to make their voices heard. In addition to his work on Post-Occupancy Evaluation (POE) and programming, and later BPE, Wolf co-wrote and co-edited books on universal design, Pueblo architecture, and architectural and esthetic criticism. He was particularly proud of contributions to his profession, including *Professional Practice in Facility Programming* for the American Institute of Architects (AIA) in 1993, and *Improving Building Performance* for the National Council of Architectural Registration Boards (NCARB) in 2003.

However, it was the three books published in the 1980s—all of which are re-issued—that together laid the conceptual, intellectual and instrumental groundwork for the cascade of studies that would be published in succeeding years and that are continuing to proliferate, as several chapters in this volume illustrate. Together, these books formalized the concept of studying occupied buildings in order to understand how well building architecture was meeting users' needs in the buildings they occupied, and, more generally, what could be learned about the psychology of space use and the needs of people who inhabit, use and respond to built space, and guiding practitioners towards the steps necessary to ensure long-term building quality and suitability.

Extensive collaborative work between Wolf and many colleagues but notably Ulrich Schramm led to the comprehensive notion of Building Performance Evaluation, the framework for which guides much of the work reported in this book. BPE goes beyond programming—consulting users about their needs and behavior patterns in order to develop a responsive and user-sensitive basis for design and space-use recommendations—and beyond POE—systematic study of the built environment in use, to provide feedback on building performance and on occupants' comfort and satisfaction and indicate needed improvements. In their introductory chapter to *Assessing Building Performance*, Preiser and Schramm present BPE as follows:

> While in the past building delivery was seen as a linear, end product-oriented process, the integrative framework is a dynamic, evolving and non-mechanical model … which can be depicted as an ever-expanding helix of knowledge on building performance … [that] … attempts to respect the complex nature of performance evaluation in the building delivery process, as well as through the entire life cycle of buildings. The BPE framework defines the building delivery and life cycle from the perspective of all parties who are involved with a building (p. 16).

In so doing, they make the innovative intellectual leap from the discrete need to include user information and perspectives in programming (pre-planning) and evaluating (post-occupancy) buildings, all the way to a systematic application of relevant user feedback to every stage of the building delivery cycle. This perspective is further developed in *Enhancing Building Performance*, and also in the Introduction to this volume, where Preiser, Hardy and Schramm elaborate on the six stages of the BPE process by indicating the multiple layers on which building performance evaluation is focused—namely, technical, functional and behavioral. These authors have long argued that the BPE approach, being performance-based, provides valuable tools to professionals engaged in building design and construction. As stated in their Introduction,

> The key aspect of performance criteria is that they constitute objective, quantifiable, and measurable 'hard' data, as opposed to 'soft' criteria, which are derived from qualitative and often subjective assessments. Thus the BPE model is based on a feedback system comparing explicitly stated performance criteria with the actual, measured performance of a building (p. 13).

The power of the BPE concept, and the extensive proliferation of work that it has inspired, emerges clearly in this volume. As subsequent chapters of the book demonstrate, evaluating building performance requires a sound theoretical framework, valid and reliable approaches to measurement, and competent and committed approaches to data analysis and interpretation. In Part II of this volume, evolving and innovative approaches to building evaluation include elements of all three.

Oseland describes the growing practice of assessing an occupied building throughout its lifecycle so that occupant feedback is routinely used as a tool for continuous improvement, to the point that practicing professionals now receive training in how to carry out POE. Loftness et al. describe tools for collecting data in field conditions that can be quickly analyzed and results applied directly to understanding how building occupants are affected by ambient interior conditions. Another type of toolkit is presented by Persky et al. that has evolved to target one specific building type, in this case, courthouses. And Senick et al. present a different approach, using dataset modeling to simulate building user populations in order to add validity and meaning to small datasets. As approaches to environmental assessment evolve, aided and abetted by information technology, the urgency increases of ensuring that building occupants themselves understand how technology that controls their environment, as explained by Schramm et al. in their research on building automation systems.

Pursuing the theory of measurement in POE and BPE, Francescato et al. raise an important question when they review the meaning of user satisfaction as an evaluation criterion. While measuring ambient conditions in an occupied environment requires certain types of instrumentation, measuring occupant responses in terms of their attitudes and behavior requires tools of a different sort and raises theoretical questions of how users' priorities are defined. The complexity of defining meaningful outcome measures increases in the BPE context where different varieties of user feedback are appropriate for each stage. In another example, Preiser and Petronis illustrate convincingly how to apply the results of in-use studies to all stages of building delivery, and most particularly, to the final stage of occupying or 'activating' a complex building such as a hospital.

Part III addresses advances in knowledge resulting from BPE by offering case study examples of the numerous and variable ways in which POE has evolved into BPE in different parts of the world. While Hodulak analyses the complex challenges of workspace change in Germany, Vischer describes how a user feedback tool first presented in *Building Evaluation* evolved in Canada, and now measures not just occupant satisfaction but the work environment's contribution to effective task performance. Nubani expands the measurement toolkit by innovatively applying the well-established method of space syntax to patterns of communications in offices in Dubai. It is clear that the studies reported in these chapters target a range of behavioral outcome variables, including Sylvest's analysis of collaboration and social interaction in the workplace.

Joiner has also updated his original contribution to *Building Evaluation*, indicating how POE has now become well-established in the New Zealand context, as has Rosenheck, who describes how numerous US organizations and government agencies have adopted POE on a systematic basis. Barker places POE and BPE in the context of large-scale institutional decision-making, explaining how feedback from building evaluation can be applied to improving technical requirements in U. S. Army buildings.

Part IV looks more closely at the evolution of study methods in the context of recent projects. Elzeyadi explains how feedback on building performance based on measures of occupants' comfort and satisfaction in the form of a continuous loop provides essential information to designers, managers and occupants. In a similar vein, Becker points out the value of using POE not just for research but as a diagnostic tool that can be used for continuous improvement, and he makes the link to evidence-based design emphasizing the importance of both instrument measures of interior conditions and of systematically collected feedback from users. Also in the context of professional practice, Parshall (a previous contributor) and Fonseca describe how collecting user feedback can be designed to fit into a demanding project schedule, supplying information directly into both programming and POE stages.

An important new direction for POE and BPE that has emerged since *Building Evaluation* was published is environmental sustainability, and Marans and Callewaert address how users are affected by sustainable building features. While Mallory-Hill and Golgolewski contribute a discussion of relevant ways of using data collected from sustainable buildings to inform and improve human comfort in sustainable environments, Bain places POE and BPE in the context of large-scale property management and asset management decision-making, referring to the challenges of large institutions such as the university campus.

Expanding the sphere of influence of BPE from building interiors to the total environment, Kato et al. describe how POE has been applied not just to facility design but also to aspects of the socio-psychological experience of patients in children's hospitals in Japan. Finally, Fay reiterates the importance of practical ways of feeding building use information forward to inform design decision-making, and describes the 'charrette' method as a collaborative way of making this happen.

The rich array of research and practice presented in this volume can be characterized as progress in three general areas. First, the chapters are notable for advances in tools and techniques of measurement. Preiser, Hardy and Schramm stress the importance and complementarity of quantitative and qualitative measurement. To be meaningful, sophisticated data collection to measure the 'objective' aspects of building performance—for example, indoor air quality, thermal comfort, lighting and noise levels—has to be applicable to changing field conditions and must avoid disturbing occupants or disrupting their behavior. In parallel, behavioral and attitudinal data from occupants on their 'subjective' experience must be reliable and complementary. When *Building Evaluation* was first published few such tools existed; this is no longer the case. Moreover, instruments that measure occupant behavior reliably have evolved to capture more sophisticated data, whether user attitudes and degrees of satisfaction, or levels and varieties of user comfort, or observable behaviors such as seat occupancy, circulation patterns, or targeted behavior specific to building types such as courthouses and shopping malls. While in the early years the two categories of data collection and analysis rarely overlapped, much of the innovation and improvement that has occurred in BPE measurement has focused on integrating quantitative and qualitative research results. Most researchers and practitioners now see the need for data that can be described as 'objective' or 'subjective' to be collected, analyzed and integrated in a meaningful way, and they are increasingly skilled at making this happen.

Second, as POE has evolved into BPE, the range and type of steps in the building delivery and occupancy process that are considered worth studying have increased. While at the outset, POE—as its name suggests—focused on occupied buildings and a previously concluded design process, BPE has expanded into defining the building delivery process in a series of 'phases' and 'loops', all of which are relevant to researchers and practitioners. From the feedback sought in POE to the feed-forward sought for programming, the accounts in these chapters address aspects of building production and management that make user feedback relevant to asset management, facilities management, planning and programming, design and construction, 'activation', and sustainability. What we have learned from BPE is that it is no longer sufficient to simply decide to plan, design and build a building because now we have the tools and data to ask: What is the problem to be solved, Whether a building is the best solution, and How it will create an appropriate and supportive environment for users.

Third, it emerges clearly from both this book and its predecessor that there is no limit to the type and number of facilities that are candidates for BPE. All buildings have users, and most facilities have several different user groups not all of whose needs are in harmony, e.g. patients and health care staff, museum visitors and curators, inmates and correctional personnel. From POE studies of social housing and institutional buildings such as libraries and psychiatric facilities—examples of building types where users are clearly defined and their range of behaviors limited —that characterized the early years, the enrichment of the BPE model has diversified study targets to include various types of health care and educational facilities, government and various types of office buildings, military facilities, buildings constructed to sustainable standards, as well as buildings that accommodate entertainment, needs of the criminal justice system, and a range of exterior natural environments. While not all facility types are referenced in this volume, the examples in these chapters indicate the expanding world of BPE and the increasing richness of the physical environments that users occupy and that can therefore be improved.

What has changed since Wolf Preiser conceived *Building Evaluation* in the 1980s is the dominant paradigm, and of this he would be proud. Moving from POE to BPE represents a major paradigm shift away from the 'one-off' approach to each individual project that professional practice accepted in the past. The importance of systematically recognizing, measuring, understanding and regularly and habitually integrating the user's point of view into planning, creating and operating buildings has a primacy that we hoped for and aspired to but could not ensure. Thanks to Wolf Preiser and his colleagues, the consistently high standard of their research and of his publications over the years between *Building Evaluation* and *Building Performance Evaluation*, the widespread influence of his inspiration and guidance, and the enthusiasm and momentum going forward, we can point to the powerful new paradigm of which this book is evidence.

Although we have lost our leader, Wolf's books and writings will continue to inspire work in this field, and a growing number of enthusiastic professionals in our industry and discipline will continue to expand not only the tools and techniques available to measure users' behavior in buildings, but also the range and types of environments which affect their occupants and how this knowledge is used. Moreover, as the construction and real estate industries evolve, and more sophisticated economic and technological tools are brought to bear on the process of creating buildings, our discipline will also evolve: identifying innovation, measuring impacts, and providing continuous feedback on use to every stage of decision-making.

I think I can speak for my fellow contributors in saying our work for this book offers an expression of our profound gratitude to Wolfgang Preiser for his lifetime contribution to creating and growing this field. We recognize and acknowledge what he has done to guide and inspire, to enable us to do work we love and place it in a meaningful context, and to publish our work in which he so much believed. Wolf: you will be missed.

References

Mallory-Hill, S., Preiser, W. F. E., & Watson, C. (2012). *Enhancing building performance.* London: Wiley-Blackwell.

Preiser, W. F. E. (1983). The habitability framework: A conceptual approach towards linking human behavior and the built environment. *Design Studies, 4*(2), 84–91.

Preiser, W. F. E. (1985). *Programming the built environment*. New York, NY: Van Nostrand Reinhold.

Preiser, W. F. E. (Ed.) (1989). *Building evaluation*. New York, NY: Plenum Press.

Preiser, W. F. E. (Ed.) (1993). *Professional practice in facility programming*. New York, NY: Van Nostrand Reinhold.

Preiser, W. F. E. (2003). *Improving building performance*. Washington, DC: National Council of Architectural Registration Boards (NCARB).

Preiser, W. F. E., Rabinowitz, H. Z., & White, E. T. (1988). *Post-occupancy evaluation*. New York: Van Nostrand Reinhold. (republished in 2015, London: Routledge).

Preiser, W. F. E., & Vischer, J. C. (Eds.) (2005). *Assessing building performance.* Elsevier Butterworth Heinemann. (republished in 2015, London: Routledge).

Preiser, W. F. E., Vischer, J. C., & White, E. T. (Eds.) (1991). *Design intervention: Toward a more humane architecture*. New York, NY: Van Nostrand Reinhold.

Vischer, J. C. (1989). *Environmental quality in offices*. New York, NY: Van Nostrand Reinhold.

Index

W.F.E. Preiser et al. (eds.), *Building Performance Evaluation*,
DOI 10.1007/978-3-319-56862-1

CPSIA information can be obtained
at www.ICGtesting.com
Printed in the USA
BVOW06*1802020917
493852BV00007BA/79/P

9 783319 568614